Topics in
Current Physics

11

Topics in Current Physics

Founded by Helmut K. V. Lotsch

Volume 1 **Beam-Foil Spectroscopy**
Editor: S. Bashkin

Volume 2 **Modern Three-Hadron Physics**
Editor: A. W. Thomas

Volume 3 **Dynamics of Solids and Liquids by Neutron Scattering**
Editors: S. W. Lovesey and T. Springer

Volume 4 **Electron Spectroscopy for Surface Analysis**
Editor: H. Ibach

Volume 5 **Structure and Collisions of Ions and Atoms**
Editor: I. A. Sellin

Volume 6 **Neutron Diffraction**
Editor: H. Dachs

Volume 7 **Monte Carlo Methods** in Statistical Physics
Editor: K. Binder

Volume 8 **Ocean Acoustics**
Editor: J. DeSanto

Volume 9 **Inverse Source Problems** in Optics
Editor: H. P. Baltes

Volume 10 **Synchrotron Radiation** Techniques and Applications
Editor: C. Kunz

Volume 11 **Raman Spectroscopy** of Gases and Liquids
Editor: A. Weber

Volume 12 **Positrons in Solids**
Editor: P. Hautojärvi

Volume 13 **Computer Processing of Electron Microscope Image**
Editor: P. W. Hawkes

Volume 14 **Excitons**
Editor: K. Cho

Volume 15 **Physics of Superionic Conductors**
Editor: M. B. Salamon

Raman Spectroscopy

of Gases and Liquids

Edited by A. Weber

With Contributions by
S. Brodersen J. M. Friedman H. W. Klöckner
G. V. Knighten J. W. Nibler D. L. Rousseau
H. W. Schrötter R. P. Srivastava A. Weber
P. F. Williams H. R. Zaidi

With 103 Figures

Springer-Verlag Berlin Heidelberg New York 1979

Professor Dr. Alfons Weber

Physics Department, Fordham University, Bronx, NY 10458, and

Molecular Spectroscopy Division, National Bureau of Standards,
Washington, DC, 20234, USA

ISBN 3-540-09036-3 Springer-Verlag Berlin Heidelberg New York
ISBN 0-387-09036-3 Springer-Verlag New York Heidelberg Berlin

Library of Congress Cataloging in Publication Data. Main entry under title: Raman spectroscopy of gases and liquids. (Topics in current physics ; v. 11). Bibliography: p. Includes index. 1. Raman spectroscopy. 2. Gases—Spectra. 3. Liquids—Spectra. I. Weber, Alfons, 1927–. II. Series.
QC454.R36R35 535′.846 78-11693

Offset printing and bookbinding: Konrad Triltsch, Graphischer Betrieb, 8700 Würzburg.
2153/3130-543210

Preface

The Raman effect is a most useful tool for the study of molecular vibrations and molecular structure. Information about the structure and symmetry of molecules, as well as about their vibrational energies can be obtained to a reasonable degree of satisfaction from their infrared and Raman vibrational spectra. The body of knowledge of the vibrational infrared and Raman spectra of molecules is immense and is now so well organized and understood that it is found to be represented in any standard upper level undergraduate curriculum in chemistry. The rotational energies of a molecule and quantitative details about its structure can only be obtained through the techniques of microwave, and high-resolution infrared and Raman spectroscopy of low pressure gases and vapors. The results of such investigations are of interest not only to the academic scientists, but also to scientists and engineers who are active in applied fields of chemistry and physics, as well as the atmospheric sciences.

This book deals with basic investigations of the Raman scattering of light by gases, with some attention also being given to liquid substances. After a brief introductory chapter that delineates the historical development of Raman spectroscopy of gases, high-resolution rotation-vibrational and pure rotational Raman spectroscopy is described in Chapters 2 and 3. The all-important intensity parameter, the Raman scattering cross section, is treated in Chapter 4, while the broadening of Raman lines due to the effects of intermolecular forces is taken up in Chapter 5. Resonance Raman scattering is presented in Chapter 6 while the new field of CARS, Coherent Anti-Stokes Raman Scattering, which owes its existence to the laser, is treated in the last chapter. The whole tone of the book reflects current research interests which are based on Raman spectra generated by laser sources.

The topics presented are arranged in such a manner that the reader is led from the study of the ideal, isolated molecule, to the effects of intermolecular forces, and finally to the considerations of the nonlinearities in the Raman scattering process. Each chapter contains an extensive list of references to the research literature whereby further details, and topics not taken up in this book, can be obtained. A supplementary list of references, compiled after the completion of the manuscripts, is added to the end of the book to insure a complete listing of the current literature. It was my intention to have this book carry the imprint date of 1978 so as to have its publication occur in the 50th anniversary year of the discovery of the Raman effect. Indeed the manuscript was well in hand but for technical reasons the publication date had to be set for 1979.

I am indebted to the authors of the various chapters for the careful preparation of their manuscripts and also for their good will and patience in seeing the results of their efforts reach the public domain. Finally I wish to thank Dr. H. Lotsch and the staff of Springer-Verlag for their cooperation and for bringing out this book in the minimum possible time.

Bronx, N.Y., and Washington, D.C. *Alfons Weber*
October, 1978

Contents

List of Contributors

BRODERSEN, SVEND
Kemisk Institut, Aarhus University, DK 8000 Aarhus C, Denmark

FRIEDMAN, JOEL
Bell Telephone Laboratories, Murray Hill, NJ 07974, USA

KLÖCKNER, HANS WOLFGANG
Sektion Physik, Universität München, Schellingstraße 4
D-8000 München 40, Fed. Rep. of Germany

KNIGHTEN, GARY
Chemistry Department, Oregon State University, Corvallis, OR 97331, USA

NIBLER, JOSEPH W.
Chemistry Department, Oregon State University, Corvallis, OR 97331, USA

ROUSSEAU, DENIS L.
Bell Telephone Laboratories, Murray Hill, NJ 07974, USA

SCHRÖTTER, HEINZ W.
Sektion Physik, Universität München, Schellingstraße 4
D-8000 München 40, Fed. Rep. of Germany

SRIVASTAVA, RAJENDRA P.
Physics Department, University of New Brunswick, Fredericton
New Brunswick, E3B 5A3, Canada

WEBER, ALFONS
Physics Department, Fordham University, Bronx, NY 10458, USA
and
Molecular Spectroscopy Division, National Bureau of Standards
Washington, DC, 20234, USA

WILLIAMS, FRAZER
Department of Electrical Engineering, Texas Tech University
Lubbock, TX 79413/USA

ZAIDI, HAIDER RAZA
Physics Department, University of New Brunswick, Fredericton
New Brunswick, E3B 5A3, Canada

1. Introduction

A. Weber

That the laser has revolutionized the field of molecular spectroscopy is a well known commonplace. This is indeed so emphatically true in the case of Raman spectroscopy, that a newcomer to this field is of the understandable impression that it was practically nonexistent prior to the invention of the laser in 1960. That such an impression is very much in error is, of course, well documented by the many research articles and books dealing with the Raman effect that have been published since its discovery just 50 years ago, in 1928. The impact made by the laser is, however, so strong that the appellation "Laser Raman Spectroscopy" is now often used to underline the break with the past.

There are now available many research monographs that deal with various specialized aspects or even whole research fields that are based on the Raman effect. To mention only a few of the more recently published ones we have the volumes edited by ANDERSON [1.1] which contain contributions that span nearly all aspects of the Raman effect, the books by KONINGSTEIN [1.2] and LONG [1.3], the series edited by CLARK and HESTER [1.4], and the books by CARDONA [1.5], and CLAUS et al. [1.6], which deal with light scattering by solids only. The proceedings of international Raman conferences are published in book form [1.7,8] and a specialized journal devoted to, among other light scattering phenomena, "all aspects of Raman spectroscopy" [1.9] began publication in 1973. The volumes on molecular spectroscopy issued by The Chemical Society [1.10] as part of their Specialist Periodical Reports contain annual reviews of the current literature on Raman spectroscopy.

The present volume grew out of a request to bring together in one book the results of contemporary research on Raman spectroscopy of gases as there have been numerous advances in this field since the appearance in 1973 of the last comprehensive review of this subject [1.11]. This restriction to gases would have been easily accomplished only a few years ago but is unwise now in view of the role played by liquid and solid substances in contributing to the understanding of resonance, nonlinear, and stimulated Raman processes. Nevertheless, the spotlight in this book is on the single molecule.

Although the Raman effect was experimentally discovered in 1928 in liquids and solids, Raman scattering by gases has played a fundamental role in the development of the basic theory of the phenomenon. The Raman spectra of gaseous hydrogen chloride

[1.12], hydrogen [1.13-15], oxygen [1.14,16,17], nitrogen [1.14,16], carbon monoxide [1.18,19], nitric oxide [1.14,19,20], carbon dioxide [1.21,22], nitrous oxide [1.22], ammonia [1.23-25], methane [1.23,26,27], ethane [1.25,27], ethylene [1.25, 27], and acetylene [1.25,27] were quickly studied under moderate resolution and using high pressures. These early investigations culminated in the PLACZEK and TELLER's [1.28] comprehensive theory of the rotational and rotation-vibrational Raman scattering by free molecules and also in the more inclusive general theory of the Raman effect of PLACZEK [1.29]. From then until the early 1950's Raman spectra of gases were studied only occasionally until new experimental techniques developed by H.L. Welsh at the University of Toronto and B.P. Stoicheff, then at the National Research Council of Canada, caused a revival of interest in this field. The new activity resulted in major contributions to our knowledge of molecular structures, Raman scattering cross sections, and intermolecular forces. These techniques, and the results that were obtained with them, have been described previously in several review articles [1.11,30-32,41].

The first laser excited Raman spectrum of a gas showing resolved rotational structure was reported in 1965 [1.33], and this was soon followed by several reports describing improvements in the experimental technique [1.34-37]. These advances were summarized in several reviews [1.11,38,40]. This early work in "Laser Raman Spectroscopy of Gases" addressed itself to the spontaneous Raman scattering and the quest for improvements in the efficiency and resolution capabilities of the method. Since that time there have been major advances in other aspects of gas-phase Raman spectroscopy, and the present book provides to the chemist and physicist a comprehensive review of this field of study.

Chapters 2 and 3 are on high-resolution studies of low pressure gases and vapors. Taken together they comprise a review of the current state of knowledge of the rotational and rotation-vibrational energies of diatomic and polyatomic molecules insofar as these have been studied by means of "Laser Raman Spectroscopy". Chapter 2 contains a modern version of the theory of rotational Raman scattering. This theory forms the basis of understanding of the results that are given in both chapters. The molecular-structure problems that are attacked by the method of high-resolution Raman spectroscopy are the same as those considered by infrared and microwave spectroscopy and an acquaintance with these topics will aid in the assimilation of the contents of these two chapters. The experimental techniques that are employed in high-resolution Raman spectroscopy of gases are still sufficiently novel and unique, and the apparatus is specially designed and built, that both chapters contain extensive and detailed descriptions of them. There is, therefore, some overlap between Chaps.2 and 3 but the duplication that does occur is believed to be more than made up by the complementary features to guide those interested in initiating their own research program in this field.

The determination of intensities and transition probabilities has been notoriously more difficult than that of finding accurate energy levels. This problem is particularly acute in Raman spectroscopy and, except for an occasional early result, reliable Raman scattering cross sections did not become available until the late 1950's (see [1.32] for a review of the pre-laser work). The availability of laser techniques has reopened this field of investigation and Chap.4 contains a treatment of the theory and experimental determination of Raman scattering cross sections from gases. The results that were obtained with pre-laser techniques only are also included in this chapter after having been reevaluated so as to place them on an equivalent basis with the newer, laser-derived results. This chapter also includes a discussion of the Raman scattering cross sections of molecules in the liquid state of aggregation.

Whereas Chaps.2 and 3, and most of Chap.4 deal with dilute gases in which molecules may be considered "free", the effects of intermolecular forces that become important with increasing gas density are taken up in Chap.5.

The Raman line widths, -shifts, and -shapes are functions of the gas pressure and density and also depend on the nature of the interaction partners. Quantitative studies of these dependencies were initiated in 1950 at the University of Toronto and the results of these investigations were summarized in a review by GRAY and WELSH [1.41] published in 1971. There have been many theoretical and experimental advances in this field since that time and Chap.5 addresses itself to these.

The phenomena described in Chaps.2-5 are those of the nonresonant Raman effect. Although strenuous efforts were made in the pre-laser era to experimentally study resonance Raman scattering it was only by means of narrow line, and also tunable lasers that sufficiently exact findings on simple molecules became available to allow quantitative evaluations of the various theoretical approaches. These questions are taken up in Chap.6.

The subject of the last chapter of this book, CARS (*C*oherent *A*nti-Stokes *R*aman *S*cattering) owes its existence to the laser. Although already observed in 1963 the CARS phenomenon could not be exploited in molecular spectroscopy until tunable lasers of sufficient power became available. This subject is currently the least settled and most actively investigated one of the various topics presented in this book.

An overview of the topics and their sequential appearance in the various chapters reveals an attempt to present a coherent coverage of Raman spectroscopy of molecules as a modern field of research. The topics covered in Chaps.2-6 were all studied prior to the invention of the laser, and indeed some of the molecules now under investigation with laser techniques were already studied as early as 1929. However, all of these topics have benefited from the laser, while that of the last chapter is a practical impossibility without laser sources. The "free" molecule is the center of attention in Chaps.2-4 while molecular interactions and Raman scattering by condensed phases are also considered in Chaps.5-7. Several subjects had to be excluded from this book due either to the desire to maintain a topical focus or because they are already adequately treated elsewhere. Among these are the various applications of gas-phase Raman spectroscopy to problems of the atmosphere or analytical chemistry (see, for example, [1.42,43]). The analysis of molecular rotation and rotation-vibration spectra has progressed way beyond the simple numerical and graphical techniques used for so long [1.30,43]. No account of the currently used statistical techniques is included here in view of the full treatment described elsewhere [1.45].

The publication of this book occurs in the 50th anniversary year of the discovery of the Raman effect. The discovery provoked a veritable stampede in the scientific world with what for those times an enormous number of over 400 publications on Raman spectroscopy by the year 1931, and already merited a full-length review in the *Handbuch der Physik* as early as 1929 [1.46]. The Kramers-Heisenberg dispersion formula was immediately recognized as giving the correct description of the scattering process but it was not until the simplification of the polarizability model was introduced by PLACZEK [1.47] that the full power of the Raman effect could be brought to bear to the study of molecular structure. Indeed the name of PLACZEK is indissolubly linked to the Raman effect, as his penetrating analysis of this phenomenon has withstodd the test of time [1.48]. Though mature and sedate, the field of Raman spectroscopy is still full of excitement and interest and promises to be so for many more years to come.

References

1.1 Anderson, A. (ed.): *The Raman Effect*, Vols.1 and 2 (Marcel Dekker, New York 1971 and 1973)
1.2 Koningstein, J.A.: *Introduction to the Theory of the Raman Effect* (D. Reidel Publishing, Dordrecht 1972)
1.3 Long, D.A.: *Raman Spectroscopy* (McGraw-Hill, New York 1977)
1.4 Clark, R.J.H., Hester, R.E. (eds.): *Advances in Infrared and Raman Spectroscopy*, Vols.1, 2 and 3 (Heyden and Son, London 1975, 1976 and 1977)
1.5 Cardona, M. (Ed.): *Light Scattering in Solids*, Topics in Applied Physics, Vol.8 (Springer, Berlin, Heidelberg, New York 1976)
1.6 Claus, R.B., Merten, L., Brandmüller, J.K.A.: *Light Scattering by Phonon-Polaritons*, Springer Tracts in Modern Physics, Vol.75 (Springer, Berlin, Heidelberg, New York 1975)
1.7 Mathieu, J.P. (ed.): *Advances in Raman Spectroscopy*, Vol.1, Proc. 3rd Intern. Conf. on Raman Spectroscopy (Heyden and Son, London 1973)
1.8 Schmid, E.D., Brandmüller, J., Kiefer, W., Schrader, B., Schrötter, H.W.: Proc. 5th Intern. Conf. on Raman Spectroscopy (H.F. Schulz Verlag, Freiburg/Br. 1976)
1.9 *Journal of Raman Spectroscopy* (Heyden and Son, London since 1973)
1.10 *Specialist Periodical Reports, Molecular Spectroscopy*, Vols.1-4, Senior Reporters Barrow, R.F., Long, D.A., Millen, D.J., Sheridan, J. (The Chemical Society, London 1973, 1974, 1975, and 1976)
1.11 Weber, A.: "High Resolution Raman Studies of Gases", in *The Raman Effect*, Vol. 2, Chap. 9, ed. by A. Anderson (Marcel Dekker, New York 1973) pp. 543-757
1.12 Wood, R.W.: Nature *123*, 166, 279 (1929); Phys. Rev. *33*, 1097 (1929); Phil. Mag. *7*, 1929 (1929)

1.13 McLennan, J.C., McLeod, J.H.: Trans. R. Soc. Can. (III) *23*, 19 (1929); Nature *123*, 160 (1929)
1.14 Rasetti, F.: Proc. Natl. Acad. Sci. *15*, 234, 515 (1929); Nuovo Cimento *6*, 356 (1929); Phys. Rev. *34*, 367, 548 (1929)
1.15 Bhagavantam, S.: Nature *128*, 70, 188, 272 (1931); Ind. J. Phys. *6*, 319, 331, 557 (1931); Ind. J. Phys. *7*, 632 (1932)
1.16 Rasetti, F.: Z. Phys. *61*, 598 (1930)
1.17 Trumpy, B.: Z. Phys. *84*, 282 (1933)
1.18 Amaldi, E.: Z. Phys. *79*, 492 (1932)
1.19 Bhagavantam, S.: Phys. Rev. *42*, 437 (1932)
1.20 Rasetti, F.: Nuovo Cimento *7*, 261 (1930); Z. Phys. *66*, 646 (1930)
1.21 Houston, W.V., Lewis, C.M.: Phys. Rev. *37*, 227 (1931); Proc. Natl. Acad. Sci. *17*, 229 (1931)
1.22 Langseth, A., Nielsen, J.R.: Phys. Rev. *44*, 326, 911 (1933)
1.23 Dickinson, R.G., Dillon, R.T., Rasetti, F.: Phys. Rev. *34*, 582 (1929)
1.24 Amaldi, E., Placzek, G.: Naturwissenschaften *20*, 521 (1932); Z. Phys. *81*, 529 (1933)
1.25 Lewis, C.M., Houston, W.V.: Phys. Rev. *44*, 903 (1933)
1.26 Bhagavantam, S.: Nature *129*, 830 (1932); *130*, 740 (1932)
1.27 Lewis, C.M., Houston, W.V.: Phys. Rev. *41*, 389 (1932)
1.28 Placzek, G., Teller, E.: Z. Phys. *81*, 209 (1933)
1.29 Placzek, G.: "Rayleigh Streuung und Raman Effekt", in *Handbuch der Radiologie*, Vol. 6, Pt.II, ed. by E. Marx (Akademische Verlagsgesellschaft, Leipzig 1934) pp.205-374. English translation by A. Werbin: "Rayleigh and Raman Scattering", UCRL Transl. No. 526L, available from National Technical Information Service, U.S. Department of Commerce, Springfield, Va
1.30 Stoicheff, B.P.: "High Resolution Raman Spectroscopy", in *Advances in Spectroscopy*, Vol. 1, ed. by H.W. Thompson(Wiley-Interscience, New York 1959) pp.91-174
1.31 Stoicheff, B.P.: "Raman Effect", in *Experimental Physics, Molecular Physics*, Vol. 3, ed. by D. Williams(Academic Press, New York 1962) pp.111-155
1.32 Murphy, W.F., Holzer, W., Bernstein, H.J.: Appl. Spectrosc. *23*, 211 (1969)
1.33 Weber, A., Porto, S.P.S.: J. Opt. Soc. Am. *55*, 1033 (1965)
1.34 Weber, A., Porto, S.P.S., Cheesman, L.E., Barrett, J.J.: J. Opt. Soc. Am. *57*, 19 (1967)
1.35 Barrett, J.J., Adams, N.I.: J. Opt. Soc. Am. *58*, 311 (1968)
1.36 Barrett, J.J., Weber, A.: J. Opt. Soc. Am. *60*, 70 (1970)
1.37 Weber, A., Schlupf, J.: J. Opt. Soc. Am. *62*, 428 (1972)
1.38 Weber, A.: "Progress in Raman Spectroscopy of Gases", in *Developments in Applied Spectroscopy*, Vol.10, ed. by Perkins,A.J., Grove,E.L., Kaelble,E.M., Westermeyer,J.E. (Plenum Press, New York 1972) pp.137-160
1.39 Edwards, H.G.M.: "High Resolution Raman Spectroscopy of Gases", in *Essays in Structural Chemistry*, ed. by Downs,A.J., Long,D.A., Staveley,L.A.K (Plenum Press, New York 1971) Chap.6, pp.135-162
1.40 Cherlow, J.M., Porto, S.P.S.: "Laser Raman Spectroscopy of Gases", in *Laser Spectroscopy of Atoms and Molecules*, Topics in Applied Physics, Vol.2, ed. by H. Walther (Springer, Berlin, Heidelberg, New York 1976) pp.255-282
1.41 Gray, C.G., Welsh, H.L.: "Intermolecular Force Effects in Raman Spectra of Gases", in *Essays of Structural Chemistry*, ed. by Downs,A.J., Long,D.A., Staveley,L.A.K. (Plenum Press, New York 1971) Chap.7, pp.163-188
1.42 Lapp, M., Penney, C.M. (Eds.): *Laser Raman Gas Diagnostics* (Plenum Press, New York 1974)
1.43 Inaba, H.: "Detection of Atoms and Molecules by Raman Scattering and Resonance Fluorescence", in *Laser Monitoring of the Atmosphere*, Topics in Applied Physics, Vol.14, ed. by D.E. Hinkley (Springer, Berlin, Heidelberg, New York 1976) Chap.5, pp.153-236
1.44 See, for example, the methods of analysis described in G. Herzberg: *Molecular Spectra and Molecular Structure*, Vol.1, *Spectra of Diatomic Molecules*, 2nd ed. (Van Nostrand, Princeton,N.J. 1950); Vol.2 *Infrared and Raman Spectra of Polyatomic Molecules* (Van Nostrand, Princeton,N.J. 1945)
1.45 Albritton, D.L., Schmeltekopf, A.L., Zare, R.N.: "An Introduction to the Least-Squares Fitting of Spectroscopic Data", in *Molecular Spectroscopy: Modern Research*, ed. by K. Narahari Rao (Academic Press, New York 1976) Chap.1, pp.1-67; see also Albritton,D.L., Schmeltekopf,A., Zare,R.N.:J. Mol. Spectrosc. *67*,132 (1977)

1.46 Pringsheim, P.: "Raman Spectra", in *Licht und Materie*, Handbuch der Physik, Band 21 (Julius Springer, Berlin 1929) pp.607-633
1.47 Placzek, G.: Z. Phys. *70*, 84 (1931); and "Raman Effekt und Molekülbau", in *Molekülstruktur* (Leipziger Vorträge), ed. by P.M. Debye (S. Hirzel, Leipzig 1931) pp.71-106. See also the comprehensive treatment given in [1.29]
1.48 See, for example, Musher, J.I., Schulman, J.M.: "On the Calculation of Raman Intensities", in Int. J. Quant. Chem., Symposium No. *5*, 183-190, 1971, for a critical evaluation of Placzek's approximations in the light of exact quantum mechanical dispersion

2. High-Resolution Rotation-Vibrational Raman Spectroscopy

S. Brodersen

With 19 Figures

The present chapter deals with rotation-vibrational Raman spectra of gases observed under high resolution. The concept of "high resolution" must not be transferred immediately from the infrared region. In the present case it means a limitation to an observed resolution better than about 0.5 cm^{-1}. The aim of such investigations is typically the same as the aim of studies within high-resolution infrared and microwave spectroscopy: to explore the fields of rotation-vibrational interaction and rotation-vibrational transitions and to determine molecular constants.

High resolution and high precision in the determination of frequencies are essentials in this kind of research. At the present state of the art, rotation-vibrational Raman spectra cannot be obtained with a resolution at all comparable to that of the best infrared instruments. But the Raman spectra are often able to give information which cannot be extracted from infrared or microwave spectra, or they may yield significantly more accurate values for molecular constants. The explanation lies in the intensity expression for Raman spectra which is very different from the expression for "intensity" in absorption. This leads primarily to more relaxed selection rules, allowing a significantly larger selection of changes in both rotational and vibrational energy to be observed. But even if certain transitions are allowed in both cases, the intensity distribution in the spectra will be quite different, and the possibility exists for extracting from the Raman spectra information which cannot be obtained from infrared or microwave spectra.

In the following it is attempted to give an up-to-date description of the state of the art. In the first part, the theory is presented using a unified nomenclature. In the second part, the presently used experimental technique is described, and in the third part, the results obtained within the last years are reviewed. Throughout, it is attempted to draw the outlines rather than to be complete. For a detailed study it is necessary to consult the original literature, to which a number of references are given. It is the hope of the author that the present chapter will help the reader to see the potentialities of rotation-vibrational Raman spectroscopy and to understand some of the general principles involved.

Comprehensive reviews on Raman spectroscopy of gases have been given in 1959 by STOICHEFF [2.1] and in 1973 by WEBER [2.2]. Both of these cover a broader range than the present chapter, including also pure rotation. The present review does not, in general, intend to cover the experimental results reviewed in those articles, which should be consulted for a more complete record of the older work. Shorter reviews of the more recent literature have been given in 1973 by EDWARDS and LONG [2.3] and in 1975 by EDWARDS [2.4].

2.1 Theory

In this section the theory of rotation-vibrational Raman spectra will be presented. This is more or less equivalent to discuss the expression for the spectral intensity, the selection rules derived from this expression, and the predictions of band structures computed from this expression. It is assumed that the reader has a basic knowledge of the treatment of the rotation-vibrational problem.

The presently used theory of Raman intensity is not essentially different from the theory developed in the early 1930's by PLACZEK and TELLER [2.5] and PLACZEK [2.6]. It assumes 1) that all transitions begin and end within the non-degenerate electronic ground state, 2) that the energy of the non-stationary transition-state is well below the energy of the first electronically excited state, and 3) that the intensity of the exciting light is sufficiently low so that the induced dipole moment is a linear function of the exciting field, as described by a 3×3 polarizability tensor.

The intensity expression derived, was in 1963 presented in a more accessible form by HERRANZ and STOICHEFF [2.7] taking into account any kind of rotation-vibrational interaction. More recently the intensity and the selection rules were discussed by RASMUSSEN and BRODERSEN [2.8] and HEGELUND et al. [2.9] in a more convenient form using the Clebsch-Gordan coefficient formulation. HEGELUND et al. [2.9] discussed only linear molecules and symmetric tops but the basic results are valid for any molecule. Many theoretical results have been published in other papers, and reference to some of these will be given below.

In the following an account will be given of the intensity expression, the more important selection rules and the calculated contours for different types of bands and molecules. All results are presented using a unified symbolism. No proofs will be given, most of them may be found in the original literature referred to above and in the following.

2.1.1 Intensity Expressions

A computation of spectral intensities must be based on a knowledge of the wave functions and energies of the initial and final states. In the present case all transitions are supposed to take place within the electronic ground state, so the proper wave functions are the rotation-vibrational wave functions in which the nuclear spin part can be accounted for, more or less, by means of a statistical nuclear spin degeneracy g_{ns} of a rotation-vibrational state. The rotation-vibrational wave function is in the general case determined by a diagonalization of the rotation-vibrational Hamiltonian matrix in terms of a suitable set of basis functions. We shall not discuss this procedure, nor the approximations involved. We shall assume, however, that the resulting wave functions may be expressed in the form

$$\psi = \sum_{V,K} c_{VK}|V\rangle|JKM\rangle \tag{2.1}$$

where $|JKM\rangle$ is a symmetric top rotational wave function, using HOUGEN's phase convention [2.10], and $|V\rangle$ is a product of harmonic one- or two-dimensional oscillator vibrational wave functions. V is thus a vector of vibrational quantum numbers, possibly including some ℓ quantum numbers. Note that K is a signed quantum number, except in special cases to be discussed below.

Raman intensities depend on the polarizability of the molecule, i.e., the relation between an external electric field described by a field vector and the induced dipole moment. If both vectors are described by their components in a molecule-fixed coordinate system (x,y,z), the polarizability is a symmetric 3×3 matrix with 6 different real elements α_{xx}, α_{yy}, α_{zz}, α_{xy}, α_{xz} and α_{yz}. The polarizability elements act as operators in the matrix elements occurring in the intensity expressions, and it is convenient to use the following linear combinations [2.8] denoted by the common symbol $T_\sigma^{(\omega)}$

$$T_0^{(0)} = -3^{-1/2}(\alpha_{xx}+\alpha_{yy}+\alpha_{zz}) \tag{2.2}$$

$$\begin{aligned}
T_2^{(2)} &= (\alpha_{xx}-\alpha_{yy})/2 + i\alpha_{xy} \\
T_1^{(2)} &= -\alpha_{xz} - i\alpha_{yz} \\
T_0^{(2)} &= (2\alpha_{zz}-\alpha_{xx}-\alpha_{yy})/6^{1/2} \\
T_{-1}^{(2)} &= \alpha_{xz} - i\alpha_{yz} \\
T_{-2}^{(2)} &= (\alpha_{xx}-\alpha_{yy})/2 - i\alpha_{xy} \quad .
\end{aligned} \tag{2.3}$$

The operator $T_0^{(0)}$ with $\omega = 0$ appears in the intensity expression for isotropic scattering and the five operators with $\omega = 2$ in the expression for anisotropic scattering.

The intensity observed in a Raman experiment [2.5-7,11,12] depends, among other factors, on the angle between the electric field vectors of the exciting light and the scattered light. The directions of propagation of the two beams are only important indirectly because they contribute in defining this angle. In the two limiting cases of a zero angle and a 90° angle, the intensity may, respectively, be written

$$I(0^\circ) \propto (10\ m_{iso}^2 + 4\ m_{an}^2)\ \exp(-E''/kT) \tag{2.4}$$

and

$$I(90^\circ) \propto 3\ m_{an}^2\ \exp(-E''/kT) \quad , \tag{2.5}$$

where m_{iso}^2 and m_{an}^2 are the squares of the transition moments for isotropic (ω=0) and anisotropic (ω=2) scattering, respectively. The common proportionality factor, left out in both expressions, contains as factors: the intensity of the exciting light, the fourth power of the frequency of the scattered light and the (infinitesimal) solid angle of observation. The Boltzmann factor stands for the number of molecules in the initial (lower) state. If the temperature is varied, the partition function should be included in the denominator.

In order to calculate the total intensity to be observed in a definite experiment, an integration should be carried out over the cone of observation, taking into account the different transmissions of the instrument for different directions of polarization. In practice it is sufficiently accurate to use the following two expressions

$$I \propto 6\ m_{an}^2\ \exp(-E''/kT) \tag{2.6}$$

and

$$I \propto (10\ m_{iso}^2 + 7\ m_{an}^2)\ \exp(-E''/kT) \quad , \tag{2.7}$$

obtained from (2.4) and (2.5) by adding the intensities for two perpendicular directions of polarization of the scattered light. Thus (2.6) applies when the direction of observation is parallel to the electric vector of the laser beam used for excitation, whereas (2.7) applies when the direction of observation is perpendicular to the electric vector of the laser beam, if the transmissions are equal for the two directions of polarization in the beam of observed light. Thus, the presence of the isotropic scattering in the observed spectrum depends on the experimental conditions, whereas the anisotropic scattering cannot be excluded.

The squares of the transition moments may both be expressed as follows [2.9]

$$m^2 = g_{ns}(2J+1)\Big|\sum_{V'K'VK} c^*_{V'K'}\ c_{VK}\ C(J\omega J';K\ K'-K)\langle V'|T^{(\omega)}_{K'-K}|V\rangle\Big|^2 \tag{2.8}$$

with $\omega = 0$ for m_{iso}^2 and $\omega = 2$ for m_{an}^2. This expression includes the effect of nuclear spin degeneracy and M-degeneracy, as suggested by the factors g_{ns} and 2J+1. Primed quantum numbers refer to the upper state, unprimed to the lower state. The star indicates a complex conjugation.

Table 2.1. Expressions for C(J2J';K K'-K)

J'-J	K'-K = 0	K'-K = 1	K'-K = 2
-2	$\sqrt{\frac{3(J-K)(J-K-1)(J+K)(J+K-1)}{2(J-1)(2J-1)J(2J+1)}}$	$-\sqrt{\frac{(J-K)(J-K-1)(J-K-2)(J+K)}{(J-1)(2J-1)J(2J+1)}}$	$\sqrt{\frac{(J-K-3)(J-K-2)(J-K-1)(J-K)}{4(J-1)(2J-1)J(2J+1)}}$
-1	$-K\sqrt{\frac{3(J-K)(J+K)}{(J-1)J(J+1)(2J+1)}}$	$(J+2K+1)\sqrt{\frac{(J-K)(J-K-1)}{2(J-1)J(2J+1)(J+1)}}$	$-\sqrt{\frac{(J+K+1)(J-K-2)(J-K-1)(J-K)}{2(J-1)J(J+1)(2J+1)}}$
0	$\frac{3K^2-J(J+1)}{\sqrt{(2J-1)J(J+1)(2J+3)}}$	$-(2K+1)\sqrt{\frac{3(J-K)(J+K+1)}{2(2J-1)J(J+1)(2J+3)}}$	$\sqrt{\frac{3(J+K+1)(J+K+2)(J-K-1)(J-K)}{2(2J-1)J(J+1)(2J+3)}}$
+1	$K\sqrt{\frac{3(J-K+1)(J+K+1)}{J(J+1)(J+2)(2J+1)}}$	$(2K-J)\sqrt{\frac{(J+K+2)(J+K+1)}{2J(2J+1)(J+1)(J+2)}}$	$-\sqrt{\frac{(J+K+1)(J+K+2)(J+K+3)(J-K)}{2J(J+1)(J+2)(2J+1)}}$
+2	$\sqrt{\frac{3(J-K+2)(J-K+1)(J+K+2)(J+K+1)}{2(J+1)(2J+3)(J+2)(2J+1)}}$	$\sqrt{\frac{(J-K+1)(J+K+3)(J+K+2)(J+K+1)}{(2J+1)(J+1)(2J+3)(J+2)}}$	$\sqrt{\frac{(J+K+1)(J+K+2)(J+K+3)(J+K+4)}{4(J+1)(2J+3)(J+2)(2J+1)}}$

The expressions for K'-K = -1 or -2 may be obtained by means of the relation $C(J2J';-K\ -K'+K) = (-1)^{J'-J}\, C(J2J';K\ K'-K)$.

The Clebsch-Gordan coefficient $C(J\omega J';K\ K'-K)$ is a well-known function of the five variables. For $\omega = 0$, all non-vanishing Clebsch-Gordan coefficients have the form $C(J0J;K\ 0\rangle$ which has the value 1 independent of J and K. For the isotropic case, (2.8) thus simplifies to

$$m^2_{iso} = g_{ns}(2J+1)\left|\sum_{V'VK} c^*_{V'K}\, c_{VK} \langle V'|T_0^{(0)}|V\rangle\right|^2 \quad . \tag{2.9}$$

For $\omega = 2$ the Clebsch-Gordan coefficient $C(J2J';K\ K'-K)$ is tabulated in Table 2.1 (rewritten from [Ref.2.13, Table 4[3]]).

Note that K'-K appears in (2.8) both in the Clebsch-Gordan coefficient and as the lower index of the operator $T^{(\omega)}_{K'-K}$.

Equation (2.8) in combination with (2.6) or (2.7) gives a general expression for the intensity of any rotation-vibrational Raman transition. It is valid for any kind of molecule and for any approximation to the rotation-vibrational interaction problem, if the resulting wave functions can be written in the form (2.1). In complicated cases the sum in (2.8) may contain a large number of terms, but if the interactions between different vibrational states may be neglected, (2.8) may be reduced considerably. A detailed discussion of these reductions is postponed to later sections where the different rotor types are discussed separately.

2.1.2 Selection Rules

The Clebsch-Gordan coefficient $C(J\omega J';K\ K'-K)$ vanishes unless the following conditions are satisfied simultaneously

$$|J'-J| \leq \omega \quad , \tag{2.10}$$

$$J+J' \geq \omega \quad , \tag{2.11}$$

$$|K'-K| \leq \omega \quad , \tag{2.12}$$

and

$$J'-J \text{ even, if } K = K' = 0 \quad . \tag{2.13}$$

Equations (2.10,11) are genuine rotational selection rules, whereas (2.12,13) in the general case only are "selection rules" for a single term in the sum appearing in (2.8). However, if the number of terms in (2.8) is reduced, both (2.12,13) may become genuine rotational selection rules as discussed in more detail below.

The vibrational matrix element $\langle V'|T^{(\omega)}_{K'-K}|V\rangle$ vanishes unless the symmetry in the molecular point group of the operator is contained in the direct product of the symmetries of the two vibrational wavefunctions. Moreover, if the irreducible representation of the operator is degenerate, the operator and the product of the two vibrational wavefunctions must transform according to identical rows in the representation. The symmetry properties of the operator is obtained from the rule [Ref. 2.11, App.XV] that a component of the polarizability transforms as the function used as an index, α_{xx} as x^2, α_{xy} as xy, etc.

As an example of the importance of this condition we may consider an F_2 fundamental band of a molecule of symmetry T_d. The two sets of functions (x,y,z) and (yz,xz,xy) transform as F_2 and according to the same three rows if taken in this order. If we choose the vibrational part of the basis functions of the F_2 state so that they transform as x, y, or z, the only non-vanishing vibrational matrix elements which can appear in (2.8) are $\langle x|\alpha_{yz}|0\rangle$, $\langle y|\alpha_{xz}|0\rangle$, and $\langle z|\alpha_{xy}|0\rangle$. Moreover, these three elements are equal in magnitude, and may be isolated outside the square as a vibrational factor in m^2.

Used in the more relaxed form, that the product of the symmetries of the two vibrational wavefunctions must contain the symmetry of some component of the polarizability, this rule is a well-known vibrational selection rule, depending on the vibrational quantum numbers only. Used as defined above, this rule is a "selection rule" on each term in (2.8), which may become a genuine rotation-vibrational selection rule if (2.8) is reduced. Again this will be discussed further below.

The rotation-vibrational wavefunctions of the two states involved in a Raman transition must have identical rotation-vibrational symmetry in the molecular point group. This leads to a number of rotation-vibrational selection rules, depending on the point group of the actual molecule. The more important of these rules will be discussed below.

2.1.3 The Vibrational Matrix Element

The polarizability depends on deformations of the nuclear framework of the molecule. To account for this, each component of the polarizability, or the operators $T_{\sigma}^{(\omega)}$, may be expressed as a series expansion at the equilibrium configuration, in exact analogy to the usual expansion of the dipole moment in the theory of infrared spectroscopy. As variables it is convenient to use the dimensionless normal coordinates.

If this expansion is inserted in the vibrational matrix element appearing in (2.8), the following series expansion results

$$\begin{aligned}\langle V'|T_{K'-K}^{(\omega)}|V\rangle &= \left(T_{K'-K}^{(\omega)}\right)_{eq} \langle V'|V\rangle + \sum_t \left(\partial T_{K'-K}^{(\omega)}/\partial q_t\right)_{eq} \langle V'|q_t|V\rangle \\ &+ \sum_{t,u} \left(\partial^2 T_{K'-K}^{(\omega)}/\partial q_t \partial q_u\right)_{eq} \langle V'|q_t q_u|V\rangle \\ &+ \text{higher order terms} \quad . \end{aligned} \tag{2.14}$$

Note that all the derivatives are taken at the equilibrium configuration. As pure rotational spectra are excluded from the present discussion, the overlap integral in the first term, and thereby the whole term, vanishes. The matrix elements in the remaining terms also vanish, except in the following cases: If q_i is the dimensionless normal coordinate of a one-dimensional (non-degenerate) factor in $|V\rangle$, and if $v_i' = v_i + 1$ and all other $v_t' = v_t$, and all $\ell_t' = \ell_t$, then

$$\langle V'|q_i|V\rangle = [(v_i+1)/2]^{1/2} \quad . \tag{2.15}$$

If $q_j^+ = q_{ja} + iq_{jb}$ and $q_j^- = q_{ja} - iq_{jb}$ are the dimensionless coordinates of a two-dimensional (two-fold degenerate) factor in $|V\rangle$, and if $v_j' = v_j + 1$ and all other $v_t' = v_t$, then

$$\langle V'|q_j^+|V\rangle = -[(v_j+\ell_j+2)/2]^{1/2} \quad \text{if} \quad \ell_j' = \ell_j + 1 \tag{2.16}$$

and

$$\langle V'|q_j^-|V\rangle = -[(v_j-\ell_j+2)/2]^{1/2} \quad \text{if} \quad \ell_j' = \ell_j - 1 \tag{2.17}$$

and all other $\ell_t' = \ell_t$. In all three cases the theoretical possibility of $v_i' = v_i - 1$, etc. has been left out as irrelevant in practice.

The corresponding expressions for the non-vanishing cases of $\langle V'|q_t q_u|V\rangle$ will not be given. It is enough to state that the corresponding "selection rules" for v_t and v_u are $v_t' = v_t \pm 1$ and $v_u' = v_u \pm 1$ (or $v_t' = v_t \pm 2$ if t and u are identical), and all other $v' = v$.

This is sufficient to show that, for a given set of quantum numbers V'V, at the most one term in (2.14) will be non-vanishing. Further, experimental evidence suggests that the expansion converges very rapidly, so that the second order terms normally are insignificant and the higher order terms safely can be neglected. Consequently, to a good approximation we may state that

$$\langle V'|T^{(\omega)}_{K'-K}|V\rangle = \left(\partial T^{(\omega)}_{K'-K}/\partial q_t\right)_{eq} \langle V'|q_t|V\rangle \quad , \tag{2.18}$$

where $\langle V'|q_t|V\rangle$ is given by (2.15), (2.16), or (2.17), depending on the changes in the vibrational quantum numbers as indicated at each expression.

The derivative appearing in (2.18) is the basic intensity parameter of vibrational and rotation-vibrational Raman spectroscopy. In principle, it can be determined experimentally from rotation-vibrational Raman spectra, but in practice only proportions between such quantities are established, using the present experimental technique. This point will be discussed further in the following sections.

2.1.4 Linear Molecules

Linear molecules are exceptional in the sense that the total angular momentum about the molecular axis always equals the vibrational angular momentum about that axis. This leads to the restriction

$$K = \ell = \sum_j \ell_j \quad , \tag{2.19}$$

so that the rotational quantum number K is fixed to a value ℓ equal to the sum of the vibrational quantum numbers ℓ_j. For linear molecules (2.1) may be written as

$$\psi = \sum_{V,L} c_{VL}|VL\rangle|J\ell M\rangle \tag{2.20}$$

where L is the vector of all the ℓ_j quantum numbers. One may use the simple basis of this expression in solving the rotation-vibrational problem, but a further good quantum number τ is obtained by using a symmetrized basis

$$|VLJ\ell M\tau\rangle = 2^{-1/2}[|VL\rangle|J\ell M\rangle + (-1)^{\rho+\tau}|V{-L}\rangle|J{-\ell}M\rangle] \quad , \tag{2.21}$$

where the sign of all the ℓ_j quantum numbers has been changed from $|VL\rangle|J\ell M\rangle$ to $|V{-L}\rangle|J{-\ell}M\rangle$. The quantity ρ is conveniently defined [2.9] as $J+\ell$ for $C_{\infty v}$ molecules and as J plus the sum of all v quantum numbers referring to a vibration of Σ_u^+ symmetry for $D_{\infty h}$ molecules. The new quantum number τ may take the values 0 and 1. In exchange for this, the same basis function (except for a possible insignificant change of sign) is obtained for the two sets of quantum numbers $VLJ\ell M$ and $V{-L}J{-\ell}M$. Only for all $\ell_j = 0$ this expression does not apply and one uses the simple basis function

$$|VOJOM\tau\rangle = |VO\rangle|JOM\rangle \quad , \tag{2.22}$$

where τ is conveniently defined as equal to ρ modulus 2. Clearly any wavefunction computed in terms of this basis may easily be expressed in the form (2.1) and the general intensity expressions may be applied.

The general "selection rules" given by (2.12,13) may in the present case be expressed as

$$|\ell'-\ell| \leqq \omega \tag{2.23}$$

and

$$J'-J \quad \text{even, if} \quad \ell = \ell' = 0 \quad . \tag{2.24}$$

For the new quantum number τ the selection rule is

$$\tau' = \tau \tag{2.25}$$

in all cases. Finally for $D_{\infty h}$ molecules there exists the rule

$$\sum v_u' = \sum v_u \quad \text{modulus } 2 \tag{2.26}$$

where v_u is the v quantum of any vibration of Σ_u^+ or Π_u symmetry. This is a well-known vibrational selection rule for molecules having a center of symmetry.

It is often a good approximation to neglect the interaction between different vibrational levels. Then the wavefunctions are given by (2.21) or (2.22). In practice we are only concerned with states having at the most one quantum number $\ell_j \neq 0$. Wavefunctions of the form (2.22) then apply to states with $\ell = 0$ (vibrational symmetry Σ) and wavefunctions of the form (2.21) apply to states with $\ell \neq 0$ (vibrational symmetry Π, Δ, etc.).

The expression (2.8) for m^2 then contains one, two or four terms, but excluding the rare case of $|\ell| = |\ell'| = 1$, (2.8) can always be reduced [2.9] to the simple form

$$m^2 = g_{ns}(2J+1)\ C(J\omega J';\ell\ \ell'-\ell)^2 |\langle V'L'|T^{(\omega)}_{\ell'-\ell}|VL\rangle|^2 \quad . \tag{2.27}$$

Strictly speaking an extra factor of 2 should be added if ℓ equals zero in one state and not in the other, but such a constant factor is normally without importance. For states with $\ell \neq 0$, the same wavefunction is indicated for both signs of ℓ. In order to use (2.27) one has to choose the signs of ℓ and ℓ' so (2.23) is satisfied. The value of m^2 as calculated from (2.27) is invariant to this choice of sign.

In this approximation the energy may be expressed in closed form [2.14], and it is no problem to compute the theoretical contour of any vibrational band. As examples we shall here discuss the structure of two fundamental bands, one with a Σ upper state ($\ell'=0$), the other with a Π upper state ($|\ell'|=1$).

The first example is thus a vibrational band with the vibrational ground state as the lower state and a fundamental vibrational state (one v=1, all other =0) of symmetry $\Sigma^+(C_{\infty v})$ or $\Sigma_g^+(D_{\infty h})$ as the upper state. Then $\ell = \ell' = 0$, and two different operators $T_0^{(0)}$ and $T_0^{(2)}$ are possible in the vibrational matrix element. Thus ω may take both of the values 0 and 2, which means that both isotropic and anisotropic scattering may be observed if the intensity expression (2.7) applies to the actual experiment.

For the isotropic component of this band the selection rule (2.10) allows $\Delta J = 0$ only, i.e., the isotropic component consists only of a Q-branch. The value of m^2 is given by

$$m^2 = g_{ns}(2J+1) \cdot (\partial T_0^{(0)}/\partial q_t)_{eq}^2/2 \tag{2.28}$$

where q_t is the dimensionless normal coordinate of the upper state vibration. As usual the energy expression indicates a small change in frequency with J, leading to a fairly sharp Q-branch.

For the anisotropic component of this band the selection rules (2.10,24) allow $\Delta J = 0, \pm 2$. The anisotropic component thus consists of an O-, a Q-, and an S-branch. With respect to frequency the Q-branch is identical to the isotropic one, but m^2 is given by a different expression

$$m^2 = g_{ns}(2J+1)J(J+1)(2J-1)^{-1}(2J+3)^{-1}(\partial T_0^{(2)}/\partial q_t)_{eq}^2/2 \tag{2.29}$$

which causes a slightly different intensity profile for low J values. The O- and S-branches give the usual double-wing pattern of the band with a spacing of approximately 4B. The expressions for m^2 for these branches is very similar to (2.29) and can easily be obtained from Table 2.1. For $D_{\infty h}$ an intensity alternation may be observed in the O- and S-branches and possibly in the Q-branch(es), because g_{ns} depends on the parity of J.

The O- and S-branches provide excellent possibilities for determining constants like ν_0, B_0, ΔB, D, etc. If the isotropic component is allowed by the experiment, it is also possible to determine the proportion between the absolute values of the two polarizability derivatives appearing in (2.28,29).

The second example is a vibrational band with the vibrational ground state as the lower state and a fundamental vibrational state of symmetry $\Pi(C_{\infty v})$ or $\Pi_g(D_{\infty h})$ as the upper state. In this case $\ell = 0$ and $|\ell'| = 1$. We may as well choose $\ell' = 1$. Then only one operator $T_1^{(2)}$ is possible in the vibrational element, and only one anisotropic (ω=2) component appears. The selection rules (2.10,24) give $\Delta J = 0, \pm 1, \pm 2$ so that five branches are possible in this case. The selection rule $\Delta\tau = 0$ is important because ℓ-type doubling splits apart the levels with different values of τ (but identical values of J).

The Q-branch forms, as usual, a sharp feature at the center of the band. However, as seen from Table 2.1, the value of m^2 is roughly proportional to J^{-1}, whereas usually m^2 is roughly proportional to J. The result is an extremely weak Q-branch which looks almost as a single line in one of the other four branches. These form two double wings with spacings roughly equal to 4B for the O- and S-branches, and 2B for the P- and R-branches. The intensity progress is practically the same within each branch, but because of the larger spacing, the O- and S-branches persist further out from the band center. The ℓ-type doubling in the upper state causes, just as in the corresponding infrared case, the apparent upper-state B-value to depend on the parity of ΔJ.

The O-, P-, R-, and S-branches may yield values for the constants ν_0, B_0, ΔB, q, D, etc. In principle the parameter $|(\partial T_1^{(2)}/\partial q_j^+)_{eq}|$ could be determined, but in practice one does not measure absolute intensities in high-resolution work. This parameter is treated as an insignificant factor in the intensity expression for all lines of the band.

2.1.5 Symmetric Top Molecules

In order to clarify the text only molecules of C_{3v} symmetry will be treated explicitly. This choice includes, however, most of all known symmetric tops. The corresponding theory for other molecular symmetries is well-known [2.9]. Sufficient details for other symmetric top molecules will be given in the result section. MILLS [2.15] has discussed the Raman selection rules for symmetric tops, based on a more general statement by HOUGEN [2.10]. BRODERSEN et al. [2.16] have described in detail the states and transitions of C_{3v} molecules.

Compared to linear molecules, symmetric tops are more complicated to discuss because K is a genuine quantum number, independent of ℓ. Thus, whereas ΔJ and J (and ω) are the only variables within a given band of a linear molecule, for a symmetric top ΔK and K are added to the list of variables, making the band structure much more complex.

The rotation-vibrational wavefunctions of a C_{3v} molecule are of two types. For each basis function $|V\rangle|JKM\rangle$ appearing in (2.1) the symmetry is defined by the quantity

$$\phi = K - \ell + 3t = -1,\ 0,\ \text{or}\ +1 \qquad (2.30)$$

where ℓ as above is the sum of all the ℓ_j quantum numbers. Each wavefunction may be written in the form (2.1) but in each linear combination occur only basis functions with one value of ϕ. Thus ϕ is a good quantum number, which may be used to characterize the rotation-vibrational wavefunction in the general case.

If in a wavefunction with $\phi = +1$, the sign is changed on K and on all the ℓ_j quantum numbers in all the basis functions in the linear combination, a wavefunction with $\phi = -1$ is obtained. These two wavefunctions are degenerate and possess E symmetry. Each of them has a spin statistical weight of $g_{ns}^{(E)}$, and the rotation-vibrational level has a total degeneracy of $2 \cdot g_{ns}^{(E)} \cdot (2J+1)$.

For wavefunctions with $\phi = 0$ a further good quantum number $\tau = 0$ or 1 is obtained (as for linear molecules with $\ell \neq 0$) by use of symmetrized basis functions

$$|VLJKM\tau> = 2^{-1/2}[|VL>|JKM> + (-1)^{\rho+\tau}|V\text{-}L>|J\text{-}KM>] \quad , \tag{2.31}$$

where ρ is defined as $J + K$, assuming no A_2 vibrations to exist for the molecule in question. Only if K and all ℓ_j are zero, is $|V0>|J0M>$ used as a basis function with $\tau = \rho = J$. A wavefunction which is a linear combination of such basis functions has A_1 symmetry if $\phi = 0$ and $\tau = 0$, but A_2 symmetry if $\phi = 0$ and $\tau = 1$. In both cases the spin degeneracy is $g_{ns}^{(A)}$. Often the splitting between levels only differing in τ is negligible. The combined level has then $A_1 + A_2$ symmetry and a total degeneracy of $2 \cdot g_{ns}^{(A)} \cdot (2J+1)$.

The spin statistical weights mentioned are the coefficients in the reduction of the symmetry of the nuclear spin wavefunctions in the C_3 rotation subgroup of the C_{3v} molecular point group

$$\Gamma_{ns} = g_{ns}^{(A)} \cdot A + g_{ns}^{(E)} \cdot E_a + g_{ns}^{(E)} \cdot E_b \quad . \tag{2.32}$$

It is now clear that any wavefunction of a C_{3v} molecule may be written in the form (2.1) so that (2.8) may be used as the expression for m^2 in all cases. The corresponding general selection rules are

$$\phi' = \phi \quad , \tag{2.33}$$

$$\tau' = \tau \quad , \tag{2.34}$$

and the general rules (2.10,11) on J and J'. In addition, the appearance of each term in (2.8) is conditioned by (2.12,13).

It is impossible to describe the structure of the Raman spectrum which may result in the completely general case. An example is shown in Fig.2.13 (Sect.2.3.3). However, in most cases the linear combination of basis functions in the wavefunction is sufficiently close to the "diagonal" approximation, discussed below, that the departures from this approximation may be described as perturbations of a given vibrational state by other vibrational states. Each rotation-vibrational state may then be described approximately by the quantum numbers of a single basis function, the intensity and frequency of each transition may be considered as perturbed from the values valid in the diagonal approximation, and the more restrictive selection rules valid for the diagonal approximation may "break down" due to the perturbation. Such a description may be very useful in the discussion of specific cases, but strictly speaking it is only an attempt to refer the general situation governed by the general intensity expression and the general selection rules given above to a limiting case where a sharper description is possible for the states and the transitions. The time has not yet come to attempt a general description of the influence of such "perturbations" on rotation-vibrational Raman spectra. Instead we shall concentrate on the diagonal approximation even though it seems clear that this approximation is sufficiently accurate for a full description of the experiment only in the minority of cases.

In the diagonal approximation the Hamiltonian matrix is supposed to be diagonal in terms of the above defined set of basis functions. Each wavefunction is then equal to a single basis function and the energy is given by a closed expression in the quantum numbers [2.14]. Again two wavefunctions with $\phi = +1$ and -1, respectively, have E symmetry and both of them may be described by one set of quantum numbers VLJKM or V-LJ-KM. If $\phi = 0$, one wavefunction is described by the set of quantum numbers VLJKMτ or by V-LJ-KMτ and two different states are described if τ is changed. In order to avoid confusion it is usually convenient to choose one set of ℓ_j values for each vibrational state or, if all $\ell_j = 0$, to use only non-negative values of K.

Just as for linear molecules, the expression for m^2 may in this approximation be reduced to the simple form

$$m^2 = g(2J+1)\ C(J\omega J';K\ K'-K)^2 |\langle V'L'|T^{(\omega)}_{K'-K}|VL\rangle|^2 \qquad (2.35)$$

where $g = 2g_{ns}$ or g_{ns} according to well-known rules [2.9].

In this approximation the conditions (2.12,13) become genuine selection rules. Similarly (2.33) may be written out as

$$\Delta K = K' - K = \ell' - \ell + 3t \qquad (2.36)$$

where t may be any integer confining ΔK to the range between $-\omega$ and $+\omega$, as required by (2.12). This selection rule indicates an important restriction on the possible values of ΔK, leading to a simplification in the structure of vibrational Raman bands.

If $\Delta\ell = 3t$, this rule requires $\Delta K = 0$, and ω may be either 0 or 2 as indicated by (2.12). In this case the band may consist of two components: one anisotropic, characterized by $\omega = 2$ and $\Delta K = 0$, and one isotropic, characterized by $\omega = 0$ and $\Delta K = 0$. The actual appearance of the isotropic component depends, as usual, on the experimental conditions.

If $\Delta\ell \neq 3t$, ΔK may have two values. for instance $\Delta K = +1$ and $\Delta K = -2$, but then only $\omega = 2$ is possible according to (2.12). In this case the band will also consist of two components, but both are anisotropic and they differ in the value of ΔK only.

Within a certain component, the vibrational matrix element factor in the intensity expression is a constant for all transitions. The variable part is

$$I' = g_{ns} \cdot (2J+1)\; C(J\omega J';K\;\Delta K)^2 \cdot \exp(-E''/kT) \quad . \tag{2.37}$$

If the band consists of two components, this expression has to be multiplied by a separate factor for each component. A calculation of the band contour will thus depend on one intensity parameter which essentially is the proportion between the absolute values of the derivatives of the two operators.

An anisotropic component will always consist of five branches with $\Delta J = 0, \pm 1$, and ± 2, whereas an isotropic component will consist of one branch only with $\Delta J = 0$. Within a certain branch each transition is characterized by a set of values for J and K of the lower state. Both the intensity and the frequency are thus functions of the two variables J and K. If wanted, the frequency expression may easily be worked out from the well-known energy expressions [2.14]. The intensity expression is obtained from (2.37) by insertion from Table 2.1 of the expression for $C(J\omega J';K\;\Delta K)$ for the values of ΔK and ΔJ of the component and branch in question, and of the relevant expression for the lower state energy E". The factor g_{ns} depends on ϕ as discussed above and thereby on K.

There is obviously no problem in computing the frequency and intensity of each transition within a given branch, and by means of suitable summation and convolution procedures to compute the contour of any given band. In order to "understand" the calculated structure, it may, however, be convenient to discuss the intensity and frequency functions of each branch in more detail.

The intensity expression given in (2.37) may be considered as consisting of three factors: the factor $(2J+1)\; C(J\omega J';K\;\Delta K)^2$ is a function of both J and K, but it depends only on the values of ω, ΔK, and ΔJ, not on the particular molecule. There is thus a limited number of such functions (in total 26). They are, in fact, indicated in Table 2.1. The Boltzmann factor is also a function of both J and K, but it depends only on the molecular constants for the lower state of the molecule, not on ω, ΔK, ΔJ or the upper state constants. The product of the last two factors in (2.37) (equal to I'/g_{ns}) is a continuous function of J and K. For given values of ω, ΔK, ΔJ and the energy constants of the lower state, this function may be visualized by plotting contour lines in the J,K-plane, as shown in Fig.2.1. To obtain the real intensity, this function has to be multiplied by $g_{ns}^{(A)}/g_{ns}^{(E)}$ for every third K value. The contour lines of the frequency function may be plotted in the same way, as also shown in Fig.2.1. From this plot one may easily visualize for which values of J and K the lines in that particular branch are most intense, and at what frequencies they will appear. Especially one may check if series of strong lines fall on top of each other, forming prominent band-heads in the spectrum, and whether an intensity alternation should be observable due to the alternation of the g_{ns} values.

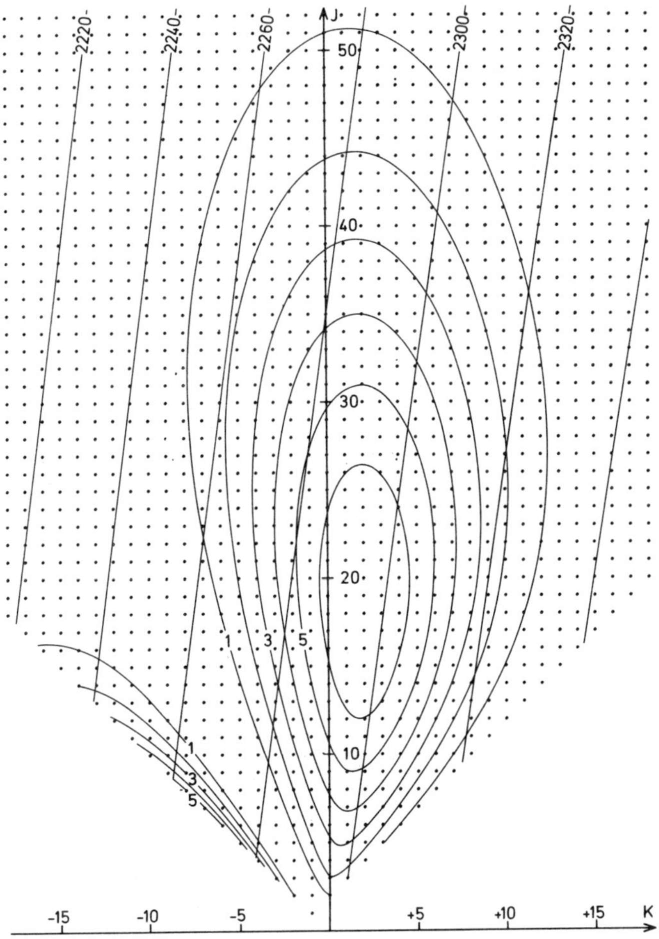

Fig. 2.1. Contour lines for the intensity function I'/g_{ns} (curved lines) and for the frequency function ν (almost straight lines), computed for the ^{R}P branch of the ν_4 fundamental band of CD_3Dι. Values of the functions are written on the contour lines. Only the whole-numbered points indicated by a dot have a physical meaning

This section will be concluded by a short description of the structure of the two types of fundamental bands in C_{3v} molecules, involving an A_1 and an E vibrational upper state, respectively.

The A_1 fundamental band is an example of the $\Delta K = 0$ type of band discussed above. Figure 2.2 shows a computed contour of the central part of the anisotropic component of such a band. In each of the five branches there is a coarse J-structure with spacing approximately equal to $2B|\Delta J|$, and a fine K-structure which is difficult to resolve.

The E fundamental band is an example of the second type of band in which two anisotropic components occur with $\Delta K = +1$ and -2, respectively. The diagram in Fig.2.3 displays the transitions of both components. For the upper vibrational state ℓ' is

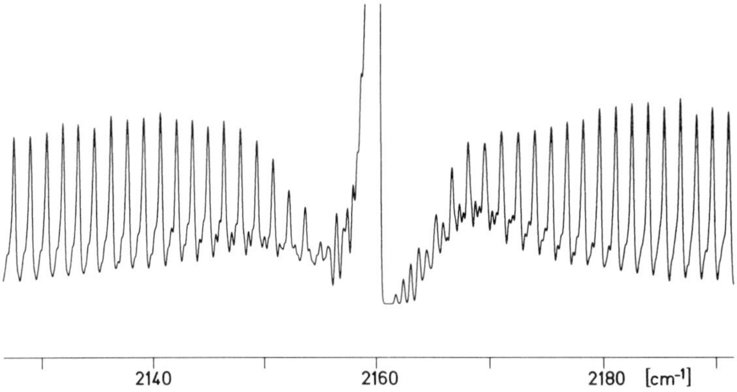

Fig. 2.2. Theoretical contour of the central part of the anisotropic component of the ν_1 vibrational band of CD_3Cl, convoluted with a Gaussian instrument function of half-width 0.25 cm^{-1}. Experimentally determined molecular constants [2.17-19] were used. All five branches start from the band origin at 2160.278 cm^{-1}. The Q-branch forms a fairly sharp feature, the P- and R-branches stop around 2135 and 2180 cm^{-1}, respectively, whereas the O- and S-branches go on to about 2090 and 2230 cm^{-1}, respectively

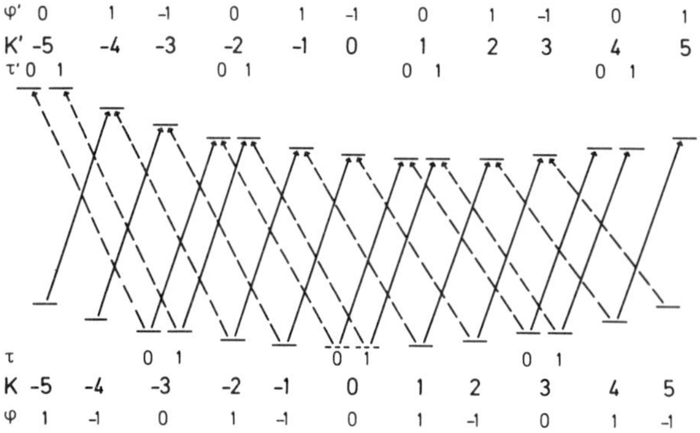

Fig. 2.3. Diagram of the transitions in an E fundamental band of a C_{3v} molecule. Each horizontal line represents a stack of levels with different values of J. The transitions of the $\Delta K = +1$ and $\Delta K = -2$ components are indicated by a full line and a dashed line, respectively. (Reproduced from BRODERSEN et al. [2.16] by permission)

chosen as +1, and K' takes all values from -J' to +J' in order to include all the rotational states. For the lower vibrational state ℓ equals zero, and all the rotational states could be described by means of the non-negative values of K. It is, however, convenient to use both signs for K in the computation as well as in the diagram, it only means that each rotational state with $K \neq 0$ appears twice in the diagramm.

A computed contour for such a band is shown in Fig.2.14 (Sect.2.3.3). In Fig.2.4 are given the contours for each component of this band, and for the $\Delta K = +1$ component also for each of the five branches separately. In the $\Delta K = -2$ component, the Q-branch

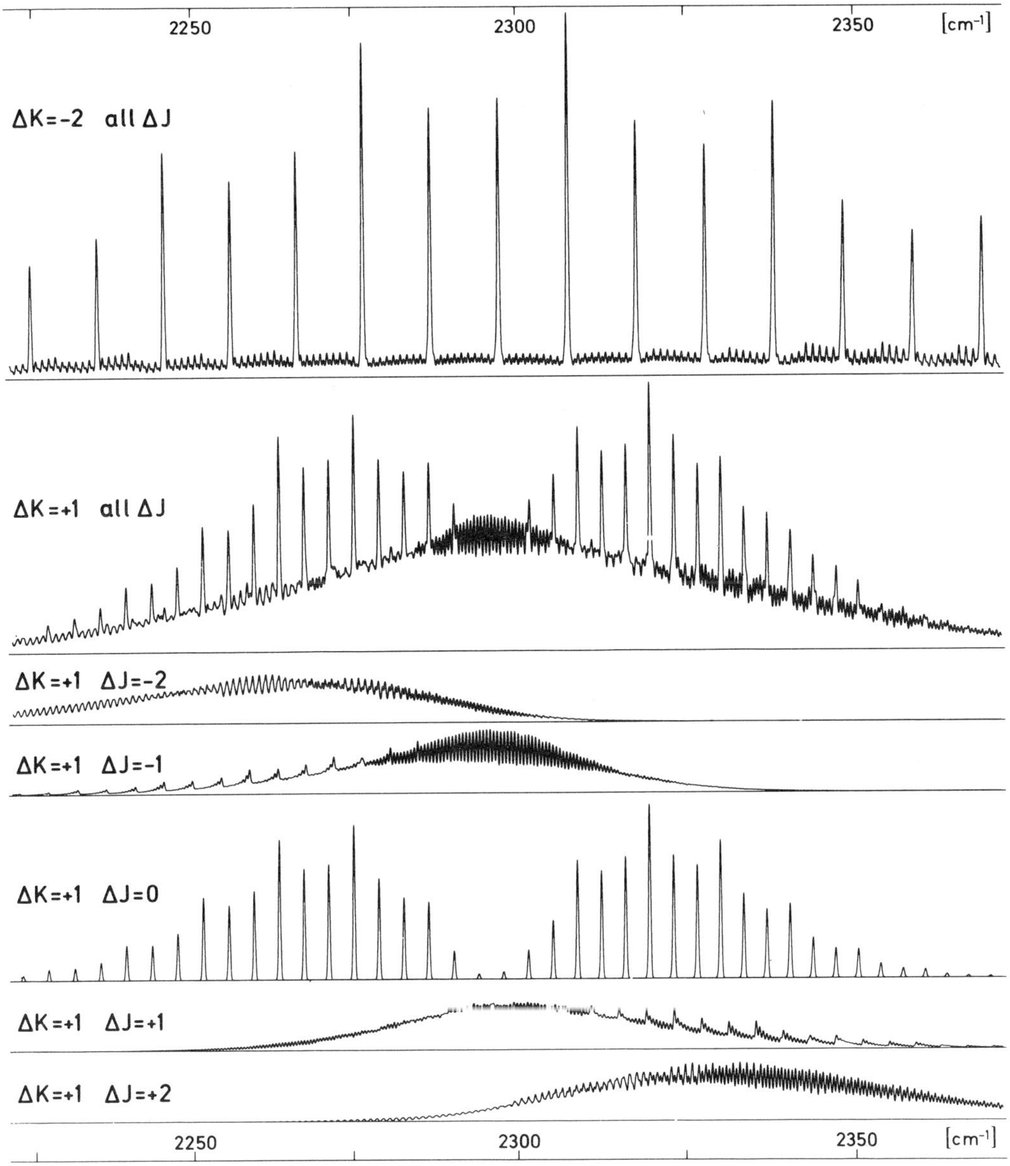

Fig. 2.4. Theoretical contours for the two components of the ν_4 band of CD_3Br (see Fig.2.14, Sect.2.3.3) and of each branch in the $\Delta K = +1$ component. (Reproduced from EDWARDS and BRODERSEN [2.80] by permission)

is the all-dominating feature, giving a series of very sharp line-shaped features. Each of these accumulations is characterized by one value of K (or K'), as indicated in Fig.2.14, and has a J fine structure. Consequently the intensity alternation due to g_{ns} is prominent. In the $\Delta K = +1$ component, the Q-branch has the same general structure but with a different spacing and an intensity minimum near the band center.

The remaining four branches of the $\Delta K = +1$ component form a semi-continuous background, in which the occurrence of discernible features depends very sharply on the values of the energy constants of the particular molecule.

In the example of Fig.2.4, the P- and R-branches for $\Delta K = +1$ show a series of unsymmetrical features due to the presence of intense transitions for values of J and K near to the lines $J = K$ or $J = -K$, indicating the limit of the physically relevant part of the J,K-plane (see Fig.2.1). When the contour lines of the frequency function cut this border line within a region of intense transitions, such an unsymmetrical structure will result. In the P-branch is seen an example of another kind of fine structure. It is a typical beat-structure with a spacing of 2B. The explanation is also given by Fig.2.1 where the slope of the contour lines of the frequency function is very close to 7 within a region of relatively intense lines. This means that transitions with $J = J_0 + 7K$ will have very nearly the same frequency, resulting in a band-head formation. The neighbouring band-heads are formed for $J = J_0 \pm 1 + 7K$, resulting in a spacing in the spectrum of nearly 2B.

The presence of the two components in a band of this type allows a determination of the A_0 and D_0^K constants as discussed in the Sect.2.3.3.

2.1.6 Asymmetric Top Molecules

If the interaction between different vibrational states cannot be neglected, the intensity within the spectrum of an asymmetric top molecule must be calculated by means of (2.8). The resulting structure is very complex and not much can be said about this general case.

Instead we shall in the following consider only those cases where the interaction between vibrational states may be neglected, and further we shall primarily discuss the molecules of symmetry C_{2v} or D_{2h}. DEROCHE et al. [2.20] have presented intensity expressions and selection rules for certain bands of a C_{2v} molecule. Recently, HILLS et al. [2.21] have discussed the intensity expressions and the selection rules for any asymmetric top.

The rotation-vibrational wavefunctions are in this approximation of the form

$$\psi = |V\rangle \cdot \sum_K c_K |JKM\rangle \quad , \tag{2.38}$$

where the rotational part and the corresponding energy must be determined by diagonalization of the Hamiltonian matrix. If a Wang basis is used, each rotational function is obtained with a definite symmetry in the D_2 point group. When rewritten in the form (2.1) the coefficients are related by

$$c_K = (-1)^{\gamma} c_{-K} \tag{2.39}$$

where γ is the sign parameter in the Wang functions. The values of the coefficients

c_K depend on the choice of correlation between the molecular-fixed axes a, b, c and the axes x, y, z used for the symmetric top basis functions.

The symmetry of the rotational wavefunctions in the D_2 group is given by

$$C_2^{(z)}|R\rangle = (-1)^K|R\rangle$$
$$C_2^{(y)}|R\rangle = (-1)^{J+\gamma+K}|R\rangle$$
$$C_2^{(x)}|R\rangle = (-1)^{J+\gamma}|R\rangle \quad . \qquad (2.40)$$

Thus each rotational wavefunction is a sum of Wang functions with a definite parity of K and $J+\gamma$.

A given vibrational transition is only allowed if the product of the vibrational symmetries in the molecular point group equals the symmetry of one of the components of the polarizability. The corresponding rotational selection rules are obtained by requiring the product of the symmetries of the two rotational wavefunctions to equal this same symmetry of the one component of the polarizability. However, this rule should be applied in the D_2 point group, and a correlation is needed to the molecular point group. D_2 is a subgroup of D_{2h} and the correlation is no problem for D_{2h} molecules. D_2 and C_{2v} are isomorphic and the correlation is obtained by replacing a σ operation by a C_2 operation about an axis perpendicular to the mirror plane. The rotational selection rule is easily converted into restrictions on the parities of $K'-K$ and of $J'-J+\gamma'-\gamma$ by means of (2.40), as indicated in Table 2.2 for the four types of bands possible for both kinds of molecules.

Table 2.2. Band types for C_{2v} and D_{2h} molecules

$\Gamma_V \times \Gamma_{V'}$ $\Gamma_R \times \Gamma_{R'}$ (in D_2)	Components of polarizability	Rotational Selection Rules	
		$K'-K$	$J'+\gamma'-J-\gamma$
A	$\alpha_{xx}, \alpha_{yy}, \alpha_{zz}$	even	even
B_z	α_{xy}	even	odd
B_x	α_{yz}	odd	even
B_y	α_{xz}	odd	odd

For the three B types of bands, only anisotropic scattering is possible (ω=2). The expression for m^2 is reduced to

$$m^2 = g_{ns}(2J+1)4|\sum_K c_K c_{K+1}'^{*} C(J2J';K\ 1)|^2|\langle V'|\alpha_{xz}|V\rangle|^2 \qquad (2.41)$$

for B_y type bands, to the analogous expression with α_{xz} replaced by α_{yz} for B_x type bands, and to

$$m^2 = g_{ns}(2J+1)4|\sum_K c_K c_{K+2}'^{*} C(J2J';K\ 2)|^2|\langle V'|\alpha_{xy}|V\rangle|^2 \qquad (2\ 42)$$

for B_z type bands. Note that all three expressions are of the simple form in which the vibrational matrix element appears as an over-all factor common for all transitions in the band. Also the sum over K and K' has been replaced by a sum over K only.

A totally symmetric (A type) band consists of two components, one isotropic with K'-K = 0 and one anisotropic with K'-K = 0 or ±2. The corresponding expressions for m^2 are

$$m^2_{iso} = g_{ns}(2J+1)|\sum_K c'^*_K c_K|^2|\langle V'|\alpha_{xx}+\alpha_{yy}+\alpha_{zz}|V\rangle|^2/3 \tag{2.43}$$

and

$$m^2_{an} = g_{ns}(2J+1)|\sum_K c'^*_K c_K \, C(J2J';K\,0)\langle V'|2\alpha_{zz}-\alpha_{xx}-\alpha_{yy}|V\rangle/6^{1/2}$$
$$+\sum_K c_K c'^*_{K+2} \, C(J2J;K\,2)\langle V'|\alpha_{xx}-\alpha_{yy}|V\rangle|^2 \quad . \tag{2.44}$$

Note that in the last expression the two vibrational matrix elements appear within the square. The intensity thus depends on the proportion between these quantities (with sign). If the experiment allows the presence of the isotropic component, the total intensity also depends on the proportion to the absolute value of the third vibrational matrix element.

It should be noted that the selection rules given in Table 2.2 depend on the choice of correlation between a, b, c and x, y, z. In principle this choice is free and no absolute meaning can be assigned to the resulting selection rules, not even to the rules on K'-K. However, many asymmetric tops are nearly symmetric, and the choice of the z axis becomes obvious. Correspondingly the selection rule on K'-K gains in absolute significance and in the limit it transforms into the usual selection rule on ΔK for a symmetric top and the band structure transforms into that of a band of a symmetric top determined by this selection rule on ΔK.

As usual each anisotropic component will consist of five branches and the isotropic component of the Q-branch only. In the general case of an asymmetric top molecule, the resulting structure of the band is very complex and cannot be described in simple terms because the energy is not given by a closed expression. HILLS et al. [2.21] have published a number of theoretical contours for molecules with $\kappa \simeq -0.9$, they show some systematic features, such as series of accumulations of Q-lines.

In the case of an asymmetric top molecule of lower symmetry than C_{2v} and D_{2h}, the symmetry condition leading to the rotational selection rules is only applicable in a subgroup of D_2. Thus, at the most, only one condition on the parity of either K'-K or J'-J+γ'-γ results, and only minor reductions are possible in the expression (2.8) for m^2. The intensity will depend on the proportion (including sign) between at least two parameters.

2.1.7 Spherical Top Molecules

Because of the high symmetry, the rotation-vibrational interaction problem of spherical top molecules is very complicated. We shall therefore only discuss the three types of fundamental bands in molecules of the methane type, neglecting interactions between vibrational states. In more general cases the Raman spectra must be computed by a direct application of (2.8). The three types of fundamentals are: the totally symmetric A_1 band, the doubly degenerate E band and the triply degenerate F_2 bands.

The basis functions used for the vibrational ground state and the A_1 upper state are simple products of the vibrational function and symmetric top rotational functions. The diagonalization leads to wavefunctions of the form (2.38), just as for asymmetric tops.

The rotational energy in these vibrational states depends mainly on J in the same way as for a linear molecule, but each J level is split in a number of closely spaced sub-levels, the so-called tetrahedral splitting. Each of the resulting sub-levels is best characterized by the symmetry of the rotation-vibrational wavefunction in T_d. To a good approximation the tetrahedral splitting follows a fixed pattern for all totally symmetric vibrational states. Only the over-all proportionality factor varies with the molecule and the vibrational state.

For a molecule of T_d symmetry, $T_0^{(0)}$ is the only totally symmetric operator, and therefore the only one possible for a totally symmetric fundamental. Consequently, in this case, only isotropic scattering occurs, and the expression for the transition moment reduces to

$$m_{iso}^2 = g_{ns}(2J+1)|\sum_K c_K'^* c_K|^2 \cdot <1|T_0^{(0)}|0>^2 \tag{2.45}$$

where the last factor because of the spherical symmetry of the polarizability reduces to

$$<1|T_0^{(0)}|0>^2 = 3(\partial\alpha_{xx}/\partial q_1)_{eq}^2/2 = 3(\partial\alpha_{yy}/\partial q_1)_{eq}^2/2 = 3(\partial\alpha_{zz}/\partial q_1)_{eq}^2/2 \quad . \tag{2.46}$$

In addition to the usual selection rules $\Delta J = 0$, the symmetries of the sub-levels are required to be the same in the upper and the lower state. The totally symmetric band should, in this approximation, consist of a single rather sharp Q-branch, possibly with some irregularities due to the tetrahedral splitting.

For the doubly degenerate E vibrational state the usual basis functions are products of a symmetric top rotational function and a product of two linear oscillator functions $|v_a>|v_b>$ for the two degenerate vibrations, with the condition $v = v_a + v_b = 1$. The wavefunctions are then again of the form (2.38) except that the sum includes a summation over the two possibilities for the vibrational function $|1>|0>$ or $|0>|1>$.

The rotational energy is again mainly as for a linear molecule, superimposed with a tetrahedral splitting. The sub-levels may again be characterized by a symmetry of the rotation-vibrational wavefunctions in T_d, but the splitting pattern is different from that of the totally symmetric vibrational states.

From the symmetry of the upper state, it is seen that only the three operators $T_2^{(2)}$, $T_0^{(2)}$, and $T_{-2}^{(2)}$ may occur in the expression for m^2. The scattering is thus fully anisotropic. If $|v_a\rangle$ and $|v_b\rangle$ are chosen so that they transform as $(2z^2-x^2-y^2)/6^{1/2}$ and $(x^2-y^2)/2^{1/2}$, respectively, all vibrational matrix elements are reduced to one, according to the rules given in Sect.2.1.2, and the expression for m^2 may be written as

$$m^2 = g_{ns}(2J+1) \cdot |\langle 01|\alpha_{xx}-\alpha_{yy}|0\rangle|^2 \cdot |\sum_K c_K\Big\{2^{1/2}\, c'^*_{aK}\, C(J2J';K\,0) + \Big[c'^*_{bK+2}+(-1)^{\Delta J}\, c'^*_{bK-2}\Big] C(J2J';K\,2)\Big\}|^2/4 \quad . \qquad (2.47)$$

Thus also in this case the intensity expression contains only one over-all intensity parameter.

In the first approximation an E fundamental band will look similar to Fig.2.2, but certain smaller changes should be expected due to the different intensity expression and to the tetrahedral splitting.

For the triply-degenerate F_2 fundamental state it is useful to consider three angular momentum vectors: $\underline{J}$ for the total, $\underline{R}$ for the rotational, and $\underline{\ell}$ for the vibrational angular momentum. These are related by

$$\underline{J} = \underline{R} + \underline{\ell} \quad . \qquad (2.48)$$

Each of these vectors is characterized by two quantum numbers for the total length and for the projection on the molecule-fixed z axis: J and K, R and K_R, ℓ and K_ℓ, respectively. In all cases we have

$$K = K_R + K_\ell \quad , \qquad (2.49)$$

and in the triply-degenerate fundamental state in which $\ell = 1$, there are the following possibilities for J and R

$$R = J-1,\ J,\ \text{or}\ \ J+1 \quad . \qquad (2.50)$$

These three cases are conventionally labeled by -, 0, and +, respectively. (In the other vibrational states discussed above $\ell = 0$, and $K_R = K$ and $R = J$ always.)

The basis functions usually used in the present case are constructed in the following way. First the vibrational function is chosen as a product of three linear oscillator functions $|v_x\rangle|v_y\rangle|v_z\rangle$ subject to the condition $v = v_x + v_y + v_z = 1$. The three degenerate functions are combined to give three new functions $|v\ell K_\ell\rangle$ as follows

$$|1\ 1\ -1\rangle = (|1\rangle|0\rangle|0\rangle - i|0\rangle|1\rangle|0\rangle)/2^{1/2}$$
$$|1\ 1\ \ 0\rangle = |0\rangle|0\rangle|1\rangle$$
$$|1\ 1\ +1\rangle = -(|1\rangle|0\rangle|0\rangle + i|0\rangle|1\rangle|0\rangle)/2^{1/2} \quad . \tag{2.51}$$

These vibrational functions may then be combined with symmetric top rotational functions to give the so-called "coupled" basis functions

$$|v\ell JRK_R\rangle = \sum_{K_\ell} (-1)^{R-J+K}\ C(J\ell R;K-K_\ell)|v\ell\rangle|JKM\rangle \tag{2.52}$$

where $v=\ell=1$ and $K=K_R+K_\ell$. The Clebsch-Gordan coefficient $C(J\ell R;K\ K_\ell)$ for $\ell=1$ is tabulated in standard textbooks, e.g., [Ref.2.13, Table 2^3].

Using these basis functions, the wavefunctions of a triply-degenerate vibrational state are obtained in the form

$$|v\ell J\rangle = \sum_{R\ K_R} c_{RK_R}\ |v\ell JRK_R\rangle \tag{2.53}$$

which may be reduced to the form (2.1) by introducing (2.51,52).

The rotational energy in this state depends primarily on J and R-J in the following way

$$E^-_{rot} = BJ(J+1) - 2B\zeta(J+1) + \ldots$$
$$E^0_{rot} = BJ(J+1) - 2B\zeta + \ldots$$
$$E^+_{rot} = BJ(J+1) + 2B\zeta J + \ldots \tag{2.54}$$

for $R-J=-1$, 0, and +1, respectively. But each level, characterized by J and R-J, is as usual split into sub-levels, labeled by the symmetry of the rotation-vibrational wavefunction in T_d.

For the F_2 fundamental band, the only polarizability elements with the proper symmetry are α_{xy}, α_{xz} and α_{yz}. Therefore only anisotropic scattering occurs. As usual the selection rule for J is $\Delta J=0, \pm1, \pm2$. There is no "selection rule" on ΔR, except what follows from (2.50), but only $\Delta K_R=\pm2$ is possible, which limits the number of terms in (2.8). As shown in Sect.2.1.2 all the vibrational matrix elements occurring in (2.8) are equal and the expression for m^2 may be reduced to

$$m^2 = g_{ns}(2J+1)\cdot|\langle x|\alpha_{yz}|0\rangle|^2$$
$$\cdot\left|\sum_{R'K}(-1)^{J'-R'+K}\left[c_K\,c'^*_{R'K+2} + (-1)^{J+R'}\ c_{-K}\,c'^*_{R'-K-2}\right]\right.$$
$$\left.\left[2^{1/2}\,C(J'1R';K+1\ 1)\cdot C(J2J';K\ 1) + C(J'1R';K+2\ 0)\cdot C(J2J';K\ 2)\right]\right|^2 \quad . \tag{2.55}$$

As usual the band consists of five branches, but each branch is split by the Coriolis term of (2.54) in three sub-branches, labeled by -, 0, and +, respectively. Further each "line" is split into an irregular pattern due to the tetrahedral splitting in both states. A calculated contour for a F_2 fundamental is shown in Fig.2.19 (Sect. 2.3.6).

The theory outlined above has recently been discussed by, among others, KATTENBERG [2.22], CHAMPION [2.23], HILICO et al. [2.24], and ROBIETTE et al. [2.25]. References to the older literature may be found in these papers.

2.1.8 Comparison with the Theory of Infrared Spectra

The theory of infrared absorption may be expressed in the same formalism as used here. The fundamental difference between the theory of Raman spectra and absorption spectra lies in the operators. In absorption the polarizability is replaced by the permanent dipole moment of the molecule, and the three operators $T_\sigma^{(\omega)}$ have $\omega = 1$. This leads to the following fundamental differences between infrared and Raman spectra:

First, the basic over-all intensity parameters in the two kinds of spectra are derivatives of the dipole-moment and the polarizability, respectively. One cannot expect these to be related in any simple way, so that the "vibrational intensity pattern" of the two kinds of spectra is expected to be quite different. For molecules with a center of symmetry the well-known "mutual exclusion rule" exemplifies this. For many other molecules there is a tendency towards an "opposite" intensity pattern.

Second, there is no analogy in absorption to isotropic and anisotropic scattering. With $\omega = 1$, absorption may rather be placed in between these two phenomena. This applies primarily to the selection rules depending on ω, such as (2.10-13,34) and those given in Table 2.2. But also the Clebsch-Gordan coefficient in (2.8) depends strongly on ω. Thus the intensity within a branch in infrared very seldom has two maxima (as in Fig.2.1). Roughly speaking, the value $\omega = 1$ leads to an infrared rotational structure in between those for isotropic and anisotropic Raman scattering, both with respect to complexity and to the amount of information which can in principle be extracted.

These differences between the two kinds of spectra is the reason why high-resolution Raman spectroscopy is a valuable supplement to high-resolution infrared spectroscopy.

2.2 Experimental Technique

In the following a general description will be given of the instrumentation and technique used at present to obtain high-resolution rotation-vibrational Raman spectra. This will be followed by a discussion of the various instrumentations used recently for this purpose. In order to clarify the text, it is not attempted to give all details, these may be found in the references.

From the outset it is important to stress that the fundamental trouble in rotation-vibrational Raman spectroscopy is the low intensity of the scattered light. Naturally the overall intensity may vary very much from one vibrational band to the other and from one molecule to the other, but in general the intensity of rotation-vibrational spectra is lower compared to the intensity of pure rotational spectra by orders of magnitude. Often the intensity of the scattered light is at the limit of what can be detected with present techniques. One can to some extent compensate for the low intensity by a sacrifice of resolution, but for a given instrument these possibilities are limited. If a certain instrument is designed for the highest resolution it may not be useful for very weak spectra, and vice versa.

The word "resolution" is unfortunately used with different meanings by different authors. Sometimes it means the slit width recalculated in wavenumbers, or the resolving power of the spectrograph alone. Here, it is always used to indicate the resolution observed in the spectrum, either as the difference in frequency between two lines observed with a minimum in between, or as the width at half height of an isolated line. Note that a high resolution is given by a small value of this width.

2.2.1 General Features of Instrumentation and Technique

A standard equipment for recording and measuring high-resolution rotation-vibrational Raman spectra may consist of the following units: a light source, a sample cell, some imaging optics, a premonochromator, a source of a calibration spectrum, a grating spectrograph, a photographic plate as a detector, some photographic processing devices, a microphotometer, a comparator, some data input device, and a computer.

In the following a short and general description will be given of each of these units.

As a *light source* the Ar^+ ion laser is now the only one in use. Usually the strong 488 nm laser line is preferred, but the 514.5 nm line may also be used in special cases. If one aims at the highest resolution, it is necessary to single-mode the laser by means of a quartz plate in the cavity acting as a Fabry-Perot. However, this about halves the power of the laser and for the long exposures a servo loop is needed for the tilt of the quartz plate in order to keep the laser adjusted at maximum power. If a free running laser is used, the natural width of the laser line is about 0.15 cm^{-1} which may contribute significantly to the observed resolution.

The *sample cell* may be placed inside or outside the cavity. The intensity of the scattered light is directly proportional to the intensity of the exciting light, and this is generally much higher inside than outside the cavity. A fairly high factor in intensity may thus be gained by placing the cell inside the cavity, provided the losses due to the cell are low. Such losses are partly compensated for by the fact that no output mirror is needed when the cell is inside the cavity. By definition (see Sect.2.1) the sample does not absorb the laser radiation and the energy loss due to the scattering is negligible. Consequently the cell windows are the only source of losses. To minimize the losses one may use Brewster-angle windows or highly anti-reflection coated windows, or the mirror in one end of the cavity may replace one window.

A further gain in intensity is obtained by either focusing or multipassing the beam within the cell. However, especially if the beam is focused inside the cavity, it is necessary to insert a filter to prevent ultraviolet radiation from the discharge section of the laser to reach the cell. Otherwise a photolysis will cause solidification of almost any sample.

The choice between focusing and multipassing of the beam is closely connected to the design of the *imaging optics* used to send the scattered light to the spectrograph. The law of conservation of the etendue (acceptance) imposes severe restrictions on what can be transmitted through the spectrograph to the photographic plate. The limiting apertures in the spectrograph (usually the slit and the grating) may be imaged in the vicinity of the cell in different ways but the etendue cannot be changed. This quantity may be defined as the area of one image times the solid angle of the other image as seen from a point in the first image. Only light passing through both images in the right direction will pass the spectrograph, but a mirror may reverse the light travelling in the wrong direction and a factor of 2 is gained.

For a given etendue of the spectrograph there are many possible arrangements of the beam inside the cell and of the imaging optics. Examples will be given below. There is no clear decision possible at the moment of the best arrangement or of how large the difference in efficiency is between the different arrangements.

It is a great advantage if the laser, the cell and the imaging optics are arranged so that two different kinds of experiments are possible, corresponding, respectively, to the two intensity expressions (2.6,7), so that either pure anisotropic scattering or a mixture of isotropic and anisotropic scattering is observed. In the case where the isotropic component is absent, or rather minimized, the intensity of the very strong Rayleigh line is also minimized, as shown by WEBER et al. [2.26].

The Rayleigh line gives rise to scattered light which may interfere with weak vibrational bands. This may be prevented by preceding the spectrograph with a *premonochromator* which also serves as an order-sorter. The highest efficiency is obtained by use of a double-monochromator of the non-dispersive type [2.27], but a single-monochromator or even an interference filter may also be used.

The spectrograph is further preceded by some arrangement to illuminate the slit by the *source of a calibration spectrum*, usually a thorium hollow-cathode lamp. Preferably the slit and the grating should be filled in exactly the same way as is done by the Raman light, in order to have the same aberrations in the two kinds of spectra. This requires that the position of the photographic plate can be shifted with great precision between exposures, allowing for instance one comparison spectrum to be exposed above the Raman spectrum and one below, the one exposed before, the other after the Raman exposure.

The *grating spectrograph* should have a large dispersion at the plate to yield high resolution (the theoretical resolving power of the grating is usually not the limiting factor) and a large etendue in order to transmit as much energy as possible. These two requirements are contradictory. By a comparison of the different instrumentations described below, one finds that the resolution obtained in practice is almost inversely proportional to the focal length of the spectrograph, but the intensity of the faintest line to be detected is roughly proportional to the focal length. Thus the focal length seems to be the parameter which distinguishes the different kinds of instrumentations mentioned above.

A large ruled grating with a blaze angle of about 63° seems preferable. The optical arrangement is either a Littrow or a Czerny-Turner type. A cylindrical lens may be placed in front of the photographic plate in order to decrease the length of the spectral lines. This does not affect the energy transmitted, but the rate of arrival of photons per unit area of the plate is increased, resulting in a higher detectability for weak lines because of the peculiar properties of the photographic process [2.28].

The main virtue of the *photographic plate* as a detector is its ability to integrate the signal over a long period of time. For rotation-vibrational spectra long exposure times are needed, from about 2 hours to 100 hours, the upper limit being set by the reciprocity failure of the photographic emulsion. The most sensitive photographic plates (for instance, Kodak spectroscopic plates type IIa-D) also have fairly large grains. This explains the need for a high dispersion in the spectrograph. The sensitivity may be increased by a factor of about 2.5 by baking the plate before use in a thermostated oven in a stream of pure nitrogen (for instance, 24 hours at 56°C) [2.29-32]. The drawback of this procedure is an increase in the overall blackening of the plate, decreasing the intensity range which can be covered in one exposure. The spectral lines may not appear exactly in the position on the developed plate where the light has fallen, especially if near to the edge of the plate. This dislocation may be reduced to 1-2 μm by use of special developers and fixers, as shown by GOLLNOW and HAGEMANN [2.33].

An automatic *photographic processing device* is convenient and ensures a reproducible handling. This is especially important if data from different plates should be subtracted, as discussed below.

A *microphotometer* is needed to convert the spectrum on the plate into a graphical representation. A *comparator* is used to measure the positions of the spectral lines in the Raman spectrum as well as in the comparison spectra. A visual comparator is only satisfactory for well-resolved spectra. In the more usual case, a significant improvement is obtained by use of a double display on an oscilloscope, as originally suggested by TOMKINS and FRED [2.34]. An even better instrument, a computer-linked photometer, may replace both the microphotometer and the comparator, delivering complete data directly to the computer.

If a more conventional comparator is used, some *data input device* is needed to transmit the data to the *computer*. This will first produce a calibration polynomial from the positions of the lines in the comparison spectra, and then compute the frequencies for each Raman line from its position.

The optimum *recording conditions* (sample pressure, slit width, exposure time, etc.) depend primarily on the intensity of the band to be recorded, but also on the substance, the instrument and the amount of details present in the spectrum. For most of the spectra reproduced in the following sections, the scanning conditions will be indicated. Here a few remarks will be given on the choice of sample pressure.

The intensity of a spectrum is proportional to the sample pressure. In order to keep down the exposure time and the slit width, it is thus important to use a high pressure. But also the pressure broadening is usually proportional to the pressure, and it must be kept under a certain level to obtain a given resolution. Unfortunately the pressure-broadening constant is only known in very few cases. In practice the sample pressure is often a more important variable than the slit width in order to obtain maximum resolution of a certain band at a given instrument. In principle, the best conditions can be found experimentally, but often this is too expensive and the choice depends on a guess. For non-polar substances it is often possible to use a pressure of one atmosphere, although considerably lower pressures may be used to get the highest resolution of intense bands. For polar substances the best compromise is often about 200 torr. A full atmosphere may give a pressure broadening in the region of 0.5 to 1 cm^{-1}. This means that for polar substances it is not possible to obtain as high a resolution or to detect as weak lines as for non-polar substances.

2.2.2 Details of Present Instruments

In this section a short description and discussion will be given of each of the instruments used in recent years to obtain high-resolution rotation-vibrational Raman spectra. For easy reference each instrument will be named by its geographical location.

Dijon Instrument. The highest resolution in such spectra has been achieved on the instrument described by MORET-BAILLY and BERGER [2.35], by BERGER et al. [2.36], and by CHAMPION and BERGER [2.37]. A resolution of about 0.05 cm^{-1} has been observed for the strong Q-branch of O_2 by LOËTE and BERGER [2.38], see Fig.2.5. This unique result is due to a focal length of 8 m in the spectrograph, yielding a dispersion of 0.6 cm^{-1} mm^{-1} on the photographic plate. However, to detect the very weak ν_4 band of CH_4, the slit was opened to 0.35 cm^{-1} [2.39], demonstrating the fact that the highest resolution can only be achieved for strong bands.

The laser runs in single mode, as required for the highest resolution. A 7.5 m long multireflection cell of 500 liter volume was previously placed inside the cav-

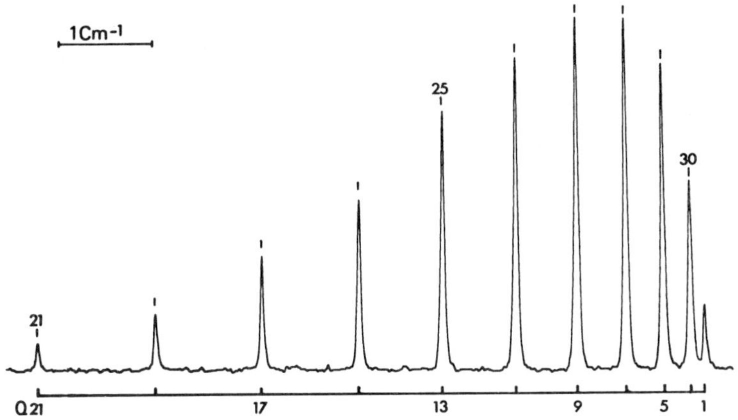

Fig. 2.5. The Q-branch of O_2 recorded by LOËTE and BERGER [2.38] on the Dijon instrument with an observed line width of about 0.05 cm^{-1}. The pressure was 80 torr and the exposure time 30 hours. The closest lines are 0.154 cm^{-1} apart. (Reproduced by permission)

ity with 5 passes of the beam but is now used outside the cavity with 60 passes. Another multireflection system is used to collect the scattered light at almost zero angle between the laser axis and the direction of observation. This implies a simultaneous observation of isotropic and anisotropic scattering, as indicated by (2.7). A single-prism order-sorter disperses the spectrum in the longitudinal direction of the slit of the spectrograph. This is equipped with a special mechanism [2.36] which automatically corrects for mechanical deformations of the spectrograph, for small changes in the laser frequency, and for displacements of the spectrum due to changes in atmospheric pressure. A cylindrical lens reduces the height of the spectrum on the plate by a factor of 3. The measurements on the plate are made by means of a microphotometer directly connected to a computer. The position of a line can be measured with a precision better than 0.01 cm^{-1}, and the Raman frequencies are accurate to 0.02 cm^{-1} [2.40].

This instrument has primarily been used to record spectra of CH_4, but recently also O_2 and SF_6 [2.41] have been studied.

Aarhus Instrument. By far the largest amount of spectra from many different compounds have been recorded on the instrumentation described by BRODERSEN and BENDTSEN [2.42,43], and by BRODERSEN et al. [2.44]. The main virtues of this equipment are: a high sensitivity to weak bands, a small cell volume and a high precision in frequency measurements. But the resolution is fairly low, although 0.21 cm^{-1} has been observed [2.45] for CHD_3, 0.3-0.4 cm^{-1} is more representative of the resolution obtained, especially for polar substances. The focal length in this spectrograph is 1.7 m, and the cylindrical lens diminishes the height of the spectrum by a factor of 10. The spectrograph in itself has a resolution of about 0.10 cm^{-1} and the linear dispersion is between 2 and 4 cm^{-1} mm^{-1}.

The laser is multi-mode and the beam is focused inside the sample cell, which also serves as the one end of the laser cavity. This leads to a cell volume of only 20 ml, which is important with expensive samples, and to a simple matching to the spectrograph, the slit image being placed inside the focused beam and the grating image inside the collecting lens. The scattered light is collected at approximately 90° to the laser axis. The plasma tube and the selector prism, and thereby the electric vector of the exciting light, can be turned by 90° about the axis, so that either of the intensity expressions (2.6,7) applies. This makes it possible experimentally to distinguish between isotropic and anisotropic scattering.

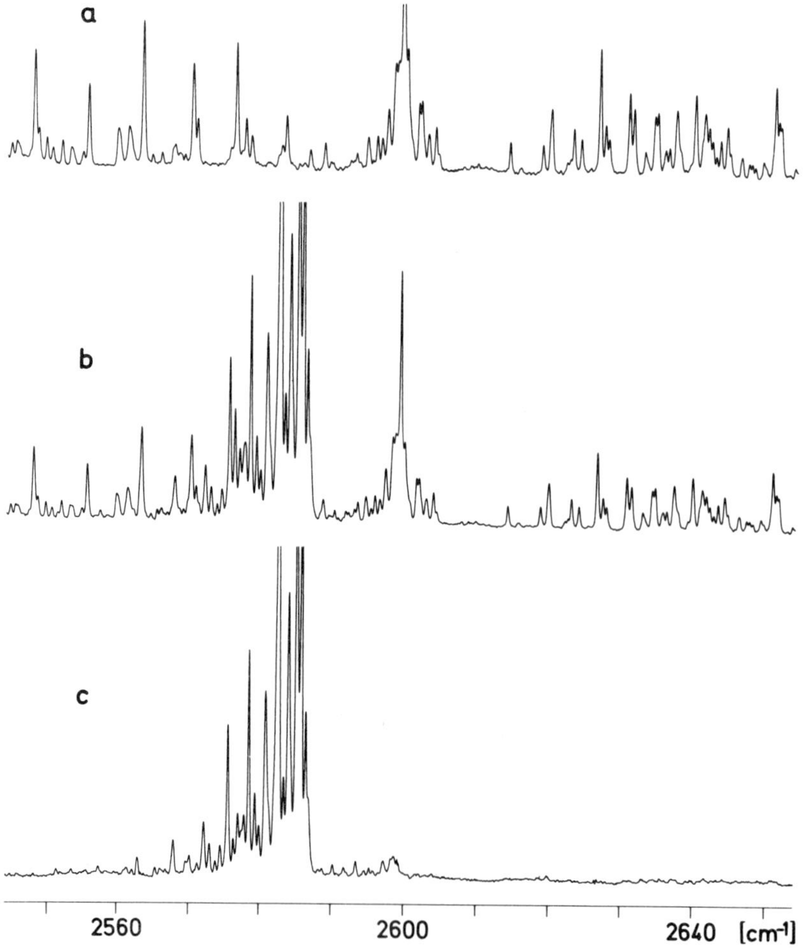

Fig. 2.6. Part of the $2\nu_4$ bands of CH_4 recorded on the Aarhus instrument. Microphotometer traces from exposures of a) anisotropic component alone and b) both components, with sample pressure 1 atm., slit width 0.15 cm^{-1} and exposure times 81 and 64 hours, respectively. c) The isotropic component computed by "subtraction" of a from b. The traces are linear in intensity and cut off to keep the figure within limits

A double-monochromator of the non-dispersive type is used as a premonochromator. The plate is processed in an automatic developing machine to ensure reproducibility. All three spectra on a plate are measured automatically in a computer-controlled photometer [2.44]. For the calibration a fourth-degree polynomial is used, in which the three highest-degree coefficients are constrained to values determined on the basis of 60 different calibration spectra [2.43]. This removes the calibration error due to the limited frequency range of a single plate. The photometer measures only the thorium lines used in the calibration, whereas 23000 points spaced by 5 μm are measured in the Raman spectrum, and the readings stored in the computer. A peak-finder program scans these data, and a microphotometer trace is plotted out linear in frequency, with a numbering of the peaks. Based on density-marks the relation between spectral intensity and photometer readings has been determined, and the microphotometer trace may be plotted on a linear intensity scale.

When each experiment is represented by a file in the computer, it is possible to make a new file representing a linear combination of spectra from different experiments. This is utilized to generate mean-value traces from different exposures of one band, or to compute the experimental trace of the isotropic component alone (see Fig.2.6) from exposures with different orientations of the laser tube.

The photometer determines the position of well-defined lines to ±0.4 μm, better than the position is defined. It seems that the frequency of the best lines can be determined with a relative precision of 0.002 to 0.005 cm^{-1}, whereas the absolute precision is 0.02 cm^{-1} or better.

Cambridge Instrument. The spectrograph used by BUTCHER and JONES [2.46] and by WILLETS et al. [2.47] has a focal length of 1 m, which implies a moderate resolution. It has, however, been applied to a relatively large variety of substances and with a number of devices to test new experimental techniques. Also a large number of pure rotational spectra have been taken at this instrument. A free-running Ar^+ laser is used, with a small multi-reflection cell inside the cavity. Recently [2.48] a White cell was used, also inside the cavity, giving a 20 times gain in intensity over the older cell. No details have been given of these cells or of the imaging optics. A filter is used as a foremonochromator, an order-sorter not being needed with the grating used in the second order. Usually a photographic film, preflashed to increase the detectability for weak lines, is used as a detector. A resolution of about 0.2 cm^{-1} has been obtained, but in practice 0.3 to 0.5 cm^{-1} is more representative.

An interesting experiment has been made by inserting an image intensifier [2.47] with the cathode in the focal plane of the spectrograph. It delivers an intensified 1:1 copy of the spectrum on a phosphor screen, which is imaged with a 1:5 magnification on a photographic plate. In principle, this set-up should increase the efficiency of the detection by a factor of about 20 and reduce the reciprocity failure. An intensity gain of 10 or 20 is claimed, but the system has only been used for one rotation-vibrational spectrum (ν_4 of CH_3I) [2.49] and no indication is given of the observed resolution.

Fordham Instrument. The use of lasers for gas Raman spectra has to a high degree been developed on this instrument, described by WEBER et al. [2.26], and WEBER and SCHLUPF [2.50]. It uses a single-mode laser in which the temperature of the Fabry-Perot is varied to change the excitation frequency in order to compensate for changes in atmospheric pressure within ±10 torr. The cell is placed inside the cavity and equipped with multireflection systems both for the laser beam and for the 90° collection of the scattered light. The focal length of 2 m has made a very high resolution possible in pure rotation (see Chap.3), for which this instrument has been used primarily. Rather few rotation-vibrational spectra have been recorded on this instrument, and no indication has been given of the resolution achieved in these cases.

Knoxville Instrument. The instrument described by FLETCHER and RAYSIDE [2.51,52] and by FLETCHER et al. [2.53] uses a photomultiplier as a detector, instead of the usual photographic plate. The spectrograph is replaced by a monochromator, the photomultiplier is cooled and followed by a photon counter, directly connected to the computer. This allows precise and convenient intensity measurements. Due to a 2 m focal length in the monochromator, a fairly good resolution of about 0.2 cm^{-1} has been obtained in the intense Q-branch of O_2 [2.51], but for the O- and S-branches the slits were opened to 0.4 cm^{-1}. A single-mode laser was used, with the beam focused in a multipass-cell placed outside the cavity.

No strict comparison has ever been carried out between the photographic plate and the photomultiplier as a detector for rotation-vibrational Raman spectra, but the results obtained for O_2 with this instrument seem to indicate that a photomultiplier may be preferred for intense spectra because of its convenience in use, whereas the photographic plate is advantageous for weaker spectra.

Toronto Instrument. Mercury excitation has, probably for the last time, been used as late as in 1971 by FAST and WELSH [2.54] to record the spectra of C_2H_2, C_2HD, and C_2D_2. The results are surprisingly good, probably due to the use of a double non-dispersive premonochromator. The focal length of the spectrograph was 1.5 m and the observed resolution about 0.3 cm^{-1}.

RICH and WELSH [2.55,56] have utilized an on-axis White cell outside the laser to multipass both the expanded laser beam and the scattered light, in connection with photoelectric recording. The observed resolution is not indicated, but the slit width was 0.35 cm^{-1}, probably due to the use of a photomultiplier as a detector.

Bradford Instrument. The instrument described by EDWARDS et al. [2.57] has primarily been used for pure rotational spectra, but also for rotation-vibrational spectra of $^3H_2$ and F_2. The laser beam is focused in the cell inside the cavity. The focal length of the spectrograph is 3 m and the dispersion 1.4 cm^{-1} mm^{-1}. The line width observed in the spectrum of $^3H_2$ is about 0.3 cm^{-1}.

2.2.3 Comparison to Infrared Experimental Techniques

The experimental technique used for rotation-vibrational Raman spectra has now reached the stage where the gamble has been taken out of the single experiment, a spectrum is just recorded and measured in much the same way as a high-resolution infrared spectrum is produced. There are, however, a number of important differences between the techniques used in the two cases.

To reach the ultimate sensitivity, the Raman experiment requires an unusual technique because a photographic plate is used as a detector. But this is really only a matter of habit. More seriously, the present technique does not allow Raman spectra to be recorded with nearly the resolution which may presently be obtained in infrared, and also for a given molecule fewer vibrational bands may in general be recorded in Raman than in infrared spectroscopy because of intensity problems. However, the Raman technique has one advantage. The experimental possibilities are virtually independent of the frequency range of the band to be studied, whereas in infrared different instruments are required for different ranges and the resolution deteriorates seriously as the frequency is decreased. In Raman spectroscopy a band at 200 cm^{-1} is recorded as easily as a band at 3000 cm^{-1} and with the same resolution. The only limitation to this is the presence of the strong pure rotational spectrum which may prevent the observation of extremely low lying bands.

2.3 Results

In this section a review will be given of the high-resolution rotation-vibrational Raman spectra obtained in recent years and of the molecular constants and other informations which have been extracted from these spectra.

The description will be arranged by molecules (and the rotor type) and preceded by a few remarks. As everywhere in this chapter it is impossible to give all details, but the references should be fairly complete, and further details may be obtained from these or directly from the authors.

2.3.1 General Remarks

The constants which can be determined from the spectra are of two kinds: 1) relative values for the derivatives of some polarizability components with respect to certain normal coordinates, and 2) constants of the molecular Hamiltonian, defining the geometry and the energy levels of the molecule in the absence of an external field. Only very few data have been obtained of the first kind and no consistent picture can be formed out of these. The second kind of constants is a relatively small contribution to the vast amount of data collected over the years from microwave, infrared and ultraviolet spectra. The main emphasis in the following will be on a demonstration of the usefulness of the Raman method to provide data which either cannot be found by other methods or with less accuracy.

The methods of analysis used in Raman spectroscopy are in principle the same as those used in infrared techniques, but the limited resolution available in Raman spectroscopy, more often than in infrared techniques, makes it necessary to use either pure curve-fitting or to correct from a peak-frequency to a subband-origin (or the like) by means of a computed contour. In such cases it is usually not possible to determine all the constants involved in a certain band, and some constants known from microwave or infrared spectra must be constrained in the analysis. On the other hand, more constants may be determined in Raman than in infrared spectroscopy, and the additional constants are often necessary to give a complete picture of the molecular states. Therefore, the ideal way to proceed is a simultaneous analysis of the two kinds of spectra. This has been done a number of times, ideally by inserting all the data in one big least-squares analysis. However, care must be taken to correct for differences in calibration errors on the two instruments.

2.3.2 Linear Molecules

One of the most obvious objects of a Raman investigation is a diatomic homonuclear molecule, because no infrared and (with the exception of O_2) no microwave spectrum exists for such molecules. The only other source of the molecular constants is the ultraviolet spectrum and in general the constants obtained in Raman spectroscopy are more accurate than those from ultraviolet spectra. High-resolution rotation-vibrational spectra of $^3H_2$, F_2, $^{14}N_2$, $^{14}N^{15}N$, $^{15}N_2$, and O_2 will be discussed below.

For other $D_{\infty h}$ molecules the normal vibrations which are inactive in infrared techniques are all active in Raman spectroscopy, and a rotation-vibrational Raman spectrum will immediately yield a number of results which can only be obtained indirectly from infrared spectra. Spectra of C_2H_2, C_2HD, C_2D_2, C_2N_2, and carbonsuboxide C_3O_2 will be discussed below. For $^{12}CO_2$ [2.58], [2.32], $^{13}CO_2$ [2.32] and CS_2 [2.58] there exist unpublished laser spectra, but they seem to give only marginal information in addition to what is known already about these molecules from other sources.

All vibrations of $C_{\infty v}$ molecules are active in both infrared and Raman techniques, and only in special cases should the Raman spectra yield otherwise unknown information. The spectra of HCN and DCN discussed below were mainly recorded to check the accuracy of the constants derived from Raman data against constants from infrared and microwave spectra.

Hydrogen. The resolved Q-branches of the fundamental vibrational bands of H_2, HD, and D_2 have been recorded by STOICHEFF [2.59] in 1957 with Hg-excitation. The corresponding Q-branch of $^3H_2$ has recently been photographed by EDWARDS [2.60] using the Bradford instrument. The constants obtained were $\nu_0 = 2464.324 \pm 0.001$ cm^{-1} (statistical error only), $B_1 - B_0 = -0.5856 \pm 0.0002$ cm^{-1} and $D_1 - D_0 = (-2.94 \pm 0.07) \cdot 10^{-4}$ cm^{-1}. Using results from pure rotation, $B_e = 20.3314 \pm 0.0006$ cm^{-1} and $D_e = (5.14 \pm 0.04) \cdot 10^{-3}$ cm^{-1} were derived, yielding $r_e = 0.74151 \pm 0.00002$ Å. This value deviates only insignificantly from the values calculated for H_2, HD, and D_2 using STOICHEFF's data. This agreement represents a convincing argument for the correctness of the data.

Fluorine. EDWARDS et al. [2.61] have photographed the unresolved Q-branch of the fundamental vibration on the Bradford instrument. They report $\nu_0 = 892.02 \pm 0.03$ cm^{-1}.

Nitrogen. BENDTSEN [2.62] has recorded the fundamental bands of $^{14}N_2$, $^{14}N^{15}N$, and $^{15}N_2$ on the Aarhus instrument. Only the anisotropic component was observed. It shows

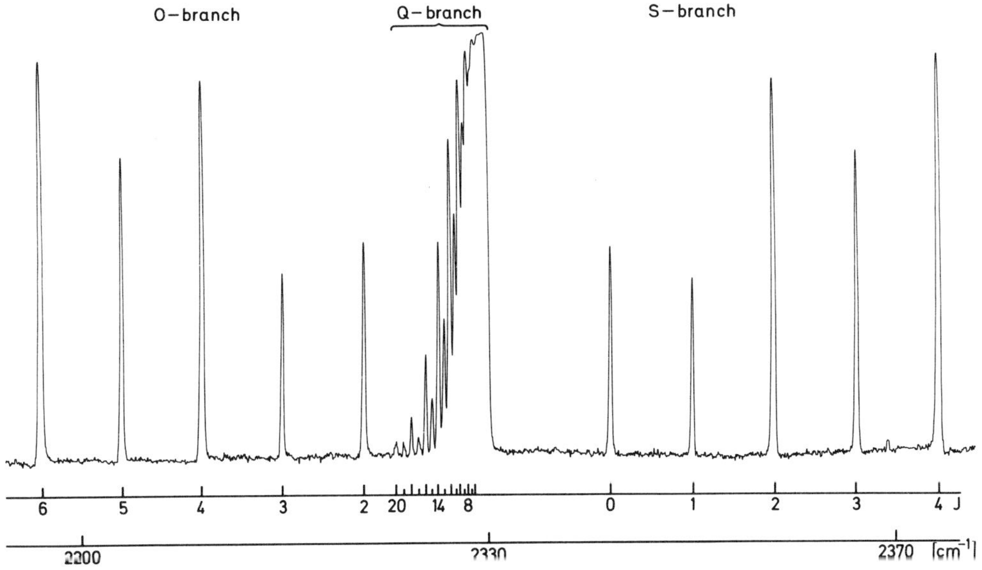

Fig. 2.7. Central part of the fundamental vibrational band of $^{14}N_2$ recorded by BENDTSEN [2.62] with a sample pressure of 1 atm., a slit width of 0.15 cm^{-1} and an exposure time of 20 hours. (Reproduced by permission)

Table 2.3. Constants for isotopic species of nitrogen, determined from Raman spectra [2.62]

	B_0[cm^{-1}]	D_0[cm^{-1}]	ν_0[cm^{-1}]	α[cm^{-1}]	r_e[Å]
$^{14}N_2$	1.989574(12)	$5.76(3)\times10^{-6}$	2329.917	0.017292(3)	1.097700(7)
$^{14}N^{15}N$	1.923596 (9)	$5.38(3)\times10^{-6}$	2291.332	0.016416(2)	1.097702(4)
$^{15}N_2$	1.857624(16)	$5.08(5)\times10^{-6}$	2252.125	0.015575(2)	1.097700(8)

Standard errors are given in parentheses in units of the last decimal. ν_0 values are accurate to ±0.01 cm^{-1}.

the structure predicted by theory (see Sect.2.1.4) including the intensity alternation due to nuclear spin. The O- and S-lines were observed for J through 16 to 21, and the Q-branches were partly resolved. The central part of the $^{14}N_2$ fundamental band is shown in Fig.2.7. Simultaneously the pure rotational spectra were measured. For two of the compounds CO was used as an internal standard to assure the highest absolute precision, since the rotational constants of this molecule are known with great accuracy from infrared measurements [2.63]. The values obtained for B_0, D_0, ν_0 and α are given in Table 2.3. The three independently derived values for r_e are included in this table. They agree well within one standard deviation. The constants given in Table 2.3 are the best known for these molecules, the B_0, α and r_e values for $^{14}N_2$ are orders of magnitude better than those obtained from ultraviolet spectra.

Oxygen. The electronic ground state of this molecule is a triplet state ($^3\Sigma_g^-$) and the electronic spin and its interaction with the nuclear rotation has to be taken into account. The energy levels may be described by the quantum numbers N of the nuclear and J of the total angular momentum. The energy is primarily determined by N in the same way as the energy is determined by J in the normal case. Due to the zero spin of ^{16}O, only states with odd N occur. For each N, three values are possible for $J = N$ or $N \pm 1$. The levels for $J = N \pm 1$ are almost coincident, and lower than the $J = N$ level by about 2 cm^{-1} for all N (except the $J = 0$ level). The constants indicating the energy both within the vibrational ground state and within the first vibrationally excited state have been determined from microwave spectra by AMANO and HIROTA [2.64].

The Raman intensity has been discussed by RENSCHLER et al. [2.65], by ALTMANN et al. [2.66], and by LOËTE and BERGER [2.38] applying various approximations. Two components occur in the fundamental band, one isotropic with $\Delta N = \Delta J = 0$, and one anisotropic with $\Delta N = 0, \pm 2$, and $\Delta J = 0, \pm 1, \pm 2$. The isotropic component consists of a Q-branch with an N structure (equal to the usual J-structure of other linear molecules) but with a very closely spaced J-structure. The anisotropic component consists of the usual O-, Q-, and S-branches ($\Delta N = -2, 0, +2$) but with so-called spin-satellites (due to $\Delta J = \pm 1$) displaced by about 2 cm^{-1} to each side of each line and a fine structure in each line due to $\Delta J = 0, \pm 2$. The intensity of the spin satellites should decrease very fast with increasing N.

The fundamental vibrational band has been observed by several authors using different instruments. RICH and LEPARD [2.67] were the first to observe the spin satellites, using the Toronto instrument described by RICH and WELSH [2.55]. FLETCHER and RAYSIDE [2.51] using the Knoxville instrument measured and analyzed the main lines of all three branches, but did not consider the spin splitting. LOËTE and BERGER [2.38] have recently recorded the Q-branch on the Dijon instrument with a resolution of about 0.05 cm^{-1}, as shown in Fig.2.5. The J-structure cannot be resolved, and the spin satellites of the anisotropic Q-branch are not seen in this

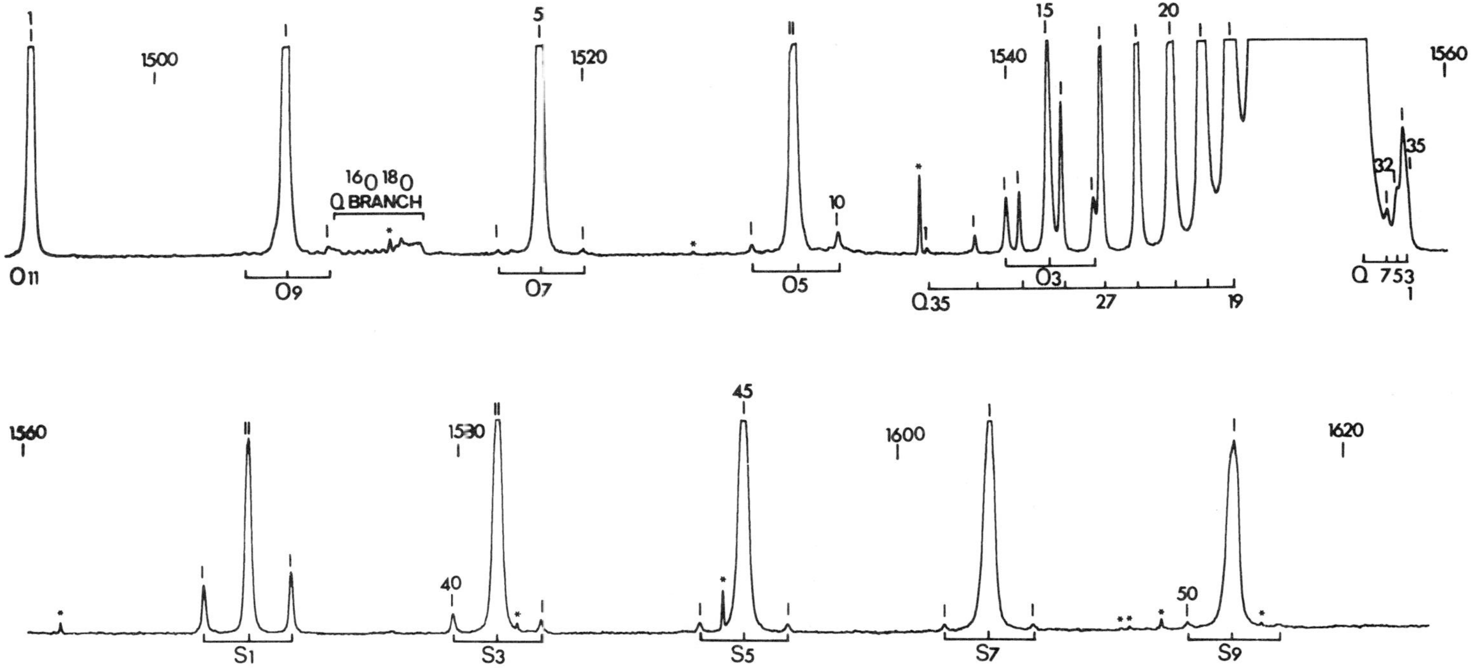

Fig. 2.8. The central part of the fundamental vibrational band of $^{16}O_2$ recorded by LOËTE and BERGÉR [2.38] with an observed resolution of about 0.15 cm^{-1}, using a sample pressure of 1 atm and an exposure time of 80 hours. The lines marked by * are Th lines used for calibration. Lines Nos. 32 through 35 are spin satellites of the anisotropic Q-branch. (Reproduced by permission)

figure because the isotropic component is very intense compared to the anisotropic one. They also recorded the O- and S-branches with a resolution of about 0.15 cm^{-1} and observed several spin satellites, as shown in Fig.2.8. Using the Aarhus instrument, the anisotropic component alone has been recorded [2.32] with less resolution but with better observations of the weak spin satellites. The rapid decrease in the intensity of the satellites is easily visualized in this spectrum.

In the analysis LOËTE and BERGER were able to determine all the constants of the upper state. It turns out that the splitting constants are determined better from microwave data [2.64], whereas their constants $B_1 = 1.421884 \pm 0.000013$ cm^{-1} and $D_1 = (-4.864 \pm 0.014) \cdot 10^{-6}$ cm^{-1} are the best known presently. The value for $\nu_0 = 1556.384 \pm 0.010$ cm^{-1} is precisely equal to the one obtained by FLETCHER and RAYSIDE [2.51] giving great confidence in the precision of this number. All constants seem more accurately determined than from ultraviolet data [2.68].

Acetylene. Using Hg excitation FAST and WELSH [2.54] have measured spectra of C_2H_2, C_2HD, and C_2D_2. The resolution is stated as about 0.3 cm^{-1}. The totally symmetric bands ν_1 and ν_2 of all three molecules and the ν_4 bands (symmetry Π_g) of C_2H_2 and C_2D_2 were observed with rotational structure, using exposure times up to 240 hours. In addition, a number of unresolved Q-branches primarily of hot bands, were recorded. The observed band structures agree with the theory given in Section 2.1.4. The constants ν_0 and α were determined and for the ν_4 bands also the ℓ-type doubling constants q_4, as well as a number of anharmonicity constants.

Recently WANG [2.69a] using the Fordham instrument has repeated the measurements for C_2H_2, and KOSTYK and WELSH [2.69b] using the cell described by RICH and WELSH [2.55,56], a single-mode laser and photographic detection have repeated the measurements for all three substances, but only some results for C_2H_2 are available at present. The values obtained for some of the constants for C_2H_2 in the two investigations are compared in Table 2.4 with infrared results. The discrepancies for the

Table 2.4. Constants for Acetylene C_2H_2 obtained from Raman and infrared spectra

	Raman		Infrared	
	[2.69a]	[2.69b]	[2.69c]	[2.69d]
ν_1	3372.887(10)	3372.838(2)	3372.851(3)	
ν_2	1974.339(2)	1974.317(2)	1974.317(3)	
ν_4	612.877(6)	612.882(3)	612.870(2)	612.871(2)
$(B_1-B_0)\times10^3$	-6.83(13)	-6.858(23)	-6.81(2)	
$(B_2-B_0)\times10^3$	-6.13(5)	-6.189(14)	-6.21(3)	
$(B_4^c-B_0)\times10^3$	-1.23(9)	-1.352(22)	-1.269(24)	-1.326(21)
$(B_4^d-B_0)\times10^3$	4.02(10)	3.968(20)	3.950(25)	3.907(21)

Standard errors are given in parentheses in units of the last decimal. For the ν_4 band c refers to ΔJ even, d to ΔJ odd.

fundamental frequencies are probably due to calibration errors, not included in the quoted errors. For the ΔB values the agreement is quite satisfactory, and the Raman values of KOSTYK and WELSH may well be more accurate than the infrared values.

Cyanogen. The C_2N_2 molecule is formally analogous to acetylene, but the ν_5 vibration of C_2N_2 is as low as 233 cm^{-1} compared to 512 cm^{-1} for the lowest vibration in C_2D_2. This causes severe problems with hot bands. WANG and WEBER [2.70] have recorded the ν_1 and ν_4 bands on the Fordham instrument, the ν_2 band being too weak to be observed. In the ν_1 band, the $\nu_5 \rightarrow \nu_1 + \nu_5$ hot band could be identified and an analysis of both bands gave among other constants $\nu_1 = 2330.512$ cm^{-1} and $X_{15} = -2.857 \pm 0.005$ cm^{-1}. The ℓ-type doubling could not be observed in the ν_4 band since the P- and R-branches could not be identified. The constants obtained from the analysis of the O- and S-branches include $\nu_4 = 502.931$ cm^{-1}. Again no estimate is given of the absolute error.

Carbon suboxide. The C_3O_2 molecule is only quasi-linear, the equilibrium configuration being slightly bend (about 17° at the central atom). Correspondingly the bending vibration ν_7 (Π_u) has an extremely low-lying fundamental state at about

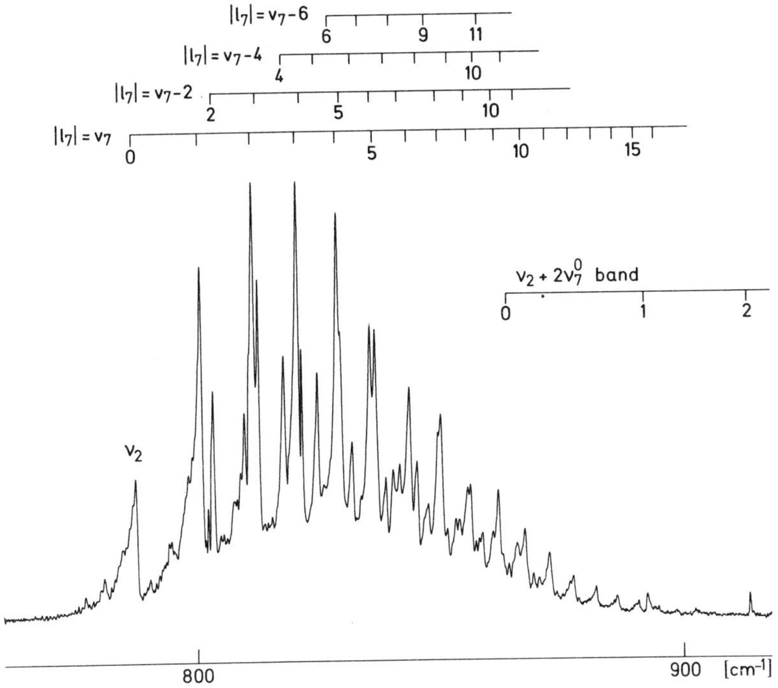

Fig. 2.9. The ν_2 band and hot bands of carbon suboxide C_3O_2 recorded by LOLCK and BRODERSEN [2.78] at a sample pressure of 50 torr, a slit width of 0.15 cm^{-1} and an exposure time of 41 hours. The assignment is given of the Q-branches in four series of hot bands $v_7\nu_7|\ell_7| \rightarrow \nu_2 + v_7\nu_7|\ell_7|$ and in a single series in the $\nu_2 + 2\nu_7^0$ band

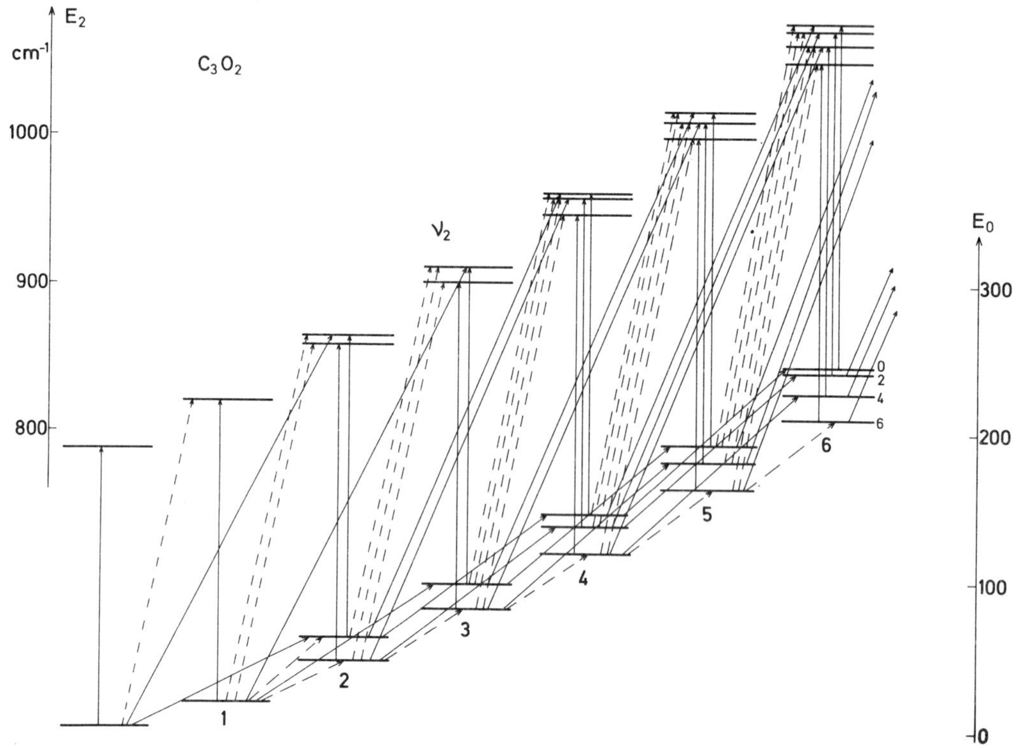

Fig. 2.10. Diagram of the $v_7\nu_7^{|\ell_7|}$ and $\nu_2 + v_7\nu_7^{|\ell_7|}$ levels of carbon suboxide and the observed transitions. Each horizontal line represents a stack of J levels, v_7 and $|\ell_7|$ being indicated. The full lines are Raman transitions, dashed lines infrared transitions

18 cm^{-1} and is highly anharmonic, the $2\nu_7^0$ state is at 60.7 cm^{-1}. Consequently any infrared or Raman band is heavily overlapped by hot bands, originating in many different $v_7\nu_7^{|\ell_7|}$ levels. This has been cleared up on the basis of several investigations of the infrared spectrum, primarily the works by CARREIRA et al. [2.71], MANTZ et al. [2.72], WEBER et al. [2.73,74], and PETERS et al. [2.75], and a low-resolution Raman spectrum recorded by CARREIRA et al. [2.71]. A theory for the molecule has been worked out by DUCKETT et al. [2.76] and by WEBER and FORD [2.77] using slightly different models.

Recently, LOLCK and BRODERSEN [2.78] have photographed the $2\nu_7^0$, ν_2, $\nu_2 + 2\nu_7^0$, ν_1, $2\nu_5^0$, and $2\nu_6^0$ bands on the Aarhus instrument. The isotropic component was included in the experiment to enhance the intensity of the Q-branches which form the dominating feature of each spectrum. Carbon suboxide polymerizes easily and it seems hopeless to keep it for extended periods of time in a cell where the laser beam is focused within the cavity. Nevertheless, it has been possible to obtain spectra of good quality by keeping the pressure at 50 torr and using only very pure samples. The ν_2 band with its hot bands is shown in Fig.2.9, which also includes the assignment of all the Q-branches. The vibrational frequencies determined from this band and the $2\nu_7^0$ and $\nu_2+2\nu_7^0$ bands, in conjunction with vibrational frequencies from the infrared ν_7^1 and $\nu_2+\nu_7^1$ bands [2.74-76], allow the construction of a level diagram for all of the $\nu_7\nu_7^{|\ell_7|}$ and $\nu_2+\nu_7\nu_7^{|\ell_7|}$ levels as shown in Fig.2.10. In fact, the assignment of most of the Q-branches in the infrared $\nu_2+\nu_7^1$ band [2.75] rests on information extracted from the Raman spectrum shown in Fig.2.9.

Hydrogen cyanide. BENDTSEN and EDWARDS [2.79] have recorded the totally symmetric ν_1 and ν_3 bands of HCN and DCN. The constants derived by the analysis agree well with constants obtained from microwave and infrared spectra, and the constants ν_0 and α are determined with similar accuracy.

2.3.3 C_{3v} Molecules

Because of the presence of a dipole moment (in most cases) and the absence of a center of symmetry, one should not immediately expect the Raman spectra of C_{3v} molecules to yield information which could not be obtained with higher precision from infrared or microwave spectra. However, if the vibrational states of a C_{3v} molecule are essentially unperturbed, the selection rule $|\Delta K| = 1$ for absorption does not allow a determination of the rotational constant "about" the figure axis (A for prolate, C for oblate tops) or the centrifugal stretching constant D_K from infrared spectra. Only if a strong (and not too localized) coupling is present between vibrational states, the so-called perturbation-allowed transitions may allow a determination of these constants with great precision. This limitation in the applicability of infrared spectra has long been a serious hindrance for a precise determination of the geometry of C_{3v} molecules.

In Raman spectroscopy a similar problem does not exist because the selection rule for ΔK for a degenerate fundamental band is $\Delta K = +1$ and -2, as discussed in Sect. 2.1.5, and the constants A (or C) and D_K are immediately determined in the analysis of such a band, if this is sufficiently resolved experimentally to allow an assignment of a reasonable number of transitions from both components. Especially for the polar molecules, the pressure broadening leads to a severely restricted resolution which imposes problems in the analysis, but as discussed below these may often be solved satisfactorily by introducing constants determined from infrared and microwave spectra in the analysis. Only if the band is heavily perturbed, is it difficult to carry through an analysis. But this is just the case where the perturbation-allowed transitions make a complete analysis of the infrared band possible. It seems thus, that the two methods supplement each other nicely. In the following the determination of A_0 (or C_0) and D_0^K from Raman spectra will be discussed for the molecules CH_3D, CHD_3, CD_3F, CD_3Cl, CH_3Br, CD_3Br, CH_3I, and $CH_3\text{-}C\equiv CH$.

Out of the spectra of these molecules only those of CH_3D and CHD_3 are sufficiently resolved to allow an independent analysis, either by means of ground state combination differences or by a direct least-squares analysis. For all the other molecules the dominating features of a degenerate fundamental band, as discussed in Sect. 2.1.5, are the two series of accumulations of Q-lines (usually called "Q-branches") whereas the four remaining branches form a more or less continuous background. In practice the analysis is based primarily on the peak-frequencies of the "Q-branches".

The procedure needed to utilize these data has been described independently by FREEDMAN and JONES [2.49], and EDWARDS and BRODERSEN [2.80]. The shape of each "Q-branch" is determined by ΔB and D_{JK}, out of which D_{JK} is always known precisely from microwave spectra, whereas a precise value for ΔB can only be obtained from a high-resolution investigation of the corresponding infrared band (or perhaps of an overtone or combination band). If the degenerate fundamental state is slightly perturbed, the variation of ΔB with K must be determined in infrared and is essential for the analysis of the Raman band. When ΔB and D_{JK} are known, the detailed shape of each "Q-branch" may be computed and utilized, either to assign a J value to the peak-frequency, or to correct this to the subband-center (J=0). Again both a ground state combination analysis and a direct analysis is possible.

For all the molecules indicated, such an analysis has been performed on the degenerate CH (or CD) stretching fundamental band ν_4 (for methylacetylene ν_6), recorded on the Cambridge instrument for CH_3I, on the Aarhus instrument for the other molecules and also on the Knoxville instrument for CD_3Cl. The constants obtained are collected in Table 2.5 which only includes those constants which are supposed to have been determined with a higher precision by this method than by any other method.

Table 2.5. Constants for C_{3v} molecules determined from rotation-vibrational Raman spectra

	A_0	$D_0^K\times10^5$	ν_4	$(A\zeta)_4$	A_4-A_0	$(B_4-B_0)\times10^3$	Ref.
CHD_3 [a]	2.6297(3)	1.8(3)	2250.88(10)	0.656(10)	-0.023(5)	-59(5)	[2.45]
CD_3F [b]	2.6033(10)	1.5(5)	2258.57(4)	0.5017(10)	-0.0122(3)	-0.46(5)	[2.32]
CD_3Cl	2.6136(5) [d]	1.99(25) [d]	2283.49(2)	0.4741(30)	-0.01291(20)	-0.169(20)	[2.19]
CH_3Br	5.1800(10)		3056.42(2)	0.3561(10)			[2.81]
CD_3Br	2.6004(10)	1.7(3)	2296.47(3)	0.4701(10)	-0.0131(2)		[2.80]
CH_3I	5.1743(6)	12.3(22)	3060.022(8)	0.3421(8)	-0.0290(3)		[2.49]
CH_3-C≡CH [b,c]	5.3102(10)	11.5(10)	2980.90(3)	0.388(5)	-0.0172(20)		[2.32]

[a] for CHD_3 replace A by C.

[b] preliminary values.

[c] for CH_3-C≡CH replace subscript 4 by 6.

[d] FLETCHER and DAUNT [2.82] have obtained the preliminary values A_0=2.6122(16) and D_0^K=1.90(44)×10^{-5}, the numbers in parentheses indicating two standard errors.

All numbers are given in cm^{-1}. The numbers in parentheses are estimated absolute errors in units of the last decimal, except for CH_3I where statistical errors are given.

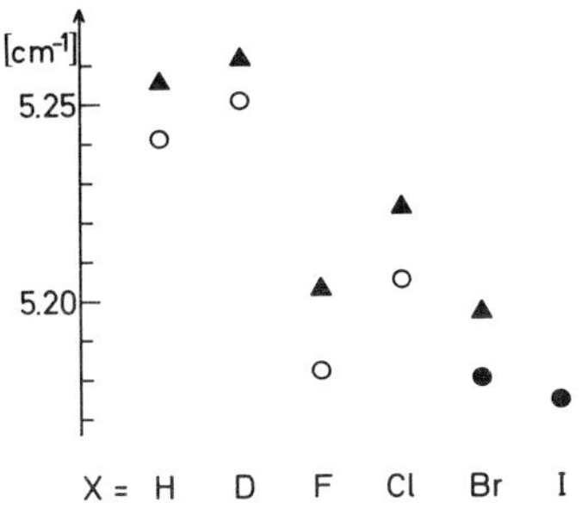

Fig. 2.11. Figure-axis rotational constants (A_0, B_0, or C_0) for the ground state of CH_3X and CD_3X molecules (the last ones multiplied by m_D/m_H). Circles are for CH_3X and triangles for CD_3X compounds. Open circles represent data from perturbation-allowed infrared transitions, filled-in symbols represent data from Raman experiments

Together with the B_0 value of CD_4 (see Sect.2.3.6) and the B_0 and A_0 values of CH_4 [2.83], CH_3D [2.87], CH_3F [2.84], and CH_3Cl [2.85] determined from infrared perturbation-allowed transitions, these values form an almost complete set of information of accurate "figure-axis" ground-state rotational constants for the series CH_3X and CD_3X with X = H, D, F, Cl, Br, and I, as shown in Fig.2.11. The perfect consistency between the two series is a strong evidence for the correctness of the data.

Some further details are given below on the Raman investigations of each molecule. For the molecules $^{12}CH_3F$ and $^{13}CH_3F$ a strong Coriolis interaction between the ν_2 and ν_5 states has been analyzed by means of the Raman spectra. Details are also given below.

Mono- and trideuteromethane. Among symmetric tops these two molecules have exceptionally large values for both rotational constants (from 2.6 to 5.2 cm^{-1}), and the rotational structure of the Raman bands may be resolved to an unusual extent. The Raman method of determining A_0 was applied for the first time in 1959 by RICHARDSON et al. [2.86] to the ν_4 band of CH_3D using Hg excitation. Their value for $A_0 = 5.243 \pm 0.002$ cm^{-1} did not fit precisely the more accurate value of $A_0 = 5.250703 \pm 0.000056$ cm^{-1} determined later by OLSON [2.87] from infrared perturbation-allowed transitions. KATTENBERG and BRODERSEN [2.88] therefore repeated the experiment and found $A_0 = 5.2502 \pm 0.0006$ cm^{-1} by an independent ground state combination difference analysis in agreement with Olson's value. If the B_0, D_J and D_{JK} (which are known precisely from microwave spectra of ordinary C_{3v} molecules) were constrained to the infrared values, $A_0 = 5.2508 \pm 0.0005$ was obtained, in perfect agreement with infrared. This seems a good proof of the absolute accuracy of the Raman method of determining A_0. Their spectrum is shown in Fig.2.12. Almost 400 lines could be measured and assigned to ν_4, the overlapping Q-branches of ν_1 and ν_5 not interfering seriously because the isotropic scattering was excluded in the experiment. Note that this band extends over more than 500 cm^{-1}.

KATTENBERG and BRODERSEN also recorded the ν_4 band of CHD_3 [2.45] and the overlapping ν_2 band. They analyzed the ν_4 band both by combination differences, by a subband analysis and by a direct least-squares analysis. Their results are given in

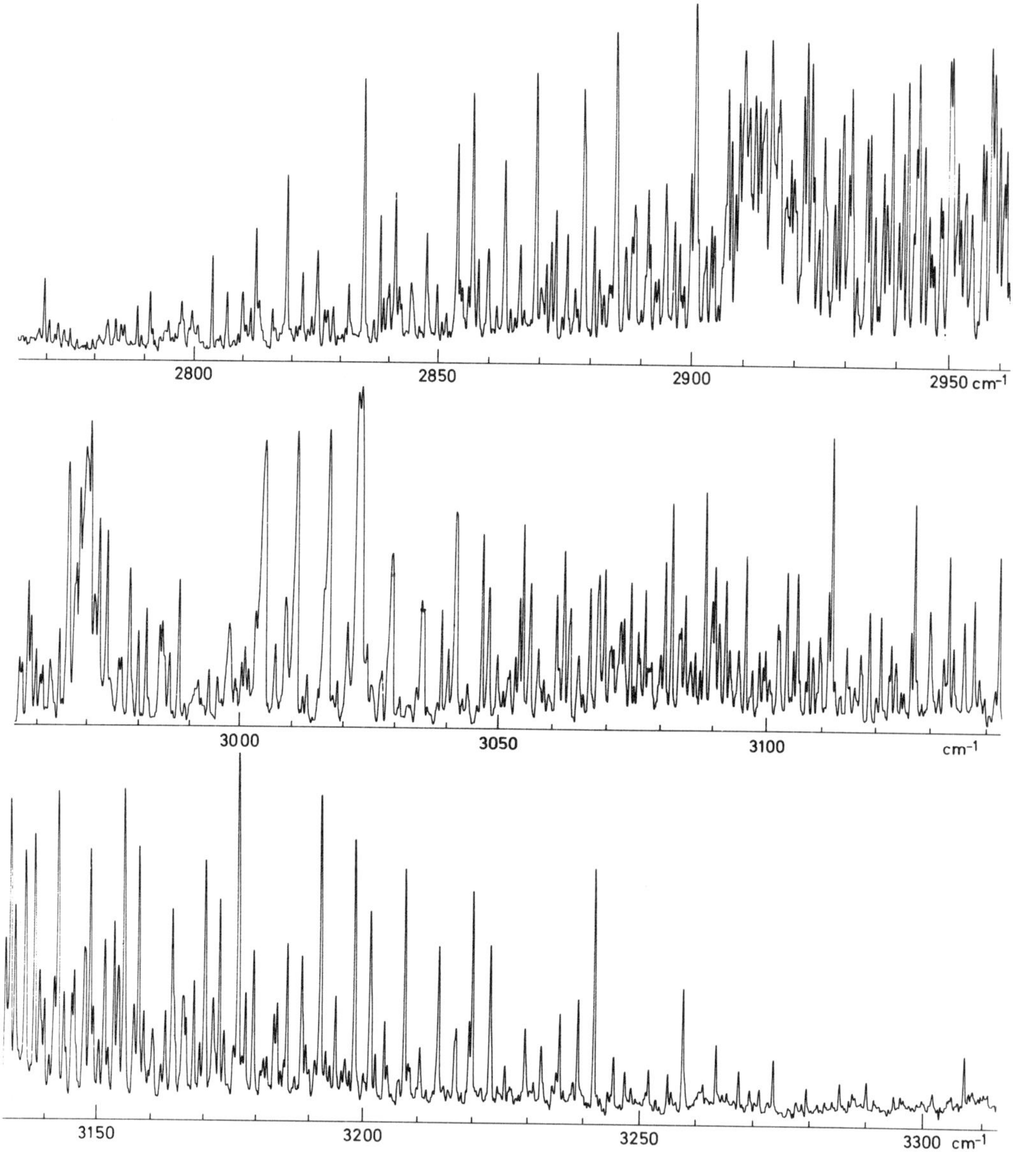

Fig. 2.12. The ν_4 band of CH_3D recorded by KATTENBERG and BRODERSEN [2.88] in three consecutive exposures. The Q-branches of $2\nu_5$ and ν_1 are seen at 2915 and 2970 cm^{-1}, respectively, but nearly all of the about 400 lines belong to the ν_4 band. (Reproduced by permission)

Table 2.5. They also analyzed the ν_2 band and found a reasonably good fit with more accurate infrared values obtained by JENNINGS and BLASS [2.89].

Methylfluoride. The ν_4 state of CH_3F is strongly perturbed and GRANER [2.84] has obtained very accurate values for A_0 and D_0^K from infrared perturbation-allowed transitions.

The ν_4 Raman band of CD_3F [2.32] is only slightly perturbed and a few fairly single O- and S-lines could be assigned. This allows the determination of ΔB, the value being in disagreement with the value $\Delta B = -0.13 \cdot 10^{-3}$ cm^{-1} from medium resolution infrared spectra by JONES et al. [2.90]. Preliminary results are included in Table 2.5. 2.5.

For CH_3F a strong Coriolis interaction between the ν_2 and ν_5 states has previously been studied in infrared by DI LAURO and MILLS [2.91]. The Raman spectrum was photographed on the Aarhus instrument and analyzed by ESCRIBANO et al. [2.92] by fitting a computed contour to the observed one, the energies being obtained by an exact diagonalization of the Hamiltonian matrix and the intensities by direct use of (2.8). A total of 14 energy constants were needed in the calculations, in addition to the intensity parameters. Ten energy constants, known from microwave [2.93,94] or infrared [2.84] techniques, were constrained and four were determined by the fitting of the curves. In fact, the Raman spectrum and the microwave spectrum of the ν_2 and ν_5 states by HIROTA et al. [2.93] were analyzed simultaneously in an iterative way. The final Raman contours are shown in Fig.2.13.

A similar investigation of the ν_2, ν_5 band of $^{13}CH_3F$ was recently carried out by HEGELUND et al. [2.95]. The spectrum is less complicated because the ν_2 and ν_5 are 12 cm^{-1} apart, compared to 8 cm^{-1} in $^{12}CH_3F$.

Methylchloride. DI LAURO et al. [2.85] have from infrared perturbation-allowed transitions determined one relation between A_0 and D_0^K for CH_3Cl, and by inserting a theoretical value for D_0^K they obtained A_0, the precision of which depends on the accuracy of the theoretical value for D_0^K. The ν_4 Raman band has been recorded [2.32] but the ν_4 state is too perturbed to allow an analysis in the usual way.

The ν_4 state of CD_3Cl is only slightly perturbed. It has been recorded independently on the Aarhus instrument [2.19] and on the Knoxville instrument by FLETCHER and DAUNT [2.82]. The first set of data, including again a few O- and S-lines, has been analyzed in conjunction with data from the corresponding infrared band recorded by PETERSON and EDWARDS [2.96], the difference in calibration being only about 0.01 cm^{-1}. The resulting constants are included in Table 2.5. Preliminary results for A_0 and D_0^K obtained by FLETCHER and DAUNT [2.82] are indicated in Table 2.5. The two sets of values for these constants are in reasonable agreement.

Methylbromide. EDWARDS and BRODERSEN have recorded the ν_4 Raman bands of CH_3Br [2.81] and CD_3Br [2.80]. The ν_4 state is almost unperturbed in both molecules, and a nice agreement was obtained between the ground state combination difference analysis and the direct analysis. The results are given in Table 2.5, and the observed and computed curves for CD_3Br are shown in Fig.2.14. The structure of this band was discussed in detail in Sect. 2.1.5. For CH_3Br all the α^A constants are known, and $A_e = 5.2442 \pm 0.0015$ cm^{-1} could be calculated. This corresponds to an equilibrium

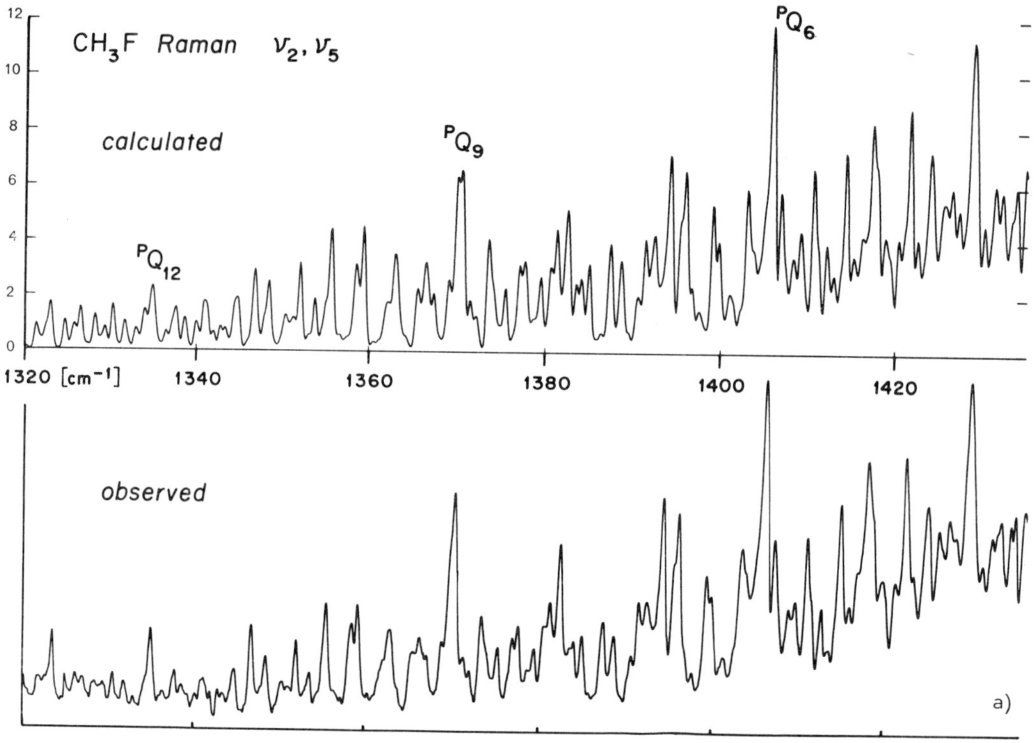

Fig. 2.13a-c. The ν_2 and ν_5 bands of CH_3F observed and calculated by ESCRIBANO et al. [2.92]. The observed spectrum was recorded at 200 torr in 64 hours. Only a few assignments are indicated because most of the prominent features are heavily blended. (Reproduced by permission)

distance of 0.32077 ± 0.00005 Å from a hydrogen atom to the figure axis. All the other five fundamental bands of CH_3Br were observed but not analyzed.

Methyliodide. MATSUURA et al. [2.97] have observed sufficient perturbation-allowed transitions in the infrared ν_5 band to get one relation between A_0 and D_0^K. FREEDMAN and JONES [2.49] recorded the ν_4 Raman band and constrained this one relation in their analysis, in addition to the constants delivered by microwave and infrared spectra. Their results are included in Table 2.5.

Methylacetylene. The ν_6 Raman band of $CH_3\text{-}C \equiv CH$ [2.32] shows the same general structure as the ν_4 bands of the methyl halides. From the infrared $2\nu_6$ band ANTILLA et al. [2.98] found $\Delta B = -2 \cdot 10^{-5}$ cm^{-1}. This is not a very accurate number, but the low value agrees with the observation of very sharp "Q-branches" in the Raman spectrum. This makes the J assignment or frequency correction less critical than in other cases. Preliminary results are given in Table 2.5.

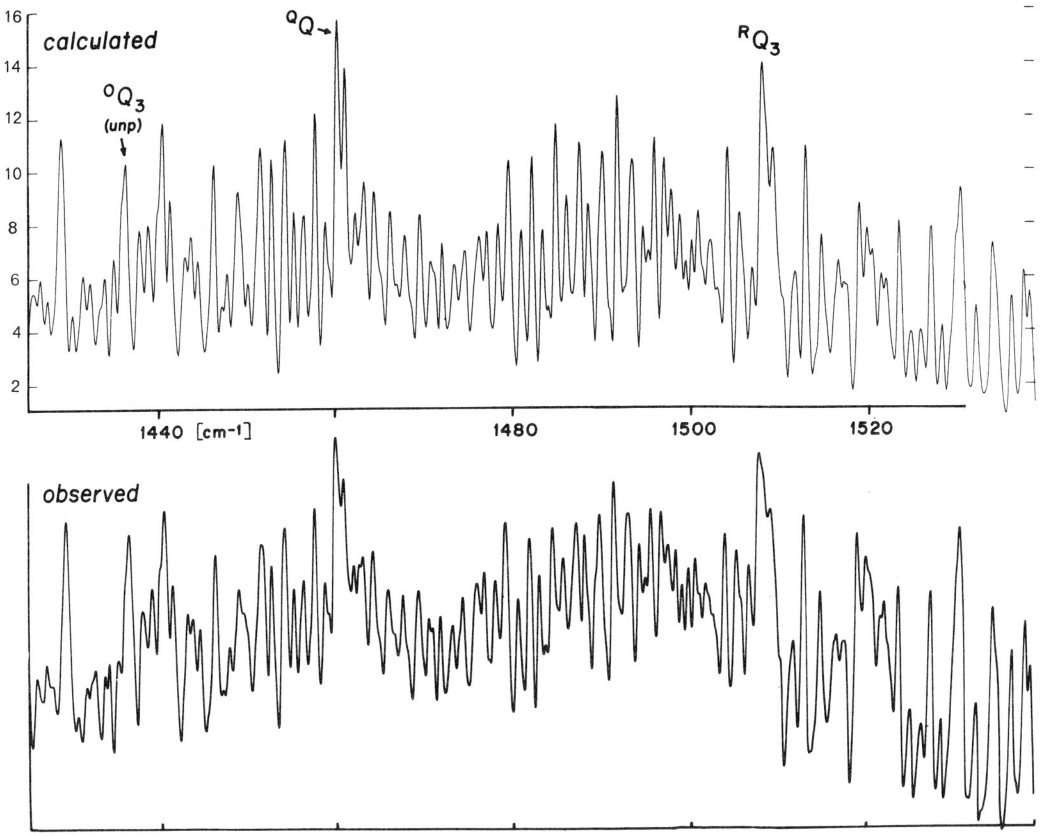

calculated
observed
16
14
12
10
8
6
4
2
1440 [cm-1]
1480
1500
1520
QQ
RQ3
OQ3
(unp)

b)

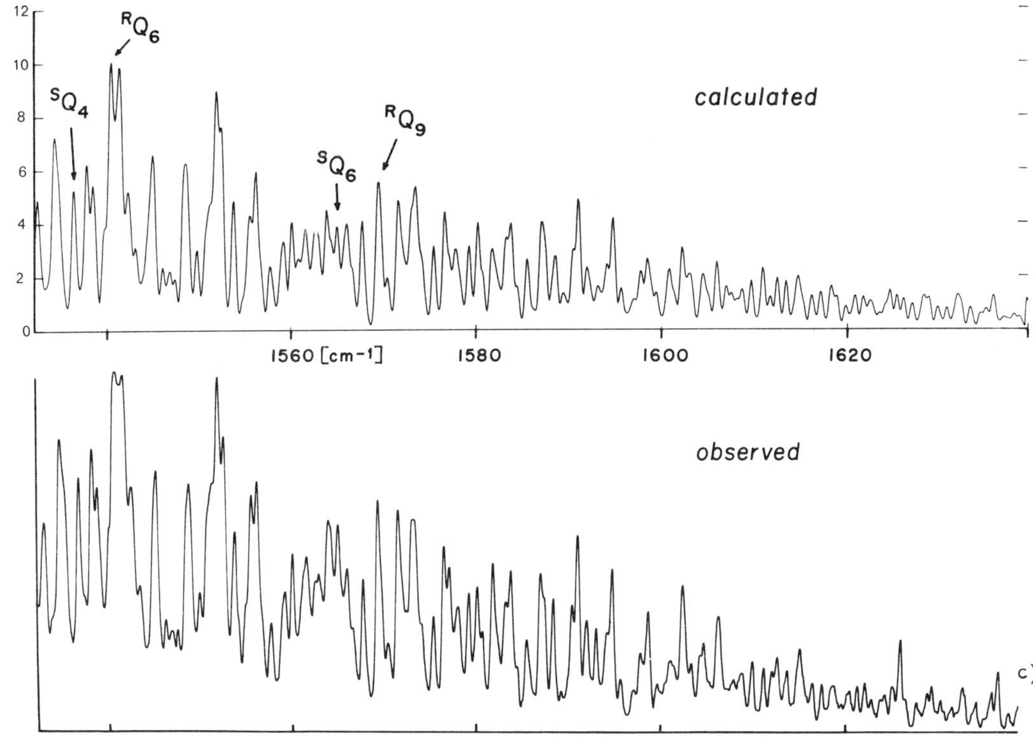

calculated
observed
12
10
8
6
4
2
0
1560 [cm-1]
1580
1600
1620
RQ6
SQ4
SQ6
RQ9

c)

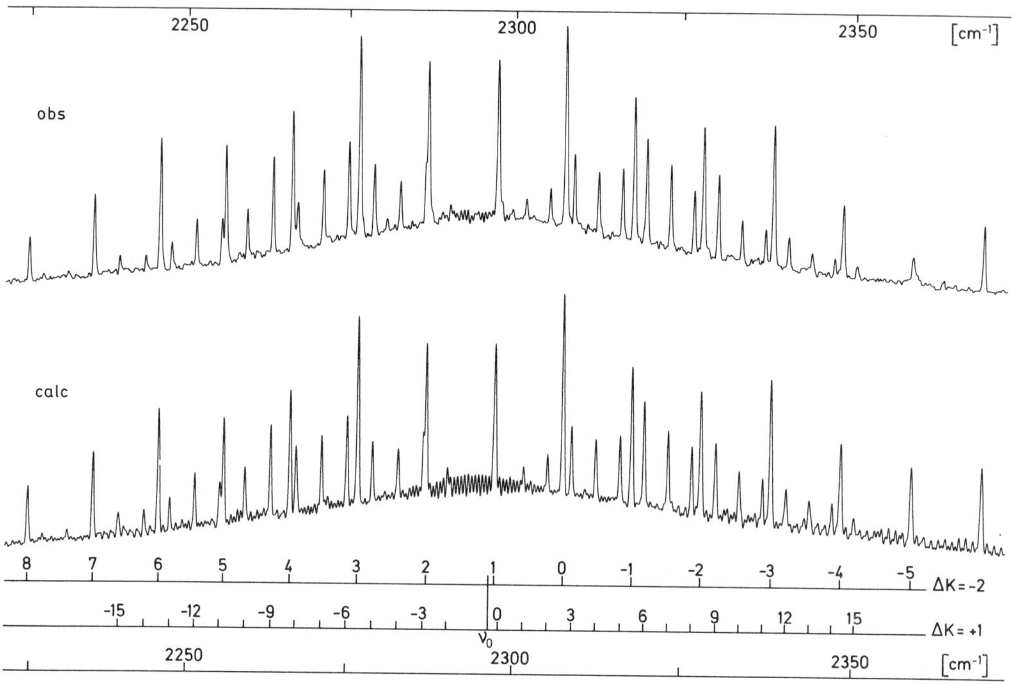

Fig. 2.14. The ν_4 band of CD_3Br as observed and calculated by EDWARDS and BRODERSEN [2.80]. The sample pressure was 200 torr, the slit width 0.15 cm^{-1} and the exposure time 16 hours. The assignment of K is given for each "Q-branch". (Reproduced by permission)

2.3.4 Other Symmetric Top Molecules

The rotation-vibrational Raman spectra of ethane, allene, cyclopropane, benzene and some of their deuterated derivatives have been studied, but those of the ethanes have not been investigated in recent years and the reader is referred to WEBER's review [2.2].

The selection rules for the D_{2d} molecule allene and the D_{3h} molecule cyclopropane are quite different from those of C_{3v} molecules. Thus the two components determined by $\Delta K = +1$ and -2 do not occur in one band for these molecules, and a determination of A_0 and D_0^K requires in practice a simultaneous analysis of a Raman band and the corresponding infrared band. The Raman spectra of $CH_2{=}C{=}CH_2$, $CD_2{=}C{=}CD_2$ and cyclopropane are discussed below. The partly deuterated molecule $CH_2{=}C{=}CD_2$ is an asymmetric top and will be discussed in the following section.

Allene. The vibrations of allene have symmetries A_1, B_1, B_2, and E. They are all Raman active, whereas only the B_2 and E vibrations are infrared active. The general selection rule for ΔK in the diagonal approximation [2.9] may be written

$$\Delta K = -\sum \Delta \ell_j - 2\sum \Delta v_B - 2\omega + 4t \quad , \tag{2.56}$$

yielding the following results for the different fundamental bands.

A_1 fundamental bands (Raman only) consist of an isotropic and an anisotropic component, both with $\Delta K = 0$. This gives the same structure as for A fundamental bands of a C_{3v} molecule (see Sect.2.1.5). For B_1 and B_2 fundamental bands in Raman, one gets $\Delta K = \pm 2$, but a simpler description is obtained by $\Delta K = +2$ and a signed K, where the same state is described for both signs of K. Such bands are dominated by one series of "Q-branches" spaced approximately by 4(A-B). The infrared B_2 fundamental band has $\Delta K = 0$. For the Raman and infrared E fundamental bands one gets $\Delta K = -1$ and +1, respectively, if the upper state ℓ is chosen as +1. The Raman band therefore looks like the $\Delta K = +1$ component of an E fundamental band of a C_{3v} molecule.

The ground state constants may in principle be determined from the "Q-branches" of a B_1 or B_2 fundamental Raman band alone, but the observations are not of sufficient quality to give accurate constants. Instead a B_2 or an E Raman band and the corresponding infrared band are analyzed simultaneously, using the same methods as for the two components of an E Raman fundamental band of a C_{3v} molecule. However, as always when data from different experiments are taken together, great care must be used to correct for differences in calibration errors.

Several Raman bands of allene were photographed in 1959 by Hg excitation by BRODERSEN and RICHARDSON [2.99]. The $\nu_1(A_1)$, $\nu_5(B_2)$ and $\nu_8(E)$ bands were recorded on

Table 2.6. Selected constants for allene and allene-d_4 obtained from rotation-vibrational Raman spectra

	Allene		Allene-d_4	
A_0	4.8124(16) 4.8110(12)	[2.100] [2.102]	2.4153(4)	[2.103]
$D_0^K \times 10^4$	1.39(15) 1.12(10)	[2.100] [2.102]	0.28(2)	[2.103]
ν_1	3015.51(4)	[2.100]	2216.63(2)	[2.103]
ν_2	1442.64(3)	[2.99]	1231.84(1)	[2.103]
ν_3	1072.582(2)	[2.99]	872.03(1)	[2.103]
ν_4	848.59(5)	[2.102]	603.65(3)	[2.103]
ν_8	3085.426(15)	[2.100]	2321.24(4)	[2.103]
$(A\zeta)_8$	-0.0313(24)	[2.100]	-0.0082(5)	[2.103]
ν_{11}	352.73(5)	[2.101]	304.40(4)	[2.103]
$(A\zeta)_{11}$	3.917(3)	[2.101]	1.6988(5)	[2.103]

Numbers in parentheses are standard errors in units of the last decimal.

the Cambridge instrument and analyzed by BUTCHER and JONES [2.100]. The $\nu_{11}(E)$ band [2.101] and the $\nu_4(B_1)$, $\nu_{10}(E)$, and $2\nu_{11}^0$ [2.102] and most of the bands of allene-d_4 [2.103] have been photographed on the Aarhus instrument and analyzed by HEGELUND et al. Some of the constants obtained are collected in Table 2.6 where primarily those constants are included which are supposed to be more accurately determined by these investigations than from infrared spectra alone.

Cyclopropane. The Raman active vibrations of cyclopropane C_3H_6 have symmetries A_1', E', and E'', and out of these only the E' vibrations are also infrared active. In this case there are two selection rules on ΔK [2.9], one identical to the rule (2.36) for C_{3v} molecules and another

$$\Delta K = \sum \Delta v'' + \omega + 2t \tag{2.57}$$

requiring the parity of ΔK to depend on ω and the changes in quantum numbers for vibrations in species A_1'', A_2'', and E''. This leads to the following selection rules for the different fundamental bands: A_1' fundamental bands have $\Delta K = 0$, giving the same structure as for C_{3v} molecules. E' fundamental bands have $\Delta K = -2$ in Raman but $\Delta K = +1$ in infrared, choosing $\ell_j' = +1$. E'' fundamental bands have $\Delta K = +1$ and are forbidden in infrared. Clearly the ground-state constants can be obtained by analyzing an E' Raman fundamental in conjunction with the corresponding infrared band. BUTCHER and JONES [2.46] have photographed the $\nu_9(E')$, $\nu_{11}(E')$, and $\nu_{12}(E'')$ bands on the Cambridge instrument. Each band shows some structure due to "Q-branches" and to band-heads as explained in the end of Sect. 2.1.5. The rotational structure of the ν_9 band was assigned and analyzed simultaneously with the infrared ν_9 band recorded by DUNCAN [2.104]. An apparent calibration discrepancy of 0.6 cm^{-1} was circumvented by using differences between ground state combination differences. They obtained $C_0 = 0.4177 \pm 0.0003$ cm^{-1} and $D_0^K = (-7.4 \pm 2) \cdot 10^{-6}$ cm^{-1}. From the ν_{11} band they got $\nu_{11} = 867.37 \pm 0.03$ cm^{-1} and $(C\zeta)_{11} = -0.399 \pm 0.003$ cm^{-1}.

Benzene. Very recently HOLLINGER and WELSH [2.105] have observed all seven Raman-active fundamentals of C_6H_6, using the cell described by RICH and WELSH [2.55,56] and photographic detection. Only partly resolved rotational structures were obtained. The E_{2g} fundamental at 1178 cm^{-1}, shown in Fig.2.15, has been analyzed by simulating the contour as shown in the figure. The fundamental frequency was found as 1177.775 $\pm$ 0.010 cm^{-1} and ζ as 0.010 ± 0.001.

2.3.5 Asymmetric Top Molecules

The ground state constants of asymmetric top molecules can usually be determined with high precision from microwave, high-resolution infrared or pure rotational Raman spectra. The main interest for rotation-vibrational Raman spectra of D_{2h} and C_{2v}

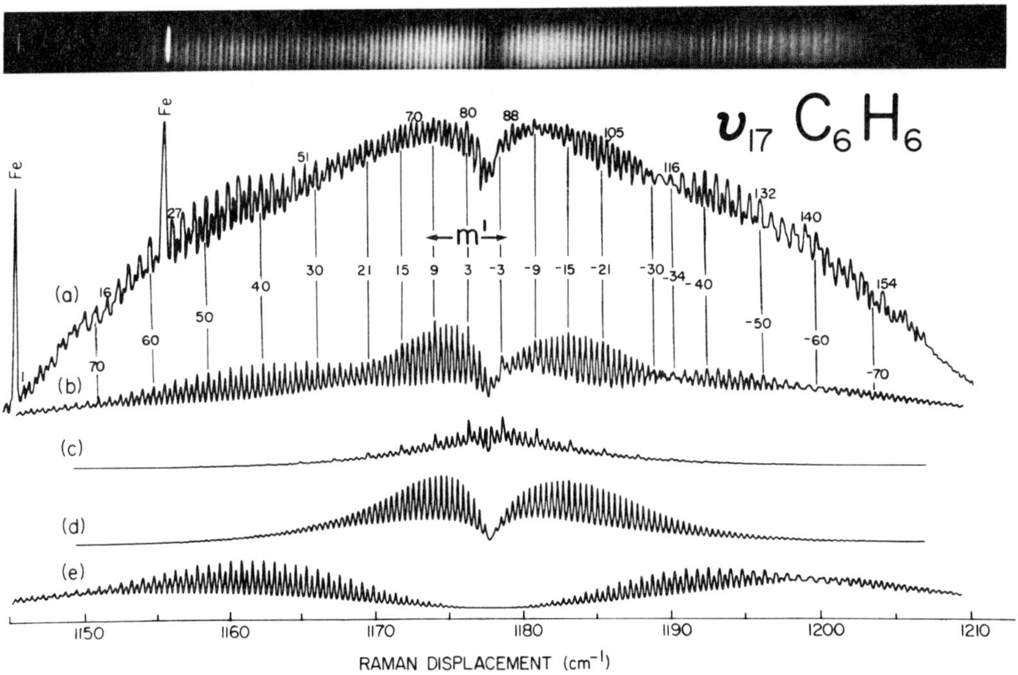

Fig. 2.15a-e. The E_{2g} fundamental of benzene at 1178 cm^{-1} as observed and calculated by HOLLINGER and WELSH [2.105]. The observed spectrum (a) is obtained by a 65 hours exposure with a sample pressure of 80 torr and a slit width of 0.15 cm^{-1}. The computed Q-branch (c), P-, and R-branches (d) and O- and S-branches (e) add up to the total computed contour (b)

molecules lies in obtaining information on those fundamental vibrational states which are inactive in infrared.

In the following the spectra of ethylene, allene-d_2, dideuteromethane and hydrazoic acid will be discussed, these being the only molecules for which rotation-vibrational Raman spectra are available.

Ethylene. Being of D_{2h} symmetry, the ethylene molecule has 6 Raman active fundamental vibrations, three of symmetry A_{1g}, two of symmetry B_{1g} and one of symmetry B_{2g}. The fundamental bands are of type A, B_c and B_b, respectively (see Sect.2.1.6).

Out of these the three totally symmetric bands ν_1, ν_2, ν_3, and the B_{1g} band ν_5 of C_2H_4 have previously been observed with Hg excitation by FELDMAN et al. [2.106]. Based on almost identical spectra from the Cambridge and Aarhus instruments DUNCAN et al. [2.107] have reanalyzed the ν_5 band. In these experiments the isotropic part of the overlapping ν_1 band was removed and the new assignment was confirmed by a simulated contour. They obtained $\nu_5 = 3083.19 \pm 0.10$ cm^{-1}.

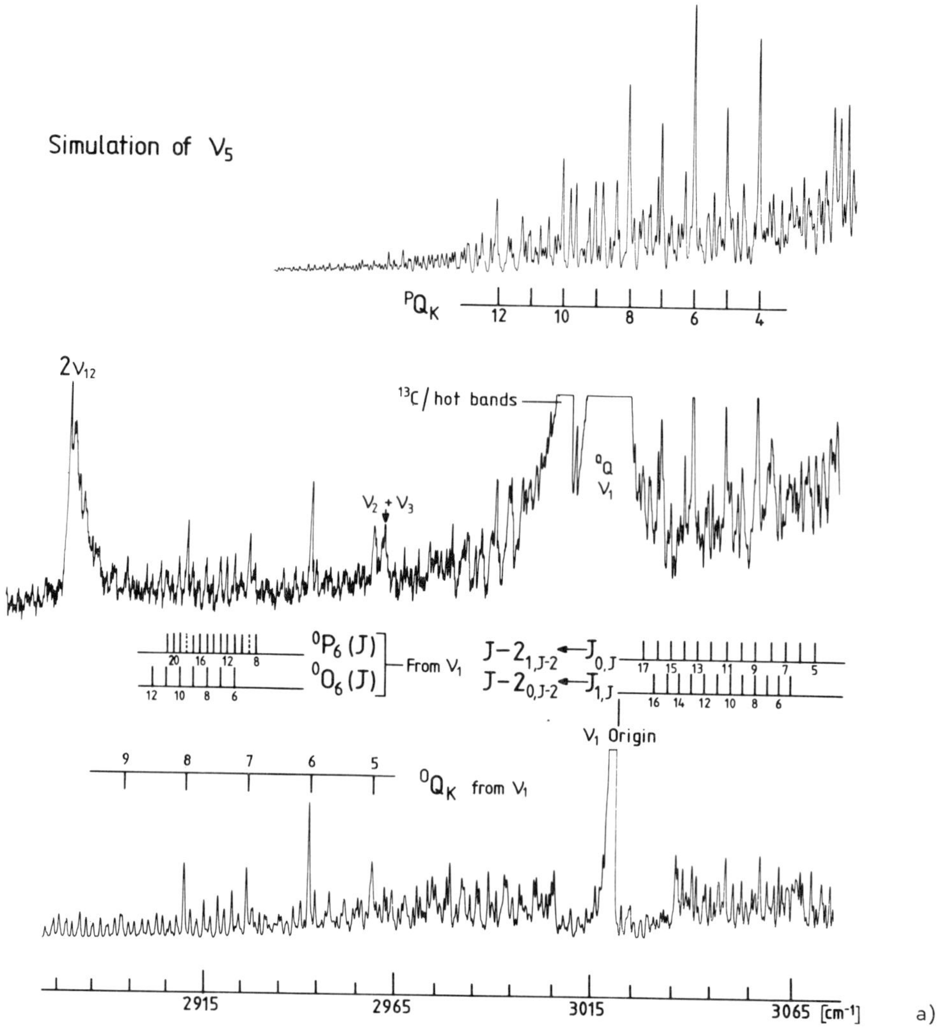

Fig. 2.16a and b. The $\nu_1(A_1)$ and $\nu_5(B_{1g})$ bands of C_2H_4 as observed and calculated by FOSTER et al. [2.48]. The sample pressure was 1 atm. and the exposure time 8 hours. The isotropic part of the spectrum was included in the experiment, but not in the calculations. (Reproduced by permission)

Recently FOSTER et al. [2.48] recorded the ν_1, ν_3, and ν_5 bands, this time including the isotropic scattering. The ν_3 band could not be satisfactorily simulated by theoretical contours, probably due to a Coriolis resonance with ν_6. A series of "Q-branches" with $\Delta K = -2$ and a spacing of approximately $4(A-\bar{B})$ was assigned in the anisotropic component of ν_1. The ν_1 and ν_5 bands were analyzed. A number of upper state constants were determined, including $\nu_1 = 3022.03 \pm 0.05$ cm^{-1}, $\nu_3 = 1343.54 \pm 0.05$ cm^{-1} and the ratios between the vibrational matrix elements of (2.44) for the ν_1 and ν_3 bands. The observed ν_1 and ν_5 bands are shown in Fig.2.16, together with computed curves for ν_5 and the anisotropic component of ν_1.

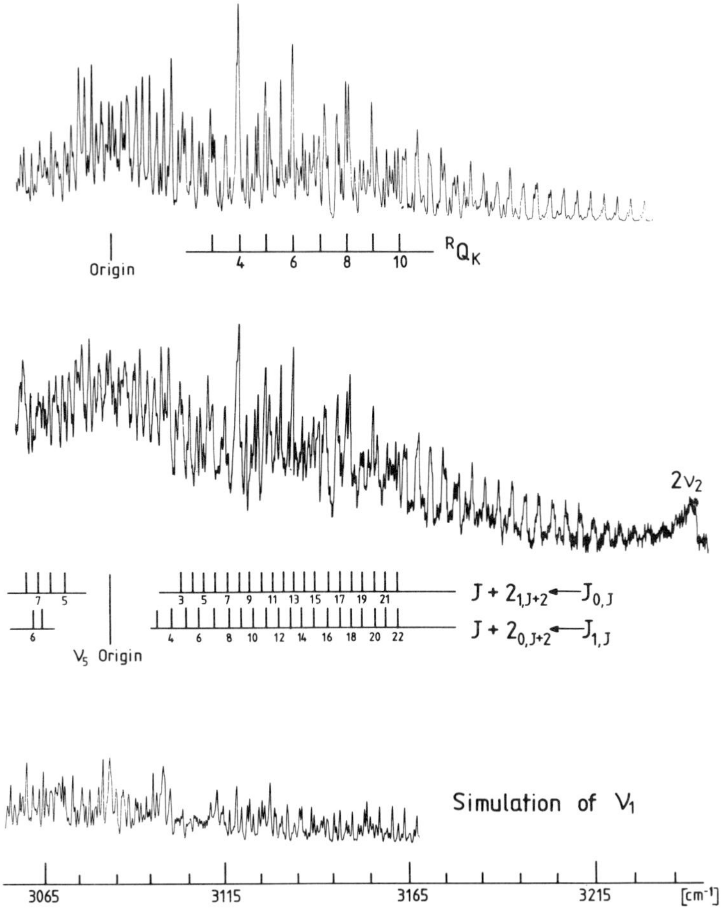

Fig. 2.16b

Figure caption see opposite page

The ν_1, ν_2, ν_3, ν_5, and ν_8 (B_{2g}) bands of C_2D_4 have been photographed on the Aarhus instrument and are presently being analyzed [2.108].

Allene-1,1-d$_2$. The substitution of two deuterium atoms in the symmetric top molecule allene makes it an asymmetric top of symmetry C_{2v} with $\kappa = -0.995$. The main effect of this is to lift the degeneracy of the E vibrations, each being split into a B_1 and a B_2 vibration, possibly far apart in frequency. The effect on the rotational energy is small and the symmetric rotor approximation may be used, except for very small values of K ($=K_a$). However, if two formerly degenerate vibrational levels are close together, a strong Coriolis resonance will be present. Further, the former A and B_2 vibrations now have A_1 symmetry and both $\Delta K = 0$ and $+2$ occur, in agreement with (2.44) and Table 2.2. A rotational structure results which is a superposition of the A and B_2 type bands discussed above for allene.

Most of the vibrational bands in allene-d_2 have been observed on the Aarhus instrument by HEGELUND [2.109]. All of the A_1 fundamental bands are dominated by a

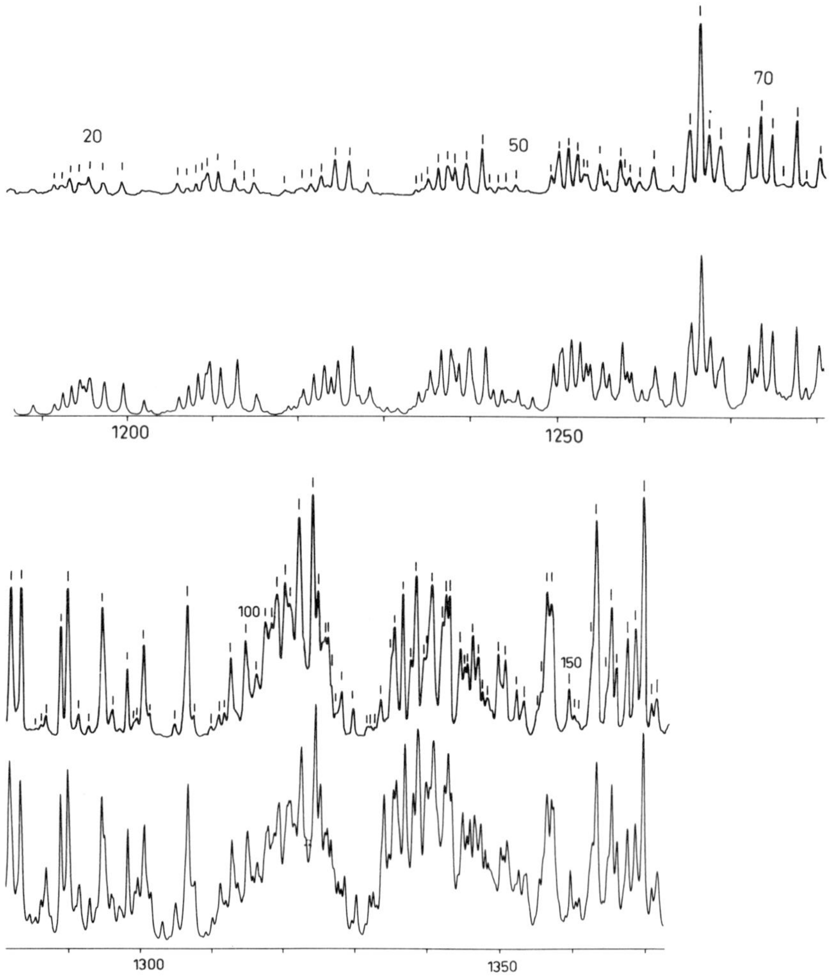

Fig. 2.17. The ν_5 band and part of the ν_3 band of CH_2D_2 observed (upper curve) and calculated (lower curve) by DEROCHE et al. [2.20]. The observed spectrum originates in three consecutive exposures of 50 to 72 hours duration at a sample pressure of 1 atm. and a slit width of 0.20 cm^{-1}. (Reproduced by permission)

strong central "Q-branch", corresponding to $\Delta K = 0$, and the ν_1, ν_5, and ν_6 bands show further a series of "Q-branches" for $\Delta K = 2$, spaced by $4(A-\bar{B})$. The formerly degenerate ν_8 and ν_{12} bands have the same structure as ν_8 in allene and allene-d_4 because they are too far apart for a Coriolis resonance, whereas the ν_{11} and ν_{15} bands are heavily perturbed.

A ground state combination difference analysis involving ν_1, ν_5, ν_8, and ν_{12} gave $A_0 = 3.2137 \pm 0.0019$ cm^{-1} and $D_0^K = (3.3 \pm 0.6) \cdot 10^{-5}$ cm^{-1}. Several upper state constants were determined, including $\nu_5 = 2227.67 \pm 0.03$ cm^{-1}, $\nu_{12} = 2323.30 \pm 0.03$ cm^{-1}, $\nu_{11} = 341.19 \pm 0.05$ cm^{-1}, and $\nu_{15} = 314.70 \pm 0.05$ cm^{-1}.

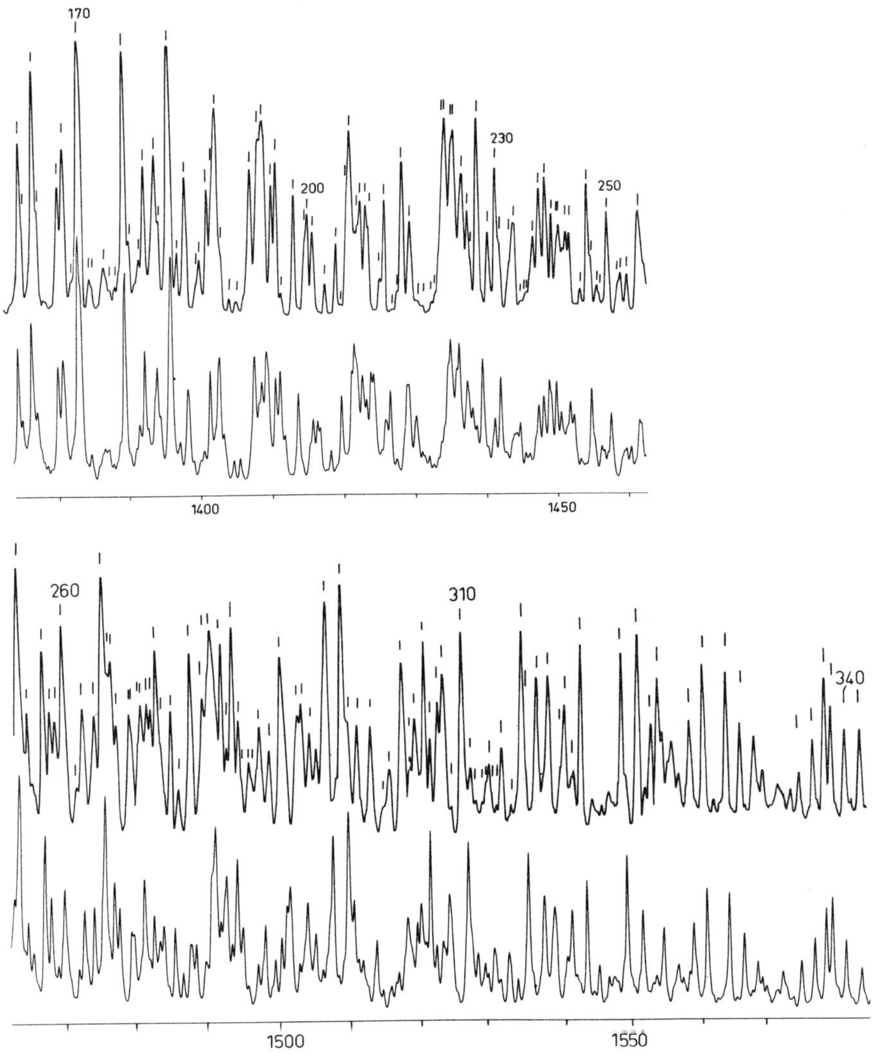

Fig. 2.17 (continued)

Dideuteromethane. This molecule is an asymmetric top of symmetry C_{2v} with $\kappa = -0.27$. All the fundamental vibrational states are quite well known from infrared spectra, except the ν_5 state, the only one of symmetry A_2 which is inactive in infrared but active in Raman. The ν_5 fundamental band has been recorded on the Aarhus instrument by DEROCHE et al. [2.20]. It is overlapped by the totally symmetric ν_3 band at high frequencies, but because both the ν_3 state and the ground state are well known, 114 ν_5 state energies could be determined from 329 lines assigned to ν_5. A least-squares analysis yielded $\nu_5 = 1331.41 \pm 0.05$ cm^{-1}, and values for A, B, C, Δ_J, Δ_{JK}, and Δ_K of the ν_5 state. In Fig.2.17 the observed spectrum is shown together with a computed contour for the ν_5 and ν_3 bands. The fitting of this contour allowed the determination of the proportion between the two vibrational matrix elements in (2.44) for the ν_3 band as well as the proportion between these and the matrix element for the ν_5 band.

Hydrazoic Acid. In addition to the pure rotational spectrum of this explosive compound (see Chap.3) it has been possible to photograph some of the vibrational bands of both HN_3 and DN_3 [2.110] on the Aarhus instrument. These molecules are slightly asymmetric, with $\kappa = -0.9992$ and -0.9977, respectively, because the hydrogen atom is out of line with the nitrogen atoms, and the ν_5 and ν_6 vibrations replace a degenerate bending vibration of a linear molecule. From infrared spectra ν_6 is known to be 607.0 cm^{-1} and 588.4 cm^{-1} for HN_3 and DN_3, respectively. The Raman spectra yield the preliminary values of $\nu_5 = 537.42 \pm 0.10$ cm^{-1} and 495.99 ± 0.10 cm^{-1}, respectively.

2.3.6 Spherical Top Molecules

The ground state constants of spherical top molecules can only be determined from infrared spectra if a sufficient number of perturbation-allowed transitions are observed. If this is not the case, as for CD_4, a simultaneous analysis of the triply degenerate ν_3 fundamental Raman band and the corresponding infrared band provides the best ground state constants.

The second point which makes the Raman spectra of spherical top molecules interesting is, as usual, the very different intensity patterns of the vibrational infrared and Raman spectra. For methane-like molecules only the two F_2 vibrations ν_3 and ν_4 are infrared active, whereas all four fundamentals are Raman sctive. However, the ν_4 Raman band happens to be very weak, whereas ν_2 is in fact visible in the infrared as a relatively weak band due to a Coriolis interaction with the nearby ν_4. Therefore, the real situation is the following: the totally symmetric ν_1 is observable in Raman spectroscopy only as an intense purely isotropic band. The doubly degenerate ν_2 may be observed in both kinds of spectra, in infrared techniques as a weak perturbation-allowed band, in Raman spectroscopy as a rather strong band, but naturally at moderate resolution. The triply degenerate ν_3 is observable with good intensity in both cases, but the selection rules are different. Finally, the triply degenerate ν_4 is in practice almost only observable by infrared spectroscopy.

As mentioned in Sect. 2.1.7., the operator $T_0^{(0)}$ of isotropic scattering is the only totally symmetric operator for spherical top molecules. Therefore, a totally symmetric band is purely isotropic,whereas all other bands as usual are purely anisotropic. This means that the experimental distinction between isotropic and anisotropic scattering is of special importance for these molecules. As demonstrated below this greatly facilitates the identification of the levels $2\nu_2$, $\nu_2+\nu_4$, and $2\nu_4$ in methane.

In the following the spectra of CH_4, CD_4, SiH_4, and SiD_4 will be discussed in some detail. The spectra of GeH_4, GeD_4, and SnH_4 have only been observed under fairly low resolution and will not be treated here. Finally some recent spectra of SF_6 (symmetry O_h) are mentioned.

Methane CH_4. The Raman spectrum of CH_4 has been photographed several times with Hg excitation, the latest study being by HERRANZ and STOICHEFF [2.7] who analyzed the ν_3 band in detail.

The ν_2 band has been recorded by CHAMPION and BERGER [2.40] on the Dijon instrument with a resolution between 0.10 and 0.13 cm^{-1}, which enabled them to resolve most of the tetrahedral splitting (see Sect.2.1.7). This is a considerable improvement compared to the corresponding infrared spectra. Constraining the ground state constants to the values determined by TARRAGO et al. [2.83] from perturbation-allowed transitions in the infrared ν_3 band, CHAMPION [2.23] assigned 126 transitions and determined 8 significant constants for the ν_2 state. The correctness of the analysis is demonstrated by an almost perfect agreement between the observed and a calculated

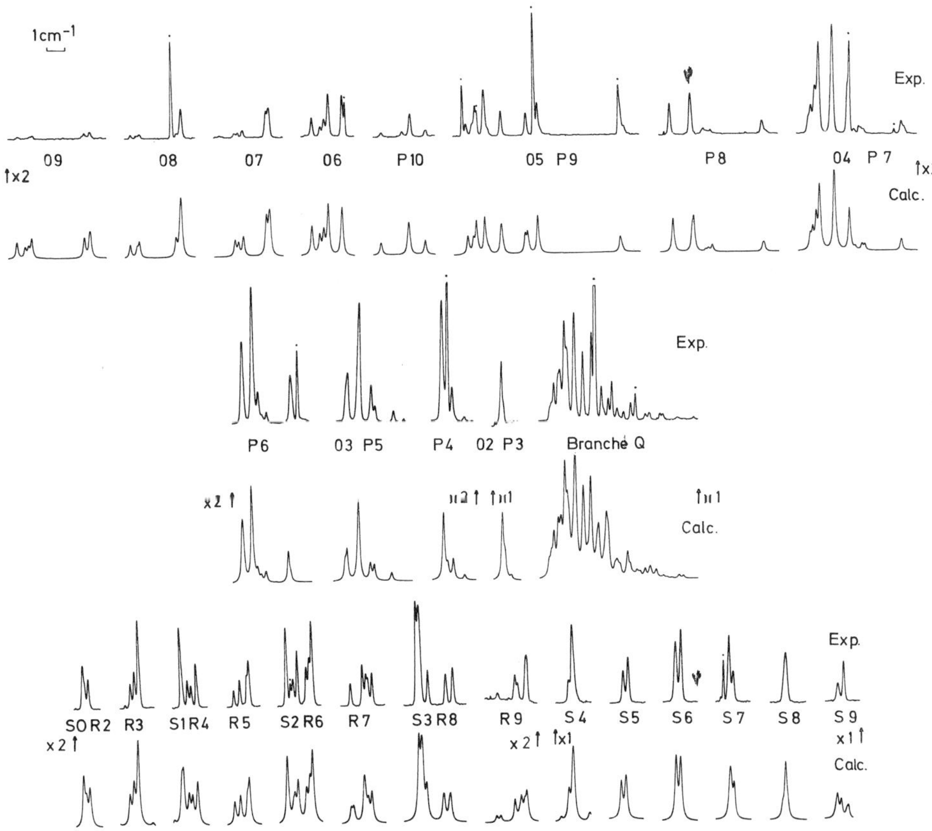

Fig. 2.18. Comparison by CHAMPION [2.23] between the observed and the computed spectrum of ν_2 of CH_4. The observed spectrum is from three exposures with sample pressure from 300 to 500 torr, slit width 0.05 to 0.057 cm^{-1} and exposure times from 30 to 87 hours. Lines marked by * are Th lines used for calibration. (Reproduced by permission)

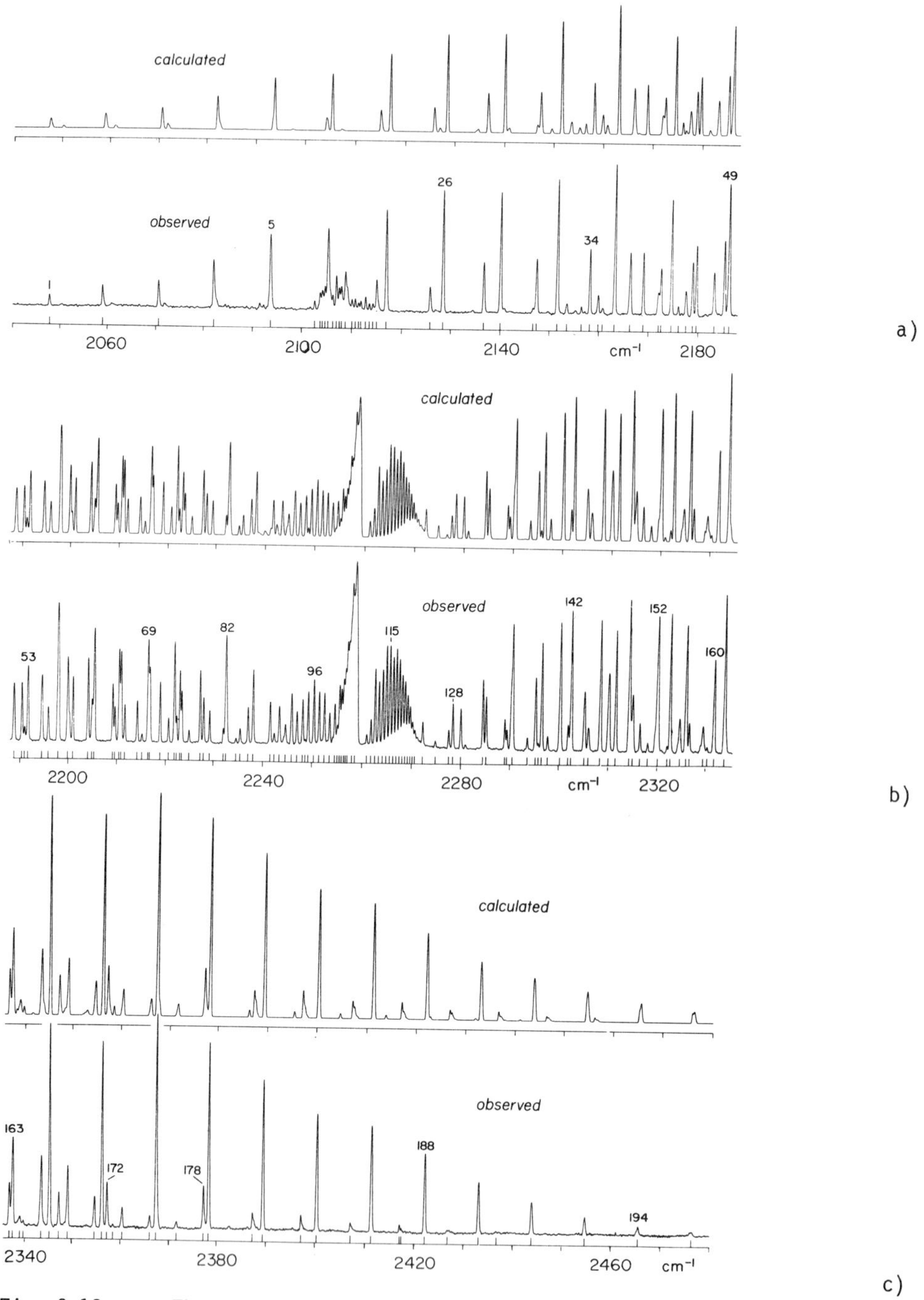

Fig. 2.19 a-c. The ν_3 band of CD_4 observed and calculated by BRODERSEN et al. [2.116]. The observed trace is from two consecutive exposures at 235 torr, 0.15 cm^{-1} slit width and 16 hours exposure time. The weak lines near 2110 cm^{-1} is the isotropic ν_1 band showing up because of the finite cone of observation. The computed curve includes ν_3 only. It is convoluted with a Gaussian slit function of 0.28 cm^{-1} width. (Reproduced by permission)

contour, as shown in Fig.2.18. Later GRAY and ROBIETTE [2.111] improved the constants by analyzing the ν_2 Raman band simultaneously with the ν_4 infrared band.

The ν_3 band has been recorded by RICH and WELSH [2.55] in Toronto without significantly improving the resolution achieved by HERRANZ and STOICHEFF [2.7], and by BERGER et al. [2.36] on the Dijon instrument with an improved resolution of about 0.15 cm^{-1}. But none of these spectra have been analyzed. Recently BERGER [2.112] has photographed the spectrum from 2850 to 3100 cm^{-1} at a resolution of 0.10 cm^{-1}. This includes a large part of the ν_3 band and assignments are given. Some lines in the very weak ν_4 band have been recorded by BERGER et al. [2.39] on the Dijon instrument, but details are not available. Probably both bands are only of marginal interest compared to the corresponding infrared bands.

With Hg excitation the ν_1 band appears as an extremely strong line at 2917 cm^{-1} without structure. Using a He-Ne laser CLEMENTS and STOICHEFF [2.113] found some structure, and recently BERGER [2.112] in the above mentioned spectrum partly resolved the strong band into 10 lines. An even higher resolution has recently been achieved by BOQUILLON et al. with the CARS technique (see Chap.7). This structure has not yet been analyzed.

In the same paper BERGER showed by comparison with a computed contour of ν_3 that a number of lines near ν_1 (2860-2920 cm^{-1}) and between 3060 and 3080 cm^{-1} did not belong to ν_3, and he assigned the first group to ν_1, the second to the A_1 component of $2\nu_2$ (which has the symmetry A_1+E). He states that all these lines are purely isotropic but without experimental evidence.

This statement has very recently been confirmed by recording the 2450 to 3125 cm^{-1} range at the Aarhus instrument [2.114] with two different orientations of the laser, making possible a complete experimental separation of the isotropic and anisotropic parts of the spectrum. This investigation has revealed a third group of isotropic lines and a very weak anisotropic band both near 2600 cm^{-1} (see Fig.2.6). The weakest of these isotropic lines are about 400,000 times less intense than ν_1. The isotropic band is due to the A_1 component of $2\nu_4$ (symmetry A_1+E+F_2), and the anisotropic band is the E or the F_2 component (or both) of $2\nu_4$. The most likely explanation of the isotropic lines near ν_1 is that they belong to the $\nu_2+\nu_4$ band (symmetry F_1+F_2), appearing as isotropic scattering because of a mixing with ν_1 due to a rotational resonance. This example demonstrates clearly the usefulness of the experimental separation between isotropic and anisotropic scattering, a feature without counterpart in infrared rotation-vibrational spectra.

Tetradeuteromethane. The complete Raman spectrum of CD_4 has been photographed with Hg excitation, the latest study being by OLAFSON et al. [2.115]. The whole spectrum has been recorded on the Aarhus instrument, using both orientations of the laser for the 1860 to 2240 cm^{-1} range.

The ν_3 band has been analyzed by BRODERSEN et al. [2.116] in conjunction with a high-resolution spectrum of the corresponding infrared band recorded by POUSSIGUE et al. [2.117]. The inclusion of the infrared data allows a calculation of the tetrahedral splitting pattern in the Raman spectrum and a correction of the observed peak-frequencies for the effect of the splitting. The analysis yielded the greatly improved ground state constants $B_0 = 2.63264 \pm 0.00013$ cm^{-1} and $D_0 = (2.74 \pm 0.15) \cdot 10^{-5}$ cm^{-1}, in addition to a large number of upper state constants. The fit between the observed and computed contour is shown in Fig.2.19.

The ν_2 band is presently being analyzed [2.118]. The tetrahedral splitting is not nearly as well resolved as for CH_4 [2.40] but the constants derived should be better than those from the infrared spectra published to date.

In contrast to the CH_4 case, the ν_1 band at about 2110 cm^{-1} appears as a very complicated bunch of isotropic lines, weakly visible on the experimental curve in Fig.2.19. Two other groups of isotropic lines, one near 1960 cm^{-1} and one round 2090 cm^{-1} were already assigned to $2\nu_4$ and $\nu_2+\nu_4$ by OLAFSON et al. [2.115]. None of these bands have been analyzed yet. The isotropic lines due to the A_1 component of $2\nu_2$ have not been observed and must be very weak in the present case.

Silane. WILLETS et al. [2.119] have recorded the ν_3 bands of SiH_4 and SiD_4 on the Cambridge instrument. In both cases the splitting into sub-branches was observed, but not the tetrahedral splitting. In the SiH_4 band a perturbation was observed, whereas the SiD_4 band appeared unperturbed. Ground state constants were derived for both molecules, including $B_0 = 2.8556 \pm 0.0015$ cm^{-1} for SiH_4 and $B_0 = 1.4338 \pm 0.0017$ cm^{-1} for SiD_4. Also a number of upper state constants were obtained for both molecules. Later it has been realized that the perturbation in the SiH_4 band is so strong that many of the assignments are incorrect [2.120]. This explains the deviation of the B_0 value for SiH_4 from the more recent determination by PIERRE et al. [2.121] of $B_0 = 2.859065 \pm 0.000010$ cm^{-1} from perturbation-allowed transitions in the infrared $2\nu_3$ band (the value given here is from a reanalysis by OZIER et al. [2.122]).

Sulphur Hexafluoride SF_6. Recently RUBIN et al. [2.123] have observed the ν_1 (A_{1g}), ν_2 (E_g), $\nu_5(F_{2g})$, and $2\nu_6$ (A_{1g} component) Q-branches of SF_6, placing the cell in a cavity outside the laser and using photoelectric detection. The focal length of the spectrometer was 2.5 m and the observed width of the calibration lines was 0.35 cm^{-1}. The observed width of the ν_2 Q-branch was only slightly larger. Estimates are given for the ν_1, ν_2, ν_5, and ν_6 fundamental frequencies. The ν_2 band has recently been studied on the Dijon instrument by Berger et al. [2.41] yielding $B_0 = 0.09111 \pm 0.00005$ cm^{-1}.

2.3.7 Conclusion

In a general sense high-resolution rotation-vibrational Raman spectroscopy is not able to compete with high-resolution infrared spectroscopy or with microwave spectroscopy. The resolution obtained in practice is by no means competitive, especially for weak bands.

However, the Raman experiment basically comprises a number of possibilities not present in an absorption experiment: the measurement of polarizability derivatives, the distinction between isotropic and anisotropic scattering, the respectively less and more relaxed selection rules for these kinds of scattering and the displacement of the observation of the low-frequency molecular transitions to an experimentally convenient frequency range.

As seen from the results indicated above, these facilities are sufficient to make rotation-vibrational Raman spectroscopy a valuable extension of infrared spectroscopy, able to yield for many different molecules informations which cannot be obtained as accurately from other sources.

Acknowledgements. I am greatly indebted to Dr. A.G. Robiette and to my colleagues F. Rasmussen, J. Bendtsen and F. Hegelund for stimulating and enlightening discussions, and to all the staff members who made the experiments in this laboratory possible.

References

2.1 Stoicheff, B.P.: "High Resolution Raman Spectroscopy", in *Advances in Spectroscopy*, ed. by H.W. Thompson, Vol.I (Interscience, New York 1959) pp.91-174
2.2 Weber, A.: "High Resolution Raman Studies of Gases", in *The Raman Effect*, ed. by A. Anderson, Vol.2 (Marcel Dekker, New York 1973) pp.543-757
2.3 Edwards, H.G.M., Long, D.A.: In *Molecular Spectroscopy*, ed. by Barrow,R.F., Long,D.A., Millen,D.J., Vol.I (Chemical Society, London 1973) pp.285-308
2.4 Edwards, H.G.M.: In *Molecular Spectroscopy*, ed. by Barrow,R.F., Long,D.A., Millen,D.J., Vol.3 (Chemical Society, London 1975) pp.383-412
2.5 Placzek, G., Teller, E.: Z. Phys. *81*, 209 (1933)
2.6 Placzek, G.: "Rayleigh-Streuung und Raman Effekt", in *Handbuch der Radiologie*, ed. by E. Marx, Vol.VI, Teil 2 (Akademische Verlagsgesellschaft, Leipzig 1934) pp.205-374. English translation by A. Werbin, "Rayleigh and Raman Scattering", UCRL Transl. No.526L, available from National Technical Information Service, U.S. Department of Commerce, Springfield, Va.
2.7 Herranz, J., Stoicheff, B.P.: J. Mol. Spectrosc. *10*, 448 (1963)
2.8 Rasmussen, F., Brodersen, S.: J. Mol. Spectrosc. *25*, 166 (1968)
2.9 Hegelund, F., Rasmussen, F., Brodersen, S.: J. Raman Spectrosc. *1*, 433 (1973)
2.10 Hougen, J.T.: J. Chem. Phys. *37*, 1433 (1962)
2.11 Wilson, E.B., Decius, J.C., Cross, P.C.: *Molecular Vibrations* (McGraw-Hill, New York 1955) Sect.3-7
2.12 Chiu, Y.-N.: J. Opt. Soc. Am. *60*, 607 (1970)
2.13 Condon, E.U., Shortley, G.H.: *The Theory of Atomic Spectra* (Cambridge University, Cambridge Press 1964)
2.14 Amat, G., Nielsen, H.H., Tarrago, G.: *Rotation-Vibration of Polyatomic Molecules* (Marcell Dekker, New York 1971)
2.15 Mills, I.M.: Mol. Phys. *8*, 363 (1964)
2.16 Brodersen, S., Rasmussen, F., Hegelund, F.: J. Mol. Spectrosc. *54*, 285 (1975)
2.17 Garrison, A.K., Simmon, J.W., Alexander, C.: J. Chem. Phys. *45*, 413 (1966)
2.18 Betrencourt-Stirnemann, C.: Thesis, Université Paris XI, 1975
2.19 Brodersen, S.: J. Mol. Spectrosc. *71*, 312 (1978)
2.20 Deroche, J.C., Graner, G., Bendtsen, J., Brodersen, S.: J. Mol. Spectrosc. *62*, 68 (1976)
2.21 Hills, G.W., Foster, R.B., Jones, W.J.: Mol. Phys. *33*, 1571 (1977)

2.22 Kattenberg, H.W.: Thesis, University of Amsterdam (1974)
2.23 Champion, J.P.: J. Physique *36*, 141 (1975)
2.24 Hilico, J.-C., Berger, H., Loëte, M.: Can. J. Phys. *54*, 1702 (1976)
2.25 Robiette, A.G., Gray, D.L., Birss, F.W.: Mol. Phys. *32*, 1591 (1976)
2.26 Weber, A., Porto, S.P.S., Cheesman, L.E., Barrett, J.J.: J. Opt. Soc. Am. *57*, 19 (1967)
2.27 van Cittert, P.H.: Physica *3*, 181 (1923)
2.28 Bird, G.R., Jones, R.C., Ames, A.E.: Appl. Opt. *8*, 2389 (1969)
2.29 Smith, A.G., Schrader, H.W., Richardson, W.W.: Appl. Opt. *10*, 1597 (1971)
2.30 Schlupf, J., Weber, A.: Appl. Opt. *10*, 2373 (1971)
2.31 Jammu, K.S., Welsh, H.L.: Appl. Opt. *15*, 423 (1976)
2.32 Unpublished work from this laboratory
2.33 Gollnow, H., Hagemann, G.: Astron. J. *61*, 399 (1956)
2.34 Tomkins, F.S., Fred, M.: J. Opt. Soc. Am. *41*, 641 (1951)
2.35 Moret-Bailly, J., Berger, H.: C.R. Acad. Sc. Paris *269*, 416 (1969)
2.36 Berger, H., Faivre, M., Champion, J.P., Moret-Bailly, J.: J. Mol. Spectrosc. *45*, 298 (1973)
2.37 Champion, J.P., Berger, H.: J. Mol. Spectrosc. *55*, 15 (1975)
2.38 Loëte, M., Berger, H.: J. Mol. Spectrosc. *68*, 317 (1977)
2.39 Berger, H., Champion, J.P., Klöckner, W., Schrötter, H.W.: In *Proc. 5th Intern. Conf. on Raman Spectroscopy*, ed. by Schmid,E.D., Brandmüller,J., Kiefer,W., Schrader,B., Schrötter,H.-W. (H.F. Schulz Verlag, Freiburg/Br 1976) p.442
2.40 Champion, J.P., Berger, H.: J. Physique *36*, 135 (1975)
2.41 Berger, H., Aboumajd, A., Saint.Loup, R.: J. Phys. Lettres *38*, 373 (1977)
2.42 Brodersen, S., Bendtsen, J.: J. Raman Spectrosc. *1*, 97 (1973)
2.43 Brodersen, S., Bendtsen, J.: J. Raman Spectrosc. *3*, 207 (1975)
2.44 Brodersen, S., Rasmussen, F., Bendtsen, J.: J. Raman Spectrosc. *6*, 67 (1977)
2.45 Kattenberg, H.W., Brodersen, S.: J. Mol. Spectrosc. *59*, 126 (1976)
2.46 Butcher, R.J., Jones, W.J.: J. Mol. Spectrosc. *47*, 64 (1973)
2.47 Willets, D.V., Freedman, P.A., Jones, W.J.: J. Raman Spectrosc. *2*, 249 (1974)
2.48 Foster, R.B., Hills, G.W., Jones, W.J.: Mol. Phys. *33*, 1589 (1977)
2.49 Freedman, P.A., Jones, W.J.: J. Chem. Soc., Faraday Trans. II, *71*, 650 (1975)
2.50 Weber, A., Schlupf, J.: J. Opt. Soc. Am. *62*, 428 (1972)
2.51 Fletcher, W.H., Rayside, J.S.: J. Raman Spectrosc. *2*, 3 (1974)
2.52 Fletcher, W.H., Rayside, J.S.: Proc. 5th Intern. Conf. Raman Spectroscopy, ed. by Schmid,E.D., Brandmüller,J., Kiefer,W., Schrader,B., Schrötter,H.W. (H.F. Schulz Verlag, Freiburg/Br 1976) p.401
2.53 Fletcher, W.H., Rayside, J.S., Wiley, W.J.: Appl. Opt. *13*, 734 (1974)
2.54 Fast, H., Welsh, H.L.: J. Mol. Spectrosc. *41*, 203 (1972)
2.55 Rich, N.H., Welsh, H.L.: J. Opt. Soc. Am. *61*, 977 (1971)
2.56 Rich, N.H., Welsh, H.L.: Indian J. Pure and Appl. Phys. *9*, 944 (1971)
2.57 Edwards, H.G.M., Good, E.A.M., Long, D.A.: J. Chem. Soc., Faraday Trans. II, *72*, 865 (1976)
2.58 Wang, C.-H.W.: Thesis, Fordham University, New York (1975)
2.59 Stoicheff, B.P.: Can. J. Phys. *35*, 730 (1957)
2.60 Edwards, H.G.M.: Private communication
2.61 Edwards, H.G.M., Long, D.A., Love, R.: Raman Conf. Reims (1972)
2.62 Bendtsen, J.: J. Raman Spectrosc. *2*, 133 (1974)
2.63 Guelachvili, G.: Optics Commun. *8*, 171 (1973)
2.64 Amano, T., Hirota, E.: J. Mol. Spectrosc. *53*, 346 (1974)
2.65 Renschler, D.L., Hunt, J.L., McCubbin, T.K., Polo, S.R.: J. Mol. Spectrosc. *31*, 173 (1969)
2.66 Altmann, K., Strey, G., Hochenbleicher, J.G., Brandmüller, J.: Z. Naturforsch. *27*a, 56 (1972)
2.67 Rich, N.H., Lepard, D.W.: J. Mol. Spectrosc. *38*, 549 (1971)
2.68 Albritton, D.L., Harrop, W.J., Schmeltekopf, A.L., Zare, R.N.: J. Mol. Spectrosc. *46*, 103 (1973)
2.69a Wang, I-Yan, Weber, A.: Indian J. Pure and Appl. Phys. *16*, 358 (1978)
b Welsh, H.L.: Private communication
c Palmer, K.F., Mickelson, M.E., Rao, K.N.: J. Mol. Spectrosc. *44*, 131 (1972)
d Pliva, J.: J. Mol. Spectrosc. *44*, 145, 165 (1972)
2.70 Wang, I.-Y., Weber, A.: J. Chem. Phys. *67*, 3084 (1977)

2.71 Carreira, L.A., Carter, R.O., Durig, J.R., Lord, R.C., Milionis, C.C.: J. Chem. Phys. *59*, 1028 (1973)
2.72 Mantz, A.W., Connes, P., Guelachvili, G., Amiot, C.: J. Mol. Spectrosc. *54*, 43 (1975)
2.73 Weber, W.H., Maker, P.D., Peters, C.W.: J. Chem. Phys. *64*, 2149 (1976)
2.74 Weber, W.H., Aldridge, J.P., Flicker, H., Nereson, N.G., Filip, H., Reisfeld, M.J.: J. Mol. Spectrosc. *65*, 474 (1977)
2.75 Peters, C.W., Weber, W.H., Maker, P.D.: J. Mol. Spectrosc. *66*, 133 (1977)
2.76 Duckett, J.A., Mills, I.M., Robiette, A.G.: J. Mol. Spectrosc. *63*, 249 (1976)
2.77 Weber, W.H., Ford, G.W.: J. Mol. Spectrosc. *63*, 445 (1976)
2.78 Lolck, J.-E., Brodersen, S.: J. Mol. Spectrosc. (in press)
2.79 Bendtsen, J., Edwards, H.G.M.: J. Raman Spectrosc. *2*, 407 (1974)
2.80 Edwards, T.H., Brodersen, S.: J. Mol. Spectrosc. *54*, 121 (1975)
2.81 Edwards, T.H., Brodersen, S.: J. Mol. Spectrosc. *56*, 376 (1975)
2.82 Fletcher, W.H., Daunt, S.J.: Personal communication from W.H. Fletcher
2.83 Tarrago, G., Dang-Nhu, M., Poussigue, G., Guelachvili, G., Amiot, C.: J. Mol. Spectrosc. *57*, 246 (1975)
2.84 Graner, G.: Mol. Phys. *31*, 1833 (1976)
2.85 di Lauro, C., Guelachvili, G., Alamichel, C.: J. Physique *37*, 355 (1976)
2.86 Richardson, E.H., Brodersen, S., Krause, L., Welsh, H.L.: J. Mol. Spectrosc. *8*, 406 (1962)
2.87 Olson, W.B.: J. Mol. Spectrosc. *43*, 190 (1972)
2.88 Kattenberg, H.W., Brodersen, S.: J. Raman Spectrosc. *4*, 323 (1976)
2.89 Jennings, D.E., Blass, W.E.: J. Mol. Spectrosc. *55*, 445 (1975)
2.90 Jones, E.W., Popplewell, R.J.L., Thompson, H.W.: Proc. R. Soc. A*290*, 490 (1966)
2.91 di Lauro, C., Mills, I.M.: J. Mol. Spectrosc. *21*, 386 (1966)
2.92 Escribano, R., Mills, I.M., Brodersen, S.: J. Mol. Spectrosc. *61*, 249 (1976)
2.93 Hirota, E., Tanaka, T., Saito, S.: J. Mol. Spectrosc. *63*, 478 (1976)
2.94 Winton, R.S., Gordy, W.: Phys. Lett. A*32*, 219 (1971)
2.95 Hegelund, F., Brodersen, S., Mills, I.M.: J. Mol. Spectrosc. *70*, 386 (1978)
2.96 Peterson, R.W., Edwards, T.H.: J. Mol. Spectrosc. *38*, 524 (1971)
2.97 Matsuura, H., Nakagawa, T., Overend, J.: J. Chem. Phys. *53*, 2540 (1970)
2.98 Antilla, R., Jaakkonen, S., Sahlström, T.: Spec. Acta *28*A, 1615 (1972)
2.99 Brodersen, S., Richardson, E.H.: J. Mol. Spectrosc. *4*, 439 (1960)
2.100 Butcher, R.J., Jones, W.J.: J. Raman Spectrosc. *1*, 393 (1973)
2.101 Hegelund, F., Bendtsen, J., Zangenberg, N.R.: J. Raman Spectrosc. *2*, 31 (1974)
2.102 Hegelund, F., Andersen, H.B.: J. Raman Spectrosc. *3*, 73 (1975)
2.103 Andersen, H.B., Hegelund, F., Zangenberg, N.R., Winther, F.: J. Raman Spectrosc. *6*, 238 (1977)
2.104 Duncan, J.L.: J. Mol. Spectrosc. *25*, 451 (1968)
2.105 Hollinger, A., Welsh, H.L.: Meeting of the Canadian Association of Physicists, June 1977, paper CE 5, and private communication from H.L. Welsh
2.106 Feldman, T., Romanko, J., Welsh, H.L.: Can. J. Phys. *34*, 737 (1956)
2.107 Duncan, J.L., Hegelund, F., Foster, R.B., Hills, G.W., Jones, W.J.: J. Mol. Spectrosc. *61*, 470 (1976)
2.108 Hegelund, F.: To be published
2.109 Hegelund, F.: J. Raman Spectrosc. *6*, 42 (1977)
2.110 Bendtsen, J.: To be published
2.111 Gray, D.L., Robiette, A.G.: Mol. Phys. *32*, 1609 (1976)
2.112 Berger, H.: J. Mol. Spectrosc. *66*, 55 (1977)
2.113 Clements, W.R.L., Stoicheff, B.P.: J. Mol. Spectrosc. *33*, 183 (1970)
2.114 Brodersen, S., Robiette, A.G.: To be published
2.115 Olafson, R.A., Thomas, M.A., Welsh, H.L.: Can. J. Phys. *39*, 419 (1961)
2.116 Brodersen, S., Gray, D.L., Robiette, A.G.: Mol. Phys. *34*, 617 (1977)
2.117 Poussigue, G., Tarrago, G., Dang-Nhu, M., Valentin, A.: J. Mol. Spectrosc. *49*, 183 (1974)
2.118 Robiette, A.G., Gray, D.L., Brodersen, S.: To be published
2.119 Willets, D.V., Jones, W.J., Robiette, A.G.: J. Mol.Spectrosc. *55*, 200 (1975)
2.120 Robiette, A.G.: Private communication
2.121 Pierre, G., Guelachvili, G., Amiot, C.: J. Physique *36*, 487 (1975)
2.122 Ozier, I., Lees, R.M., Gerry, M.C.L.: Can. J. Phys. *54*, 1094 (1976)
2.123 Rubin, B., McCubbin, T.K., Polo, S.R.: J. Mol. Spectrosc. *69*, 254 (1978)

3. High-Resolution Rotational Raman Spectra of Gases

A. Weber

With 19 Figures

One of the oldest applications of the Raman effect is to the study of molecular structure. By far the greatest amount of work has been and is currently being devoted to the study of liquid substances where the low-resolution Raman spectra give evidence for favoring one of several possible molecular geometries. Such work is done in conjunction with the technique of infrared absorption spectroscopy. For small molecules with an established geometry the technique of high-resolution Raman spectroscopy of low-pressure gases and vapors has furnished a substantial amount of empirical data giving quantitative knowledge about the moments of inertia of these molecules. In a few favorable cases even internuclear distances have been derived from these data. The results thus obtained take their place alongside the structural information derived from high-resolution infrared rotation-vibration spectra, microwave pure rotation spectra, and electron diffraction.

In a previous review [3.1] the whole field of spontaneous Raman spectroscopy of gases was surveyed, including the work that was done prior to the invention of the laser. In addition to dealing with the problem of deducing structural parameters of molecules, this review also dealt with problems of inner molecular dynamics, rotation-vibration spectra, intensities, and molecular interactions. In the present chapter the pure rotational Raman spectra of those molecules that have been studied with the laser technique will be discussed. For completeness of presentation older results, obtained by means of the mercury arc technique, will be included in the tabulations if no newer results based on the laser technique are available. The theory of the spontaneous Raman effect for free molecules has been described in various publications and a sufficiently detailed and complete modern formulation is contained in Chap.2.

Although the number of laboratories in which Raman spectra of gases are studied under high resolution is very small there exists considerable interest in both the experimental techniques employed and in the results obtained from such studies, and a fair number of review articles have been published in recent years on this topic [3.2-6]. The demands placed on the accuracy of the Raman results are such, however, that it is not out of place to initially present a rather detailed discussion of those experimental techniques that have been found suitable in the hope that they may be of use to those investigators who may also wish to engage in the study of high-resolution Raman spectroscopy of gases.[1]

[1]See also the discussion of experimental techniques by S. Brodersen in Chap.2.

3.1 Instrumentation and Techniques

The apparatus for generating highly resolved Raman spectra of gases consists of three essential parts, the laser, the spectrograph, and the scattering cell together with the associated optics. Each of these principal components requires special attention.

3.1.1 The Raman Source Unit

a) *The Laser System*

Except for the initial experiments with He-Ne lasers [3.7,8] which served to demonstrate the feasibility of using cw lasers as a source for Raman spectroscopy of gases, and a few others [3.9-11], nearly all subsequent work in this field has been done by means of the argon ion laser. Various commercially available units have been employed and Table 3.1 lists the wavelengths and wavenumbers of the various laser lines produced by an argon laser [3.12a]. A fairly complete listing of laser lines produced by various other lasers was recently published [3.12b]. The laser output power is not the same for all the lines listed in Table 3.1 and generally only the strongest ones are employed to generate Raman spectra. Little information is available on the spectral widths of these lines since they depend on the laser tube current and the output power. For a free running laser operating in multimode the width FWHM (Full Width at Half Maximum) of the 4880 Å line is at least 0.15 cm^{-1} [3.13a] for low laser powers and may reach the value of 0.3 cm^{-1} for a power level of 5 W. For high resolution studies such laser line widths are excessive and various techniques have been proposed for achieving a narrow laser line by having the laser operate in a single longitudinal mode. The most convenient of these techniques is the use of an intracavity ětalon [3.13b] by means of which a laser line of width (FWHM) $\simeq 0.001$ cm^{-1} is produced. In actual practice such narrow width of the exciting line is of no particular use in Raman spectroscopy of gases since the instrumental resolving power of available grating spectrographs are $R = \sigma/\Delta\sigma \leq 10^6$. At $\lambda = 5000$ Å this implies an instrumental limit of resolution of 0.02 cm^{-1}.

Table 3.1. Argon laser lines [3.12a]

λ_{air}[Å]	λ_{vac}[Å]	σ[cm^{-1}]
5145.3083	5146.7417	19 429.7686
5017.1628	5018.5621	19 926.0261
4965.0795	4966.4650	20 135.0456
4879.8635	4881.2264	20 486.6548
4764.8646	4766.1969	20 981.0886
4726.8683	4728.1906	21 149.7396
4657.9012	4659.2053	21 462.8879
4579.3495	4580.6327	21 831.0451
4545.0579	4546.3261	21 995.7826

The wavelengths given in Table 3.1 are those for lines produced in a hollow cathode discharge lamp [3.12a] and may be assumed to be identical to the center wavelengths produced by a free-running multimode laser of sufficient cavity length so that line displacements due to cavity mode pulling are absent. This requirement is fulfilled in practice by the free-running commercial laser units employed in Raman spectroscopy. However, significant deviations from the wavelength values cited in Table 3.1 may occur with lasers operating in single mode (see discussion below).

b) *The Scattering Cell*

A variety of scattering cells have been employed and some of the simplest arrangements for the 90° scattering configuration are shown in Fig.3.1. In these arrangements the cell may be located outside the laser cavity or within it. In the extracavity configurations the laser beam makes a single unfocused (a), or focused (b) pass through the cell. Lens L_2 collects the scattered radiation from the scattering volume and transfers it to the spectrograph. In arrangement (c) the addition of a

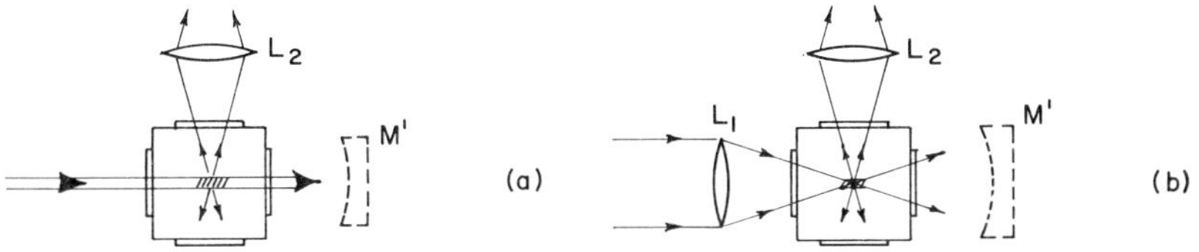

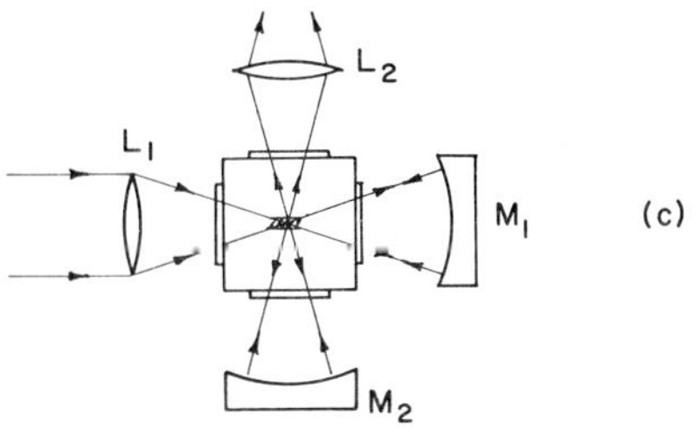

Fig. 3.1a-c. Simple Raman scattering cell arrangements for extra- and intracavity work

retroreflecting mirror M_1 enhances the intensity of the laser radiation in the scattering volume and a backing mirror M_2 serves to nearly double the collection efficiency for the scattered radiation. Intracavity arrangements result by removing the output mirror from the laser unit and replacing it by a fully reflecting mirror M' shown in dashed outline in (a) and (b). A simple removal of the laser output mirror converts (c) into an intracavity arrangement. The advantage of intracavity work is that the intensity of the exciting radiation is substantially enhanced over that

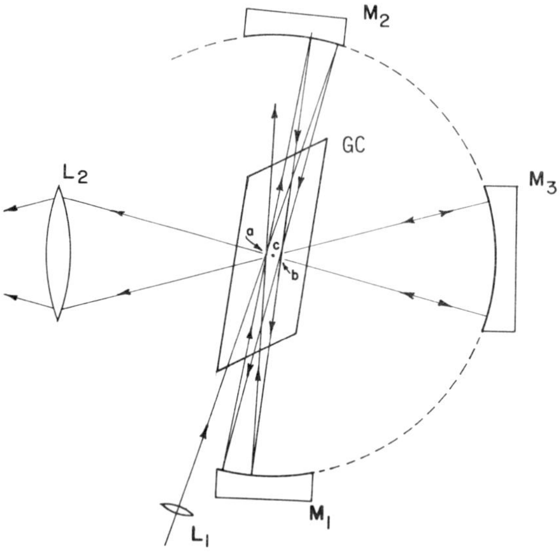

Fig. 3.2. Simple multiple-pass cell arrangement [3.18a]

possible in extracavity work. Lower gas pressures may then be used to record the Raman spectrum in a given amount of time or, conversely, for a given gas pressure the recording time may be reduced. For high resolution work it is preferable to work with low gas pressures in order to minimize the effects of pressure broadening of the Raman lines. It should be noted, however, that intracavity work is feasible only as long as the losses introduced into the laser cavity by the optical parts of the cell (i.e. windows and perhaps other components), the gas filling, and other devices placed into the laser cavity do not cause quenching of the laser oscillation. The efficiency of the system of Fig.3.1c was considered by PILZ and KRIEGSMANN [3.13c] and references cited therein.

A more complicated system is that described by RICH and WELSH [3.14] in which the volume of a typical multiple reflection mirror cell developed originally for use with the mercury arc excitation technique is efficiently illuminated by the output radiation of an argon laser. In this system the scattering configuration is longitudinal, scattered radiation formed at 0° and 180° with respect to the exciting radiation being collected by the mirror system of the cell. The volume of such a cell is, however, rather substantial. The cells needed for the schemes shown in Fig.3.1 are generally quite small, having a volume of only a few cm^3. They are thus more conservative in the use of rare and expensive gas samples.

The simple arrangements shown schematically in Fig.3.1 have been used by several investigators [3.7,11,15-17].

A very compact multiple pass system for extracavity work has been designed by KIEFER et al. [3.18a] and is shown in Fig.3.2 (see also [3.18b] for the use of such system in flame absorption spectroscopy). The two mirrors M_1 and M_2 are concentric spheres coated to maximum reflectance at 4880 Å and have their common center of curvature C located in the middle of the gas cell GC. The incident laser beam is focused by lens L_1 to a point slightly to one side of C and falls onto mirror M_2. After reflection the beam is reimaged to a point on the other side of C and then falls onto mirror M_1. This process repeats itself and is limited in the main by the aperture of the gas cell and the aberrations inherent in the off-axis operation of the mirrors M_1 and M_2. The scattered light is collected by the lens L_2 and the backing mirror. A gain factor of about 20 over the single focused beam arrangement has been realized by this technique.

A special high temperature cell for intracavity work has been developed by FREEDMAN et al. [3.19]. The cell is made of quartz and has been operated at a temperature of 900-950 K for generating the rotational spectrum of the diatomic sulphur molecule S_2.

A simple focused, multiple pass cell employing ellipsoidal mirrors has been developed by HILL and coworkers for extracavity work [3.20-22]. Gains of up to 100 in the Raman signal strength over that due to a single pass focused laser beam have been attained [3.21].

One of the most complex optical systems used in generating Raman spectra of low pressure gases incorporates a multiple-pass multiple-reflection cell placed inside the laser cavity [3.2,8,23]. This system is shown schematically in Fig.3.3. The Raman cell in the form of a cross is located inside the laser cavity which is defined by the fully reflecting laser resonating mirror LM and the Littrow prism + mirror combination LP+M. The laser beam is multipassed through the cell by means of a pair of plane beam folding mirrors FM which are adjusted to give 9 or 11 traversals of

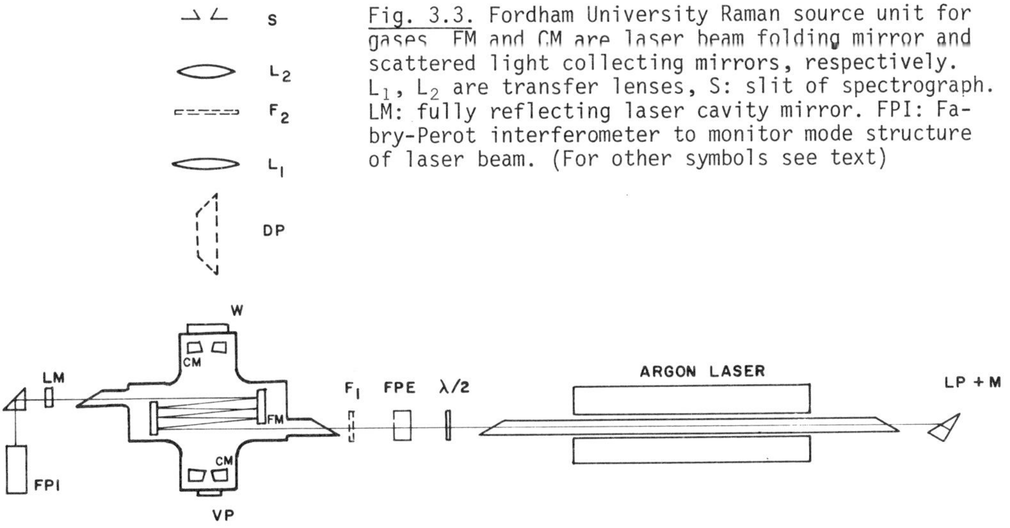

Fig. 3.3. Fordham University Raman source unit for gases. FM and CM are laser beam folding mirror and scattered light collecting mirrors, respectively. L_1, L_2 are transfer lenses, S: slit of spectrograph. LM: fully reflecting laser cavity mirror. FPI: Fabry-Perot interferometer to monitor mode structure of laser beam. (For other symbols see text)

the cell. The scattered radiation generated within the cell is collected by means of a typical 4-mirror multiple reflection mirror system CM [3.24] and emerges from the cell through the window W. The laser beam enters and leaves the cell through Brewster angle windows. In addition to the scattering cell the following components of the illuminating optics are also shown in the figure as located with the laser cavity: a halfwave plate $\lambda/2$, a solid Fabry-Perot étalon FPE, and a filter F_1.

c) *The Illuminating Optics*

Commercial lasers come usually equipped such that the beam polarization is linear and vertical whereas in a laboratory experiment the scattering plane is conveniently horizontal. In Fig.3.3 the halfwave plate, $\lambda/2$, allows the change of the laser beam polarization from the vertical to the horizontal. This is of advantage when studying pure rotation spectra since thereby the intensity of the Rayleigh scattered radiation is minimized while that of the Raman radiation is reduced only slightly compared to that produced when the exciting light is vertically polarized. The effectiveness of this procedure has been amply demonstrated for pure rotation spectra [3.1,3,8]. The control of the laser beam polarization is also of advantage in the study of rotation-vibration spectra.

For high-resolution studies it is desirable that the line width of the laser radiation be reduced to that of a single cavity mode. This is accomplished by placing a solid Fabry-Perot étalon FPE within the laser cavity. As shown by HERCHER [3.13b], the effect of a tilted intracavity étalon is to introduce a wavelength-dependent loss into the cavity. Only those axial-mode wavelengths that will resonate in both the laser and étalon cavity will be present in the output spectrum of the laser. If the free spectral range of the étalon is greater than the linewidth of the free-running laser, it is possible to have only one axial mode wavelength in the laser output whose linewidth is determined by a number of factors. Among these are the power coupled out of the cavity and the separation of the mirrors that make up the cavity [3.25,26]. However, if the wavelength of the single mode is allowed to shift randomly within the amplifying bandwidth of the laser medium, the effective linewidth of the output is increased. It is therefore imperative that the étalon cavity be stabilized against any change in its resonant wavelength if a narrow line with a single-mode laser is to be realized. The insertion of an étalon causes a reduction of the beam power to 50-75% of its pre-insertion value. However, the power per unit bandwith is actually enhanced from, say, 1 W/0.15 cm^{-1} = 6.7 W/cm^{-1} to 0.50×1 W/0.001 cm^{-1} = 500 W/cm^{-1}.

Also shown (in dashed outline) in Fig.3.3 is a filter F_1 on the input side of the Raman cell. Such a filter may on occasion be necessary and its use in intracavity work is to be avoided if possible. In addition to the emission of the laser radiation the laser discharge is also a strong source of incoherent ultraviolet radiation. With the Raman cell located inside the laser cavity this ultraviolet radiation (with wavelengths down to 2000 Å) will invariably also enter the Raman cell and interact with the gas sample. Possible photochemical reactions in the gas may result in the formation of particles. Any laser light scattered by these particles (Mie's scattering) will totally mask the weak Raman radiation that is of interest. The purpose of the filter F_1 is to absorb the incoherent ultraviolet radiation emitted by the laser discharge tube. A dispersion form of filter F_1 has been designed by CLAASSEN et al. [3.11]. It is very effective in its filtering action and also possesses the advantage of being tunable to the wavelengths of different exciting lines.

d) *The Transfer Optics*

In general the transfer optics consists of a simple lens that images the scattering regions (Fig.3.1,3) onto the slit of the spectrograph while a more complicated set of transfer optics is used with the multiple reflection cells of RICH and WELSH [3.14] and WEBER et al. (see Fig.3.3). Additional components might be an image rotator, shown as a Dove prism DP in Fig.3.3, to transform the horizontal scattering region into a vertical image on the spectrograph slit, a polarization analyzer, and a filter F_2. Since the rotational transitions are depolarized, polarization analyzers are not needed in the study of rotational Raman spectra.

3.1.2 Filters

The need for filters to prevent unwanted incoherent radiation emitted by the laser plasma from entering the Raman cell was described in the previous section. To prevent unwanted Rayleigh scattered radiation from entering the spectrograph and also to avoid the overlap of orders another filter, such as F_2 shown in Fig.3.3, is often inserted in the optical path ahead of the spectrograph. This filter is generally a glass absorption filter or an interference filter of appropriate band pass. In the cell design of RICH and WELSH [3.14], however, a polarization filter is used to suppress the strong Rayleigh scattered and also stray radiation that would otherwise enter the spectrograph.

Molecular iodine I_2 has been suggested as an absorption filter for those experiments in which the 5145 Å single-mode argon laser line is used to excite the Raman spectrum [3.27,28]. Such a filter is indeed very effective in absorbing the 5145 Å line but suffers from the disadvantage in that the absorption spectrum of molecular iodine is not restricted to this wavelength alone, but covers very densely the whole green and yellow portion of the visible spectrum. Two atlases of this absorption spectrum giving highly accurate (to ~0.004 cm^{-1}) line positions are now available [3.29,30]. While for moderate resolution spectroscopy line distortions due to the overlapping iodine absorptions may be small and tolerable, the iodine filter is definitely not recommended for use in high-resolution Raman spectroscopy of gases. The iodine absorption spectrum has, however, two other very useful applications namely, for convenient frequency stabilization of single mode argon lasers and as a source of a standard reference spectrum for the determination of unknown wavelengths or wavenumbers. These applications are described in Sect.3.1.5.

A very effective and flexible manner of removing the exciting radiation and at the same time have variable band pass filtering and order sorting is by means of a dispersion filter premonochromator. A variety of single [3.31-35] and double [3.15, 36-41] monochromators have been suggested and are in use, employing prisms or gratings as the dispersive elements.

The theory of the double monochromator has been developed by VAN CITTERT [3.36] and again more recently by LECLERCQ and WALLART [3.37] for both the subtractive (zero dispersion) and additive (double dispersion) arrangements. These investigators show that the zero-dispersion form of the double monochromator exhibits greater rejection of stray light than does the additive form. Several designs of the zero-dispersion double monochromator have been proposed employing prisms [3.15,38-40] or gratings [3.37,41] and their usefulness in Raman spectroscopy has been amply demonstrated.

Of the two types of dispersion filters, the single and the zero-dispersion double monochromator, it is the latter that is preferred for Raman spectroscopy. Not only is it a superior filter, but errors in wavelength determination which can be of great severity [3.34,42] are less likely to occur with its use. Other schemes of filtering are of lesser usefulness and are summarized elsewhere [3.41].

3.1.3 The Spectrograph

The demands placed on the resolving power and the accuracy of wavenumber determination in Raman spectroscopy of gases point to the photographic technique of recording the spectra as the method of choice. Moreover, due to the low intensities of the Raman scattered radiation, specially built spectrographs are generally used in preference over commercially available instruments [3.15,23,39a,43-48]. The characteristics of these instruments are summarized in Sect.2.2.2.

The Fordham instrument is patterned after the large coudé spectrographs used in astronomical research. Figure 3.4 shows schematically the optical arrangement. The collimator and camera mirrors are spheres of focal length 4 m and 2 m, respectively. The camera is of the Schmidt type but, since it operates with a focal ratio of $f/D \approx 14$, the customary Schmidt corrector plate is not used. The grating is a 300 groove/mm Bausch and Lomb echelle replicated from a master ruled under interferometric control. Both collimator and camera operate slightly off-axis yet the aberrations due to this arrangement are small enough to not materially affect the required resolving power. The limiting resolution of this instrument was experimentally determined to be 0.03 cm^{-1} at 4880 Å in the 12th order of diffraction [3.23]. This resolution is adequate for the study of the Raman spectra of most gases as long as the Doppler broadening of the spectral lines at room temperature defines the attain-

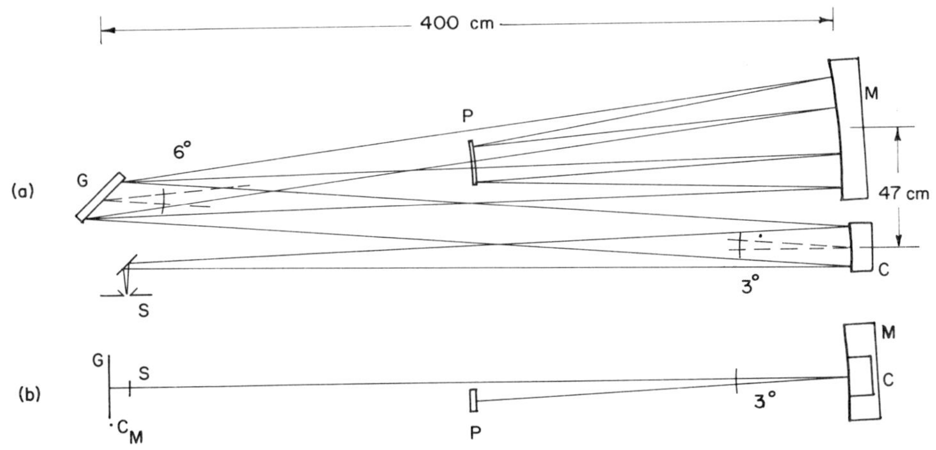

Fig. 3.4a and b. The Fordham Spectrograph (schematic) (a) plan, (b) elevation. S: slit of spectrograph. C and M are collimator and camera mirrors, respectively. G: diffraction grating. P: plate holder. The camera operates 3° off axis in the vertical plane to avoid vignetting of the diffracted beams by the plate holder. The center of curvature of the camera minor M is located in the plane of the grating and is shown by the dot labeled C_M

able limit of resolution. The curved focal field of two meters radius of curvature presents no practical problems as photographic plates are easily bent to this radius without breakage. The components of the spectrograph are independently mounted on three concrete piers which stand isolated from the floor of the laboratory in the basement of the building. The instrument is housed in a thermostatted room whose temperature is held constant to better than 0.1°C over the duration of a Raman exposure.

As is well known, spectrographs must be stable against mechanical and environmental changes. This requirement is judged to be especially severe in the case of Raman spectroscopy of gases where, as a routine matter, photographic exposure times of up to 100 hours and more are encountered. The required overall stability is generally difficult to achieve and its lack can give rise to systematic plate-to-plate shifts in the measured Raman frequencies [3.49] and can even give rise to a loss of resolution. Compensation of the distortion due to mechanical and other instabilities can be achieved by the scheme devised by CHAMPION and BERGER [3.46].

Although the need for a temperature stabilized spectrograph room is well understood and provided for [3.50], the need to also avoid variable thermal gradients is more difficult to satisfy for long focal length instruments.

Unless the spectrograph is a vacuum instrument the changes in the barometric pressure as well as in the relative humidity affect its dispersion since the index of refraction of air is a function of the pressure, temperature and the relative humidity [3.51,52].

The changes in the relative humidity of laboratory air are generally small, at least in modern, well air conditioned laboratories, and thus have a negligibly small effect on the dispersion. Changes in the barometric pressure are, however, best compensated [3.53,54]. Several effective compensation schemes [3.23,46,55-57] are now available and controlling the pressure inside the spectrograph room [3.15] is not advised. In a different scheme BRODERSEN and BENDTSEN [3.58] rely on the use of the (thorium) calibration spectra taken at the beginning and the end of the Raman exposure, and the assumption that the pressure change during the Raman exposure is such that its center of gravity is located at the midpoint of the exposure (linear pressure change). This method thus accounts for, but does not compensate barometric pressure fluctuations and lack of the assumed linearity of the pressure change can cause systematic plate-to-plate variations in the measured Raman displacements [3.49]. The equipment for the afore-mentioned compensation schemes is, however, simple to fabricate and install and use of one of these schemes is strongly recommended in high-resolution Raman spectroscopy of gases.

To overcome the low efficiency of the photographic emulsion an image intensifier may be incorporated in the spectrograph [3.44]. Although image intensifiers have been employed in astronomical spectroscopy for many years [3.59] there are a number of limitations and disadvantages that have prevented their wide use in high-resolution spectroscopy covering a broad spectral region. However, the usefulness of the image intensifier tube in Raman spectroscopy of gases has been amply demonstrated by obtaining the rotational spectrum of the sulphur diatomic molecule [3.19]. It is unlikely that the Raman spectrum of this molecule could have been photographed with a conventional spectrograph.

3.1.4 Recording Materials

For rotational Raman spectra excited by the argon laser lines listed in Table 3.1 the following photographic plates have been found to be suitable. Kodak spectroscopic plates 124-01 (formerly known as the 103a-J plate) IIa-D, and IIIa-J for spectra excited by the green and blue-green lines, and 103a-O and IIa-O for spectra excited by the blue lines. Other emulsion types, sensitive in the green, yellow and red regions may also be used. For spectra excited by the red 6328 Å He-Ne laser line the Kodak Ia-E and IIIa-F (formerly known as 127-04 and 127-05) plates are employed [3.60].

The type IIIa-J emulsion has a fine grain and is particularly well suited for high-resolution studies where a good signal/noise ratio is especially important. The sensitivity of the untreated IIIa-emulsion is very inferior to that of the 103a- or IIa-emulsions. However, by hypersensitizing the IIIa-emulsion its speed can be raised to a level comparable to that of the untreated 103a-emulsion. Of the various methods of hypersensitization that have been proposed [3.61] the method of "baking" in nitrogen or hydrogen gas at temperatures varying from 50-60°C [3.62-70] or simply "soaking" the plates for extended periods of time in a nitrogen or hydrogen atmosphere at room temperature [3.71-78] have been described in the recent literature. Although the baking in nitrogen technique does not yield the speed gains that can be attained with the baking or soaking in hydrogen technique, it is certainly safer and more convenient to employ. These methods of hypersensitizing photographic plates are not restricted to the IIIa-emulsion and comparative studies demonstrate their usefulness as well for the 103a-, IIa-, and other emulsions [3.67,72,73,79]. Disadvantages of the baking technique are in the batch-to-batch variations in the attainable gain in speed, the increase and non-uniformity of the fog level, and the poorer keeping qualities of the hypersensitized plates. There is also a slight increase in plate grain that results from baking in nitrogen. However, the overall advantage of hypersensitizing by baking in nitrogen gas has made the Kodak IIIa-J plate the plate of choice for recording rotational Raman spectra of gases excited by the radiations emitted by the argon-ion laser. Most attention has so far been paid to the increase in photographic speed by hypersensitization. A gain in speed does not, however, necessarily imply a gain in the more important detective quantum efficiency (DQE). The relationship between "speed" and "DQE" and the respective gains that result from hypersensitization has been the subject of several recent investigations [3.80].

The standard developers for the Kodak spectroscopic plates are D-19 and D-76. A specially formulated developer, MWP-2, possesses considerable advantages over the standard developers in terms of producing higher emulsion speed, lower granularity and improved edge effects [3.81].

It is generally tacitly assumed that the developed image on the processed photographic plate is a faithful record of the spectrum that forms the latent image in the emulsion. This is, however, not the case owing to the swelling and non-uniform shrinkage that is experienced by the emulsion in the developing and drying process. This emulsion shrinkage effect can cause non-uniform distortions in the positions of the spectrum lines. Three quantitative studies of this effect have recently been reported for a variety of spectroscopic and photogrammetric emulsions [3.82-84] and the Kodak Tropical Developer has been found to minimize the shrinkage effect [3.82]. The shifts in the positions of spectrum lines are of the order of 1 μm and may be as great as 20 μm near the edge of a plate developed in the standard manner. With a nominal plate factor of 2.0 cm^{-1} mm^{-1} this translates into a minimum wavenumber inaccuracy of 0.002 cm^{-1}. The shrinkage effect may be minimized by drying the plates in a horizontal position, or by placing them on edge with the direction of dispersion along the horizontal. Wavelength errors may be minimized by using a large number of standard wavelengths and employing a high order polynomial for the wavelength calibration of the plate.

3.1.5 Wavelength Standards

The iron and thorium hollow cathode lamps are now widely employed to furnish highly accurate wavelength standards for the calibration of spectrographs operating in the visible and ultraviolet regions of the spectrum. The wavelengths of the strong lines of the iron [3.85] and thorium [3.86,87] spectra are known to within a few 10^{-4} Å and, depending on the spectral region, the wavenumbers of the lines are accurate to better than 10^{-2} cm^{-1}. The thorium spectrum is preferred over the iron spectrum owing to the more numerous and sharper lines in the region of the Raman bands. However, it has been noticed by several investigators [3.88-90] that some of the thorium lines are ill suited for calibration purposes even though their published wavelengths are given to interferometric accuracy. As an alternative to either the Th or Fe wavelengths the rotational Raman spectrum of carbon monoxide may be used as an internal standard, this gas being present in the Raman scattering cell concurrently with the unknown sample gas [3.89,91].

The hollow cathode emission spectrum of uranium, and the molecular absorption spectrum of iodine vapor at room temperatures have recently been proposed to serve as secondary standards [3.92].

Fourier transform spectroscopy in the visible region was used to determine the wavenumbers, which have an accuracy of ± 0.004 cm^{-1} at 20,000 cm^{-1} (5000 Å). The iodine absorption spectrum is of particular usefulness since a very large number of sharp lines of width ≈ 0.025 cm^{-1} are densely spaced in the region of Raman bands located at $\lambda > 5000$ Å ($< 20{,}000$ cm^{-1}). Two atlases of the iodine absorption spectrum obtained with a grating spectrograph calibrated by means of microwave-excited thorium spectra [3.29] and with the Fourier transform technique [3.30] are now available and greater accuracy in the measured Raman shifts is to be expected through the use of the iodine absorption spectrum [3.94].

3.1.6 Determination of Raman Shifts

The photoelectric setting principle invented by TOMKINS and FRED [3.95] is the basis of nearly all modern research quality measuring engines used for the measurement of spectroscopic plates. Commercial instruments based on this principle and operating fully automatic and under computer control are now available, albeit at prohibitive cost. The classical spectrum measuring engine (spectrum comparator) based on the visual setting technique is, however, easily modified into a photoelectric setting comparator as described in numerous publications [3.95-104]. For sharp, symmetrical spectrum lines a setting precision of 0.5 μm or less is easily attained, but the accuracy of the measurement depends largely on the mechanical quality of the basic engine and that of the optical and electronic systems employed. The testing of such comparators for screw errors is described in [3.105,106] while errors due to improper phasing of the optical and electronic sweeps and the non-linearity of the sweep are discussed in [3.104]. Automatic and computer controlled comparators and microdensitometers have recently been developed to deal with the greatly increased amount of available plate material with improved precision and accuracy [3.107]. Two recent reports dealing with "in house" modification and construction are particularly noteworthy [3.108,109]. (See also the description of the Aarhus instrument in Sect.2.2.2).

Since all Raman spectroscopic work has been done so far with spectrographs operating in ambient air the determination of the Raman shifts requires knowledge of the index of refraction of air at the time of the exposure of the reference spectrum [3.51,52]. Several techniques for frequency or wavelength calibration are available [3.58,89,110]. The most commonly used method is the fitting of the wavelengths of the reference lines (Th, Fe, etc.) to a polynomial in the positions of these lines on the plate and interpolating the wavelengths of the Raman lines by means of this polynomial. The method employed at the Fordham University laboratory for deriving the wavenumbers of a given Raman band is based on the polynomial method also described in [3.110]. Only those thorium wavelengths determined with interferometric accuracy and recommended as secondary standards [3.86] and that also define a smooth correction curve to an assumed linear dispersion over the region of the band are employed as calibration wavelengths. The thorium lines are then further selected by fitting their wavelengths by means of an iterative least squares procedure to a fourth or higher-order polynomial, depending on the extent of the region to be calibrated. In the initial step all lines are used to determine the first set of polynomial coefficients. The wavelengths of the thorium lines and the standard deviation of the fit are then calculated. That reference line that has the largest deviation (obs-calc) in the wavelength is then rejected from the set and a new set of polynomial coefficients in computed with the remaining lines. This process is repeated and finally terminated either by retaining a predetermined number of reference lines in the calibrations set, by observing the behavior of the polynomial coefficients through the various iterations and terminating the refinement when the values of the coefficients begin to oscillate, or by simply requiring that the largest deviation (obs-calc) be less than three standard deviations. The final set of polynomial coefficients is then used to compute the wavelengths of the Raman lines and their vacuum wavenumbers using EDLÉN's dispersion formula [3.51].

For single-mode operation of the laser the wavenumber of the exciting line is generally not equal to the value given in Table 3.1. For pure rotational Raman spectra in which the average of the Stokes and anti-Stokes displacements gives the observed Raman shift the error due to an uncertain wavenumber value of the exciting line averages out to zero, provided that this uncertainty is less than the Raman line spacing. This is not the case for rotation-vibration bands and an independent determination of the wavenumber of the exciting line must be made. Exciting line wavenumbers different from those listed in Table 3.1 may also arise as the result of wavelength stabilization of the laser radiation. Thus, for the green 5145 Å lines of a low power argon laser stabilized by means of saturated absorption in an intracavity $^{129}I_2$-cell the vacuum wavelength was measured by SPIEWECK [3.111] as λ_{vac} =(5146.7465±0.0001) Å, giving a wavenumber σ = (19429.7504±0.0004) cm^{-1}, while with an extracavity $^{129}I_2$-cell the wavelength was found to be λ_{vac} = (5146.74206±0.00010) Å giving σ = (19429.76718±0.00038) cm^{-1} [3.112]. These wavenumbers differ from the one given in Table 3.1 by -0.0182 cm^{-1} and -0.00142 cm^{-1} respectively. There is reason to believe that similarly variable output wavelengths/wavenumbers will be had with stabilized high-power argon lasers [3.113].

With a straight spectrograph slit the spectrum lines are curved and wavelength dependent corrections due to the off-set between the reference spectrum and the Raman spectrum must be applied. This well known effect has been treated in quantitative detail [3.114,115]. The off-set error may be minimized by using a curved spectrograph slit, its radius of curvature being chosen for the central wavelength of the Raman region.

The accuracy of wavelength determination depends not only on the factors described above, as well as the quality of the diffraction grating and the degree to which the light from the Raman source and from the wavelength calibration source traverse identical paths in the spectrograph, but also on the quality of the camera mirror. Rather severe tolerances must be satisfied by even a spherical camera mirror if wavelength errors due to a variable focal length resulting from surface irregularities on the mirror are to be avoided [3.116]. Here again, the use of a large number of calibration lines and a high order polynomial will minimize the errors in the measured wavelengths.

3.1.7 Photoelectric Spectrometers

The photoelectric method of recording high-resolution spectra is used at the University of Tennessee [3.47]. Moreover, in a few special cases resolved rotational spectra can also be observed by means of commercial grating double monochromators. In view of the required accuracy of the measured Raman shifts the photographic method is preferred, and indeed the bulk of the work in high-resolution Raman spectroscopy of gases has been done with this technique. The unquestioned advantage of the photoelectric technique is in the realm of intensity and linewidth determinations. See, for example, Chap.4 for the determination of Raman scattering cross sections and also Sect.3.2.4 for the determination of the components of the polarizability tensor by using photoelectrically recorded spectra.

3.1.8 Interferometric Techniques

The Fabry-Perot êtalon used in the classical manner of crossing it with a stigmatic spectrograph has been employed to good advantage in the study of the rotational spectra of some diatomic and simple polyatomic molecules. The technique has been fully described by BUTCHER et al. [3.117]. A novel use of the scanning Fabry-Perot interferometer has been invented by BARRETT et al. [3.118] and has been further developed by others [3.119-122]. However, its use has so far been limited to the study of some simple and rather well known linear and symmetric top molecules.

The method of Fourier transform spectroscopy based upon the Michelson interferometer has until recently been limited to the study of infrared absorption spectra. This method has now been shown to be also applicable to the study of emission spectra in the visible region [3.92] but so far there have been no reports of any Raman spectra obtained with this technique. Several problems associated with the application of the Fourier transform technique to Raman spectroscopy have been discussed by SCHILDKRAUT and HIRSCHFELD [3.93].

3.2 Observed Rotational Spectra

3.2.1 Diatomic Molecules

Figure 3.5 shows the Stokes side of the pure rotational Raman spectrum of a typical, stable, homonuclear diatomic molecule, N_2 in the ground Σ_g^+ electronic state. The spectrum consists of a single series of lines of alternating intensity. This alternation is due to the spins of the nuclei and the statistics obeyed by them [3.123].

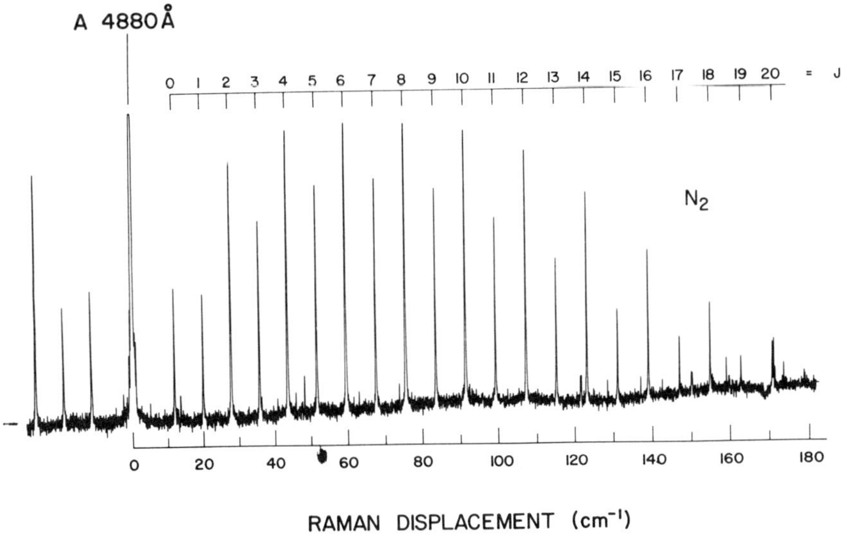

Fig. 3.5. Rotational Raman spectrum of nitrogen (Stokes side). The intensity ratio of 2:1 for the J = even:odd lines is distorted due to the nonlinear response of the photographic emulsion

This alternation is absent in the spectra of heteronuclear diatomic molecules. If the nuclear spin is zero, i.e. for ^{16}O and ^{32}S, then every alternate line is missing. For a given vibrational state v the rotational term values are represented by the expression

$$F_v(J) = B_vJ(J+1) - D_v[J(J+1)]^2 + H_v[J(J+1)]^3 \qquad (3.1)$$

where J is the quantum number of the total angular momentum, exclusive of nuclear spin. The dependence of the rotational constant B_v and the centrifugal distortion constant D_v upon the vibrational level v is given to lowest order by

$$B_v = B_e - \alpha_e(v+\frac{1}{2}) \qquad (3.2)$$

$$D_v = D_e + \beta_e(v+\frac{1}{2}) \qquad (3.3)$$

where α_e and β_e are rotation-vibration interaction constants. The dependence of H_v on v is given by a formula similar to (3.3). The subscript e denotes the equilibrium state of the molecule. The moment of inertia of the molecule in the equilibrium configuration is

$$I_e = \mu r_e^2 \qquad (3.4)$$

where μ is the reduced mass, and r_e is the equilibrium separation of the nuclei. The moment of inertia I_e and the equilibrium rotation constant B_e are related by

$$B_e = \frac{h}{8\pi^2 c\ I_e} \qquad (3.5)$$

when B_e is expressed in cm^{-1}. For the vibrational state v an "effective" moment of inertia, I_v is defined by the relation

$$I_v = \frac{h}{8\pi^2 c\ B_v} \qquad (3.6)$$

where B_v is the rotation constant for the state v. Further, B_e and D_e are connected by Kratzer's formula

$$D_e = \frac{4B_e^3}{\omega_e^2} \qquad (3.7)$$

where ω_e is the (harmonic) vibrational frequency. The higher order centrifugal distortion constant H_v is usually approximated by

$$H_v \approx H_e = \frac{2D_e}{3\omega_e^2}(12B_e^2-\alpha_e\omega_e) \quad . \tag{3.8}$$

For molecules in a $\sum$ electronic state the selection rule for the pure rotational Raman effect is $\Delta J = +2$. Transitions with $\Delta J = +2$ form the S-branch of a band, and the Raman shift for both the Stokes and anti-Stokes transitions are represented by the equation

$$\Delta\nu_S = F_v(J+2) - F_v(J) = (4B_v-6D_v+\frac{27}{4}H_v)(J+3/2)$$
$$- (8D_v-34H_v)(J+3/2)^3 + 12H_v(J+3/2)^5 \quad . \tag{3.9}$$

For nearly all Raman spectra studied so far the simpler formula

$$\Delta\nu_S = (4B_v-6D_v)(J+3/2) - 8D_v(J+3/2)^3 \tag{3.10}$$

has been found to be adequate.

Table 3.2 lists the molecular constants of diatomic molecules obtained from the study of their rotational Raman spectra. Only the most recent results are presented and citations to earlier work may be found in the references given in Table 3.2. The rotational spectra of chlorine, bromine and sulphur were obtained only with the laser technique while those of the other molecules listed, at least for their more abundant isotopic variants, were also generated by means of the mercury arc technique. The results obtained with the laser technique are decidedly superior to those obtained with the mercury arc method.

The rotational spectra of the three isotopic variants of nitrogen, $^{14}N_2$, $^{14}N^{15}N$, and $^{15}N_2$, were investigated independently by BUTCHER et al. [3.17] by means of the Fabry-Perot interferometer, and by BENDTSEN [3.89] who used a grating spectrograph. The results for B_0 and D_0 for $^{14}N_2$ obtained by BUTCHER et al. [3.17a] on the basis of (3.10) are slightly lower than those obtained by previous workers who studied the electronic emission and absorption spectra and also the rotational Raman spectrum. While for B_0 this difference was judged to be acceptable, the value gotten for D_0 ($=5.48\times10^{-8}$ cm^{-1}) is decidedly less than the previous experimental values, and is also below the value predicted theoretically (5.74×10^{-8} cm^{-1}). A subsequent fit of the data to the three constants equation (3.9) gave improved values for B_0 and D_0 ($=5.723\times10^{-8}$ cm^{-1}) but the value found for H_0 [$=(3.7\pm11.0)\times10^{-10}$ cm^{-1}] is two orders of magnitude too large in comparison to the theoretical value of 5.243×10^{-12} cm^{-1}. The discrepancy in the H_0 value indicates that this constant cannot be determined from the available data and that with these data (3.9) is inappropriate as a physical model to represent the rotational transitions. Equation (3.9) must therefore be viewed as a computational device whereby the inaccuracies of the values of B_0 and D_0 determined from (3.10) are in part removed and transferred to the H_0 term of (3.9). The

Table 3.2. Molecular constants of diatomic molecules determined from pure rotational Raman spectra

Molecule		B_0 [cm^{-1}]	D_0 [cm^{-1}]	Ref.
Hydrogen	H_2	[56.3392][a]	[0.04599][a]	[3.136]
	HD	[44.6678][a]	[0.02592][a]	[3.136]
	HT	39.783 ±0.0008	$(2.130 \pm 0.070) \times 10^{-3}$	[3.137]
	D_2	[29.9105][a]	[0.01134][a]	[3.136]
	DT	24.995 ±0.005	$(8.000 \pm 4.000) \times 10^{-3}$	[3.137]
	T_2	20.0386±0.0005	$(4.990 \pm 0.040) \times 10^{-3}$	[3.137]
Fluorine	F_2	0.88331±0.00004	$(3.48 \pm 0.06) \times 10^{-6}$	[3.138]
Chlorine	$^{35}Cl_2$	0.24310±0.00002[a]	$(1.43 \pm 0.08) \times 10^{-7}$	[3.139]
	$^{35}Cl^{37}Cl$	0.23653±0.00002[b]	$(1.44 \pm 0.05) \times 10^{-7}$	[3.139]
	$^{37}Cl_2$	0.22999±0.00002	$(1.48 \pm 0.06) \times 10^{-7}$	[3.137]
Bromine	$^{79}Br_2$	0.0819 ±0.0005[a]	2.077×10^{-8} [b]	[3.140]
	$^{81}Br_2$	0.0801 ±0.0005[a]	1.992×10^{-8} [b]	[3.140]
Nitrogen	$^{14}N_2$	1.989506±0.000027[a]	$(5.48 \pm 0.06) \times 10^{-6}$ [a]	[3.17a]
		1.989548±0.000028[b]	$(5.72 \pm 0.06) \times 10^{-6}$ [b]	[3.17a]
		1.989574±0.000012[c]	$(5.76 \pm 0.03) \times 10^{-6}$	[3.89]
	$^{14}N^{15}N$	1.923604±0.000020	$(5.297 \pm 0.038) \times 10^{-6}$	[3.17b]
		1.923596±0.000009[d]	$(5.83 \pm 0.03) \times 10^{-6}$	[3.89]
	$^{15}N_2$	1.857672±0.000027	$(5.201 \pm 0.048) \times 10^{-6}$	[3.17b]
		1.857624±0.000016[e]	$(5.08 \pm 0.05) \times 10^{-6}$	[3.89]
Oxygen	$^{16}O_2$	1.437682±0.000009	$(4.852 \pm 0.012) \times 10^{-6}$	[3.17a]
		1.437685±0.000050[a]	$(4.61 \pm 0.09) \times 10^{-6}$	[3.43]
	$^{16}O^{17}O$	1.39538 ±0.00005	$(4.55 \pm 0.12) \times 10^{-6}$	[3.137]
	$^{16}O^{18}O$	1.35781 ±0.00008[b]	$(4.7 \pm 0.2) \times 10^{-6}$	[3.43]
	$^{17}O_2$	1.35319 ±0.00009	$(4.81 \pm 0.24) \times 10^{-6}$	[3.137]
	$^{18}O_2$	1.27803 ±0.00005[c]	$(3.84 \pm 0.05) \times 10^{-6}$	[3.43]
Sulfur	S_2	0.29443 ±0.00005[a]	$(19.0 \pm 0.5) \times 10^{-8}$ [a]	[3.19]
Nitric Oxide	NO	1.69614 ±0.00010	$(5.46 \pm 0.20) \times 10^{-6}$	[3.133]

Hydrogen: [a] Values in square brackets are based on pre-laser techniques. The H_0 constant was determined to be for H_2: 5.2×10^{-5} cm^{-1}; for HD: 2.2×10^{-5} cm^{-1}; for D_2: 0.36×10^{-5} cm^{-1}. The B_e-value, for H_2: 60.841 cm^{-1}; for HD: 45.638 cm^{-1}; and for D_2: 30.442 cm^{-1}.

Fluorine: B_e = 0.889±0.001 cm^{-1}

Chlorine: [a] B_e = 0.24385±0.00002 cm^{-1}
[b] B_e = 0.23725±0.00002 cm^{-1}

Bromine: [a] Determined with the photoelectric scanning technique. Values quoted are B_e-values.
[b] Computed using Kratzer's formula (3.7).

(*Continued opposite page*)

B_0 and D_0 values determined from (3.9) are then closer to the true values of these constants. This behavior has also been observed in the analysis of the rotational Raman spectrum of carbon monoxide [3.124a]. The statistical error quotes for the constants of $^{14}N_2$ (± 0.000027 cm^{-1} or ± 0.000028 cm^{-1} for B_0 and $\pm 0.055\times 10^{-6}$ cm^{-1} or $\pm 0.056\times 10^{-6}$ cm^{-1} for D_0) are only a measure of the precision with which the data can be represented by the two frequency formulas.

Using a totally different experimental technique BENDTSEN [3.89] obtained for $^{14}N_2$ the values $B_0 = 1.989574 \pm 0.000012$ cm^{-1} and $D_0 = (5.76\pm 0.03)\times 10^{-6}$ cm^{-1} by fitting the Raman shifts to (3.10). The agreement of the D_0 value with the theoretical one is excellent. Furthermore, through the results obtained from the fundamental rotation-vibration band additional data were obtained for all three isotopic species (see Table 3.2) which then allowed an internal check on the accuracy of the experimental data. Such checks were not available to BUTCHER et al. since only the pure rotational spectra were studied by them.

The values for B_0 and D_0 reported by BUTCHER et al. and by BENDTSEN on the basis of (3.10) disagree and a comparison of the Raman shifts reported for the three molecules by these workers reveals systematic differences between them. In view of the calibration procedure used and the greater amount of data available the values for B_0 and D_0 obtained for nitrogen ($^{14}N_2$, $^{14}N^{15}N$, and $^{15}N_2$) by means of the grating spectrograph [3.89] are recommended as the best determined so far. These results were incorporated in a grand fit of all available data for the rotational constants of the electronic ground state ($X^1\Sigma_g^+$) of $^{14}N_2$ to furnish the equilibrium constants B_e, α_e, γ_e, and r_e [3.125].

Two special cases for which the rotational energy levels cannot be represented by the simple equation (3.1) need to be mentioned. The ground electronic state of oxygen is $^3\Sigma_g^-$ and, since the ^{16}O and ^{18}O nuclei have zero spin, only the odd rotational levels of $^{16}O_2$ and $^{18}O_2$ are populated in the ground state while every rotational level is populated for the mixed variants $^{16}O^{18}O$, $^{16}O^{17}O$, and $^{17}O^{18}O$. Due to various interactions of the spins of the unpaired electrons the rotational levels are split

---

Nitrogen: [a] Values determined from a two constant formula (3.10).
[b] Values determined from a three constant formula (3.9).
$H_0 = (3.7\pm 11.0)\ 10^{-10}$ cm^{-1}.
[c] $B_e = 1.998232\pm 0.000012$ cm^{-1}.
[d] $B_e = 1.931816\pm 0.000009$ cm^{-1}.
[e] $B_e = 1.865424\pm 0.000016$ cm^{-1}.

Oxygen: [a] $B_e = 1.4458\pm 0.0003$ cm^{-1}.
[b] $B_e = 1.3653\pm 0.0003$ cm^{-1}.
[c] $B_e = 1.2847\pm 0.0003$ cm^{-1}.

Sulphur: [a] Quoted errors are five standard deviations.

into triplets [3.123] and (3.1) represents only one of the three levels with the proviso that J is replaced by the quantum number of the angular momentum of the nuclei, N. The first observations of the triplet splitting in the rotational Raman spectrum were reported by LEPARD [3.39b] and JAMMU et al. [3.126]. Each rotational line is attended by a pair of weak satellites located approximately 2 cm^{-1} on each side of the line. Quantitatively useful measurements of this triplet splitting were however possible only by means of the laser technique and the photoelectric recording of the spectra [3.127-129], (see Fig.3.6). An exact treatment of the intensities of the various transitions has been worked out by LEPARD [3.130], while ALTMANN et al. [3.129] have computed the theoretical intensities of the various transitions and have compared the computer generated intensity profile to the observed spectrum. The effect of the presence of the isotopic species $^{16}O^{18}O$ as well as of impurity amounts of nitrogen and also the contribution to the observed intensity by molecules in the first vibrationally excited state is accounted for. The agreement between the observed and calculated intensities for the first five triplets $N=1 \rightarrow N=3$ to $N=9 \rightarrow N=11$ which comprises 15 lines with intensities spanning a ratio of 1:100 is excellent. The spin splitting of the rotational lines has also been observed in the rotation-vibration band [3.128].

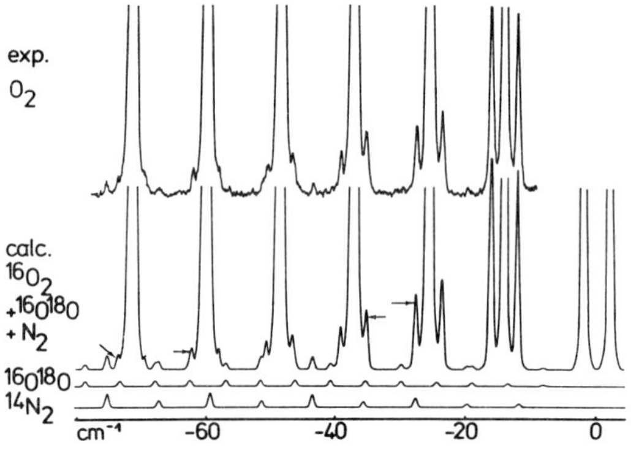

Fig. 3.6. Rotational Raman spectrum of oxygen (Stokes side) showing the spin satellites. The arrows in the calculated spectrum indicate the intensity of the spin satellites computed by neglecting the contributions due to the $^{16}O^{18}O$ molecule (natural abundance) and trace amounts of nitrogen in the gas sample. Features of the $^{16}O^{18}O$ and $^{14}N_2$ rotation spectra are visible in the observed spectrum (top). (From [3.129] with permission)

The ground state rotational constants of oxygen are well known from the study of the microwave [3.131a], laser magnetic resonance [3.131b], and electronic band spectra [3.132]. The study of the rotational Raman spectrum was therefore mainly concerned with the intensity problem.

The second special case is that of nitric oxide, NO. This is the only stable diatomic molecule known to have a $^2\Pi$ electronic ground state. Owing to spin-orbit interaction this state is split into $^2\Pi_{1/2}$ and $^2\Pi_{3/2}$ substates which are separated by about 120 cm^{-1}, the $^2\Pi_{3/2}$ substate having the higher energy. Each electronic substate has its own set of vibrational and rotational levels. For a Π electronic state the selection rule for pure rotational Raman transitions ist that $\Delta J = 0$, +1, and +2.

An R-branch ($\Delta J=+1$) is therefore present in the pure rotation spectrum in addition to the usual S-branch ($\Delta J=+2$). For this molecule also, the ground state molecular constants are well known from infrared and microwave spectroscopy and the focus of the Raman studies was the intensity distribution in the pure rotation as well as in the electronic Raman spectrum of the transition between the $^2\Pi_{1/2}$ and $^2\Pi_{3/2}$ electronic substates.

Laser generated pure rotational Raman spectra showing resolution of the R-branch lines were obtained by SHOTTON and JONES [3.133] and RENSCHLER et al. [3.134] but the rotational spectrum showing the most extensive detail was obtained by FAST et al. [3.135] using the mercury arc technique. Figure 3.7 shows the spectrum recorded by them with the photographic technique using exposures of up to 120 h duration. Due to the use of a zero-dispersion band pass filter [3.39] which was set to exclude the exciting line from the grating spectrograph they were unable to observe the R-branch structure of the pure rotation spectrum but did observe the R-branch transitions of the electronic spectrum [3.14,135]. The theoretical intensities shown in the figure were computed by LEPARD [3.130]. They are shown again in Fig.3.8 to present more clearly the complexity of the whole spectrum and to point out the weakness of the electronic transition. A repetition of the experimental work with single mode laser excitation would be of interest to verify the theoretical intensity distribution.

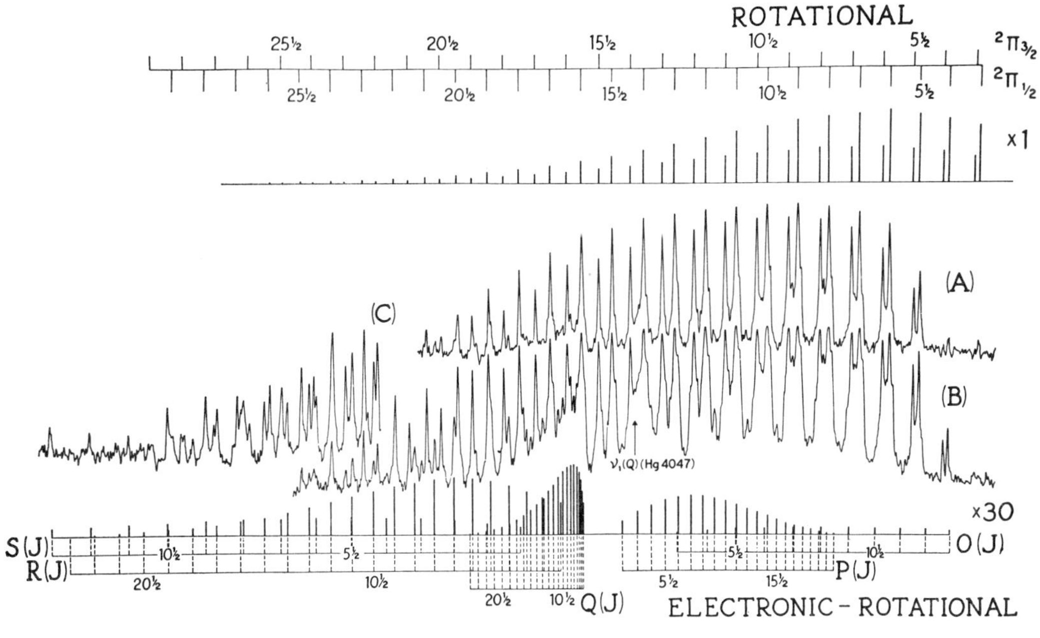

Fig. 3.7. Pure rotational Raman spectrum of nitric oxide (Stokes side) excited by the 4358 Å line of a mercury arc. Exposure times are (A) 3 hr, (B) 12 hr, and (C) 120 hr. (From [1.135] with permission)

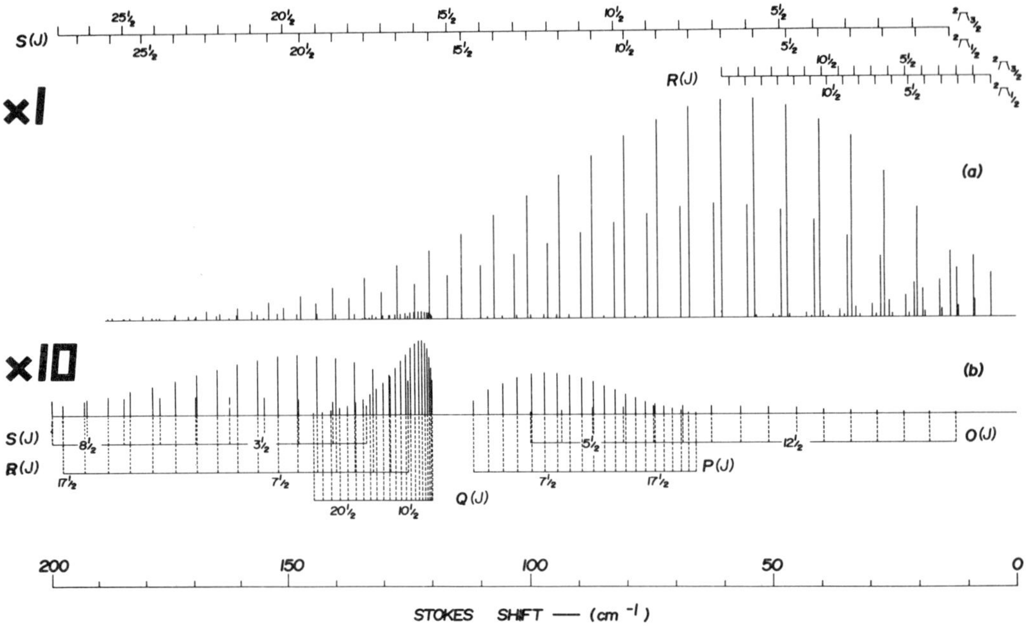

Fig. 8a and b. Raman spectrum of nitric oxide calculated for (a) both the pure rotational and the electronic transitions, and (b) the electronic transition only. The intensity scale for (b) is expanded by a factor of 10 with respect to that shown in (a). (From [1.130] with permission)

3.2.2 Linear Polyatomic Molecules

Rotational Raman spectra of polyatomic molecules have been studied only for the Σ_g^+ ground electronic state. The vibrations of a linear polyatomic molecule divide themselves into two groups, the parallel (bond stretching) vibrations and the perpendicular (bending) vibrations. The fundamental vibrations of the latter type (π-species) are always doubly degenerate and their rotational energy levels are split into doublets due to the rotation-vibration interaction (ℓ-type doubling). The states of parallel vibrations as well as the ground vibrational state are always non-degenerate (σ-species) and their rotational energy levels are single. For linear polyatomic molecules in their ground vibrational state the rotational Raman effect is described by the same selection rules and relations that are appropriate for diatomic molecules in a Σ_g^+ electronic state, namely (3.1,5,6,9,10), while the remaining equations assume a more general form. The rotational Raman spectra of nearly all linear polyatomic molecules studied so far have been adequately described by this model. Table 3.3 lists the molecular constants obtained so far from pure rotational Raman spectra.

Pure rotational Raman scattering by molecules in excited vibrational states gives rise to a rotational Raman spectrum (rotational hot band) that overlaps that due to molecules in the ground state. The ground state rotation spectrum of $C^{16}O_2$ and $C^{32}S_2$ consists of an S-branch only, with every line with $J = \text{odd}$ missing owing to the zero value of the spin of the ^{16}O and ^{32}S nuclei. The lowest excited vibrational state

Table 3.3. Molecular constants of linear polyatomic molecules determined from pure rotational Raman spectra

Molecule		B_0 [cm^{-1}]	D_0 [cm^{-1}]	Ref.
Carbon Dioxide	CO_2			
	$(00^00)^a$	0.3902634±0.0000044	$(1.270±0.035)\times10^{-7}$	[3.141]
	$(01^10)^b$	0.3906398±0.0000082	$(1.419±0.048)\times10^{-7}$	[3.141]
Carbon Disulfide	CS_2			
	$(00^00)^a$	0.1091200±0.0000024	$(1.057±0.075)\times10^{-8}$	[3.141]
	$(01^10)^b$	0.1093639±0.0000086	$(1.783±0.121)\times10^{-8}$	[3.141]
Hydrogen Cyanide	HCN	1.478277 ±0.000029	$(2.97 ±0.04)\times10^{-6}$	[3.91]
	DCN	1.207986 ±0.000023	$(2.31 ±0.04)\times10^{-6}$	[3.91]
Xenon Fluoride	XeF_2	0.11327 ±0.00001	$(2.1 ±0.2)\times10^{-8}$	[3.137]
Acetylene	$^{12}C_2H_2$	1.176628 ±0.000016	$(1.59 ±0.02)\times10^{-6}$	[3.142]
Iodoacetylene	C_2HI	[0.10622 ±0.00010]^a	$[(3.9 ±1.0)\times10^{-8}]^a$	[3.143]
Diacetylene	C_4H_2	[0.14689 ±0.00004]^a	$[(3.2 ±0.6)\times10^{-8}]^a$	[3.144]
	C_4D_2	[0.12767 ±0.00005]^a	$[(2.3 ±0.8)\times10^{-8}]^a$	[3.144]
Cyanogen	$^{12}C_2{}^{14}N_2$			
	(a)	0.157135 ±0.000010	$(2.50 ±0.08)\times10^{-8}$	[3.145]
	(b)	0.157479 ±0.000028	$(1.15 ±0.12)\times10^{-8}$	[3.145]
	(c)	0.157721 ±0.000038	$(-1.14 ±0.32)\times10^{-8}$	[3.145]
	(d)	0.158589 ±0.000052	---	[3.145]

Carbon Dioxide: [a] Ground state constants.
[b] Constants for the c-levels of (01^10)-state.

Carbon Disulfide: [a] Ground state constants.
[b] Constants for the c-levels of (01^10)-state.

Iodoacetylene: [a] Results obtained with pre-laser technique.

Diacetylene: [a] Results obtained with pre-laser technique.

Cyanogen: [a] Ground state constants.
[b] Constants of $(00001^1)^c$-state.
[c] Constants of $(00001^1)^d$-state.
[d] (00002)-state constant.

of these molecules is the double degenerate $\nu_2(\pi)$ state describing the bending vibration with $\nu_2^1 = 667$ cm^{-1} for CO_2 (room temperature population ~7%) and $\nu_2^1 = 397$ cm^{-1} for CS_2 (room temperature population ~21%). The selection rule for the rotational Raman effect for molecules in a Π-vibrational state is $\Delta J = 0, +1, +2$ so that an R-branch ($\Delta J=+1$) occurs in addition to the usual S-branch. The intensities of the R-branch transitions decrease as J^{-1} so that only a few lines of this branch are expected to be observed. The rotational hot bands showing these R- and S-branches were observed in CO_2 [3.9,141], CS_2 [3.10,141], acetylene, C_2H_2 [3.142] and cyanogen, C_2N_2 [3.145].

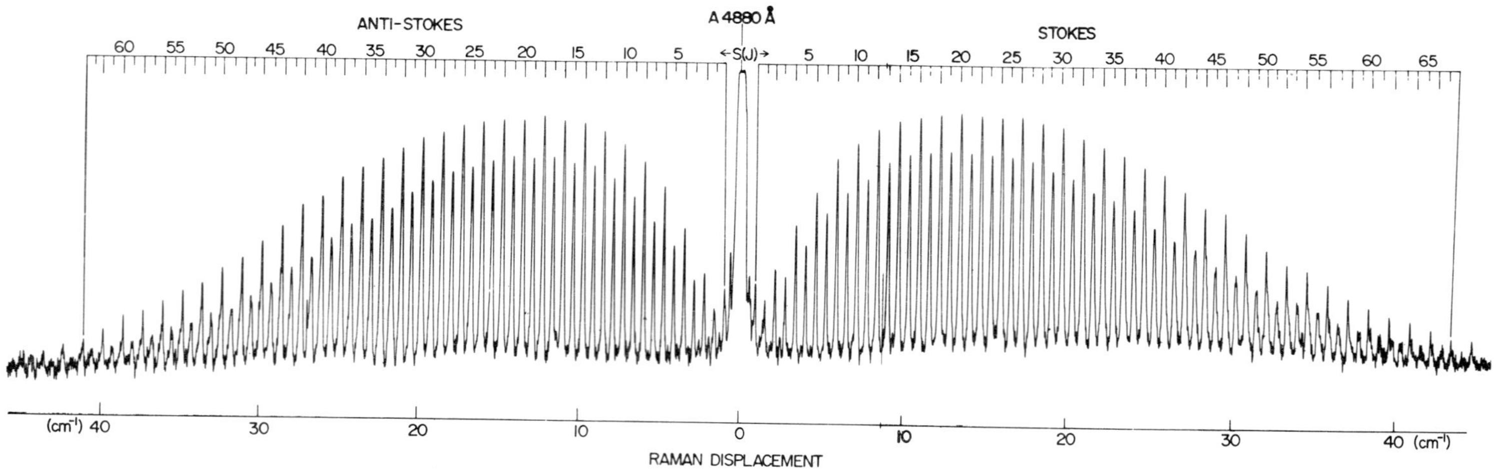

Fig. 3.9. Rotational Raman spectrum of cyanogen, C_2N_2 at a pressure of 100 torr, photographed in 10 min on a hypersensitized Kodak IIIa-J plate. The intensity ratio of 2:1 for the J = even:odd lines is distorted due to the non-linear response of the photographic emulsion [3.145]

For CO_2 and CS_2 the observation of the hot bands was facilitated by the absence of ground state rotation lines with J = odd. This assistance in their observability does not exist for C_2N_2 where the actual resolution of the hot band lines from the ground state spectrum was achieved and the line splitting due to the ℓ-type doubling was also observed [3.145]. Figure 3.9 shows the pure rotation spectrum of C_2N_2 photographed in 10 min using a gas pressure of 100 torr. Even though this spectrum appears to be perfect in that the lines are well resolved from one another, are quite sharp and symmetric and occur with a virtually absent underlying continuum, the molecular constants derived from it differ markedly from those obtained by means of infrared spectroscopy. This discrepancy is due to the presence of rotational hot bands in the Raman spectrum. Figure 3.10 shows part of the Stokes side of the cyanogen spectrum photographed in 6-1/2 h, using a gas pressure of 50 torr. The hot band structure due to molecules in the $\nu_5^1 = 233\ \mathrm{cm}^{-1}$ state is sufficiently well resolved from the ground state lines, as is the ℓ-type doubling. The analysis of this portion of the rotation spectrum gave the values of the constants listed in Table 3.3. These are in excellent agreement with the infrared determined values.

The presence of rotational hot bands in the observed pure rotation spectrum limits the accuracy with which the ground state constants B_0 and D_J can be determined from pure rotation spectra. The severity of this problem depends on the frequency of the low lying modes and the form of these vibrations.

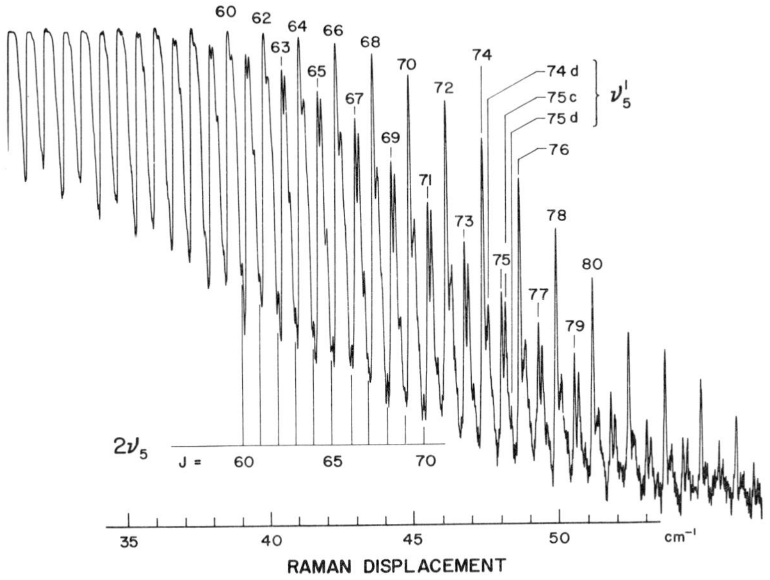

Fig. 3.10. High-J region of the rotation spectrum of cyanogen at a gas pressure of 50 torr, recorded in $6\frac{1}{2}$ h on a hypersensitized Kodak IIIa-J plate. The ν_5^1 and $2\nu_5$ rotational hot bands are resolved from the ground state spectrum. The ℓ-type doubling in the ν_5^1-state hot band is partially resolved [3.145]

3.2.3 Symmetric Top Molecules

The rotational spectra of a few select symmetric top molecules are shown in Figs. 3.12-16. This selection covers the range from the simple unambiguous case of BF_3 (Fig.3.12) to heavy molecules whose observed rotation spectra are unresolved composites of rotation spectra in many vibrational states (Figs.3.15,16). All spectra consist of an R- and an S-branch, and the intensity alternation in the observed lines is due to the different intensity distributions in the R- and S-branches, as well as the fact that every R-branch line with J = even nearly coincides with an S-branch line. Each observed S-branch line is also an unresolved composite of (J+1) individual transition for each value of J (K-structure), while each R-branch line has J-components.

The rotational term values for an oblate symmetric top molecule with principal moments of inertia $I_A = I_B < I_C$ are, to lowest order in the centrifugal distortion effect,

$$F_v(J,K) = B_vJ(J+1) + (C_v-B_v)K^2 - D_JJ^2(J+1)^2 - D_{JK}J(J+1)K^2 - D_KK^4 \quad . \tag{3.11}$$

For prolate tops $I_A < I_B = I_C$ and the coefficient (C_v-B_v) becomes (A_v-B_v). The selection rule for pure rotational transitions in a totally symmetric vibrational state is that $\Delta J = 0, +1, +2$, and $\Delta K = 0$. Accordingly the rotational Raman displacements of the R- and S-branches are respectively given by

$$\Delta\nu_R = 2(B_v-D_{JK}K^2)(J+1) - 4D_J(J+1)^3 \tag{3.12}$$

where J=1,2,3,..., and |K|=1,2,3,...,J, and

$$\Delta\nu_S = (4B_v-6D_J-4D_{JK}K^2)(J+3/2) - 8D_J(J+3/2)^3 \tag{3.13}$$

where J = 0,1,2,... and |K| = 0,1,2,...,J.

Although not explicitly so labeled, the centrifugal distortion constants D_J, D_{JK}, and D_K depend on the vibrational state and are functions of the vibrational quantum number just as is true of the rotation constants A_v, B_v, and C_v. Expressions for the term values and the R- and S-branch Raman shifts that include the next higher order centrifugal distortion contribution involving the constants H_J, H_{JK}, H_{KJ}, and H_K have been given elsewhere [3.164].

The intensities $I(J,K\to J',K')$ of the individual Stokes transitions for an oblate symmetric top molecule in a totally symmetric vibrational state are given for the R-branch by

$$I_R(J,K;J+1,K) = C(\nu_0-\Delta\nu_R)^4(2J+1)g_{ns}\, b_{J,K;J+1,K} \times \exp\{-[B_vJ(J+1)+(C_v-B_v)K^2]hc/kT\} \tag{3.14}$$

and for the S-branch

$$I_S(J,K;J+2,K) = C(\nu_0-\Delta\nu_S)^4(2J+1)g_{ns}\, b_{J,K;J+2,K} \times \exp\{-[B_vJ(J+1)+(C_v-B_v)K^2]hc/kT\} \quad (3.15)$$

where $\Delta\nu_R$ and $\Delta\nu_S$ are the Raman shifts given by (3.12,13). Numerical and physical constant coefficients are lumped together in the coefficient C and only the rigid rotor term value is used in the Boltzmann factor. The statistical weight g_{ns} depends on the value of K and is determined by the values of the spins of, and the spin statistics obeyed by the equivalent nuclei, as well as the point group symmetry of the molecule. Tabulations of the nuclear spin statistical weights g_{ns} for a variety of molecules may be found in [3.165-172].

Several planar symmetric top molecules of symmetry D_{3h} have recently been studied. For the general case of the planar symmetric top molecule $W_3X_3Y_3Z_3$ belonging to the D_{3h} point group (Fig.3.11) the nuclear spin statistical weights are listed in Table 3.4 for the various fundamental vibrational species. Provided that they are planar and belong to the D_{3h} point group, these weights are also appropriate for molecules of the type $X_3Y_3Z_3$ (e.g., s-triazine, $C_3N_3H_3$; cyanuric fluoride, $C_3N_3F_3$), X_3Y_3, and also Y_3 by replacing, respectively, the factors $(2I_W+1)$, $(2I_W+1)(2I_Z+1)$, and $(2I_W+1)(2I_X+1)(2I_Z+1)$ by unity. Further, for the symmetrically substituted benzenes of the type $W_6X_3Y_3$ the formulas are applicable with $\Omega = (2I_W+1)^2(2I_X+1)(2I_Y+1)$ while for molecules of the type AX_3 with the A-atom located at the center of the X_3-triangle (e.g., SO_3, BF_3) the formulas are valid with $\Omega = (2I_X+1)$ and multiplying *all* entries in the table by $(2I_A+1)$, where I_A is the nuclear spin of the central atom. The W, X, Y, and Z atoms are located on the corners of equilateral triangles. If the

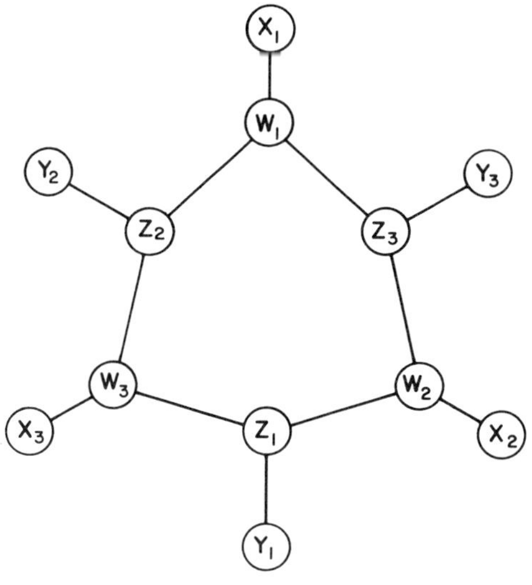

Fig. 3.11. The planar $W_3X_3Y_3Z_3$ molecule with D_{3h} symmetry

Table 3.4. Nuclear spin statistical weight factors for sym-$W_3X_3Y_3Z_3$ having D_{3h} symmetry[a]

	A_1	A_2	E
K = 0, J = even	$\frac{\Omega}{6}(\Omega^2\mp 3\Omega+2)$	$\frac{\Omega}{6}(\Omega^2\pm 3\Omega+2)$	$\frac{\Omega}{3}(\Omega^2-1)$
K = 0, J = odd	$\frac{\Omega}{6}(\Omega^2\pm 3\ +2)$	$\frac{\Omega}{6}(\Omega^2\mp 3\Omega+2)$	$\frac{\Omega}{3}(\Omega^2-1)$
K = 3p	$\frac{\Omega}{3}(\Omega^2+2)$	$\frac{\Omega}{3}(\Omega^2+2)$	$\frac{2\Omega}{3}(\Omega^2-1)$
K = 3p±1	$\frac{\Omega}{3}(\Omega^2-1)$	$\frac{\Omega}{3}(\Omega^2-1)$	$\frac{\Omega}{3}(2\Omega^2+1)$

$\Omega = (2I_W+1)(2I_X+1)(2I_Y+1)(2I_Z+1)$

For K = 0 upper (lower) sign applies for FD(BE) statistics.
For K ≠ 0 weights apply to both FD and BE statistics.

[a] I_W, I_X, I_Y, I_Z = nuclear spin of the W, X, Y, Z-atom.

total number of such triangles formed by fermions is even, then the overall species of the total wavefunction must be A_1 and BE-statistics applies, whereas if the total number of such triangles is odd, FD-statistics applies and the overall species is A_2. For the case of the AX_3 molecule this rule is independent of the nature of the A-atom.

For the Stokes transitions the PLACZEK-TELLER factor [3.168] $b_{J,K;J',K}$ is, for the R-branch

$$b_{J,K;J+1,K} = \frac{3K^2[(J+1)^2-K^2]}{J(J+1)(J+2)(2J+1)} \quad , \tag{3.16}$$

and for the S-branch

$$b_{J,K;J+2,K} = \frac{3[(J+1)^2-K^2][(J+2)^2-K^2]}{2(J+1)(J+2)(2J+1)(2J+3)} \quad . \tag{3.17}$$

For the anti-Stokes transitions the factors $b_{J+1,K;J,K}$ and $b_{J+2,K;J,K}$ can be computed from the relation [3.168]

$$(2J+1)b_{J,K;J',K'} = (2J'+1)b_{J',K';J,K} \quad . \tag{3.18}$$

Tables of these factors are given in [3.1]. Since the K-structure is not resolved in the observed spectra, each observed Stokes Raman line free of overlap has an intensity given by

$$I_R(J,J+1) = \sum_{K=1}^{J} I_R(J,K;J+1,K) \tag{3.19}$$

and

$$I_S(J,J+2) = \sum_{K=0}^{J} I_S(J,K;J+2,K) \quad . \tag{3.20}$$

However, since the observed rotational Raman spectrum is a composite of the ground state spectrum and the hot bands, the intensity distribution in the individually observed Raman lines (see Figs.3.12-16) is determined not only by the unresolved K-structure but also by the presence of the hot bands.

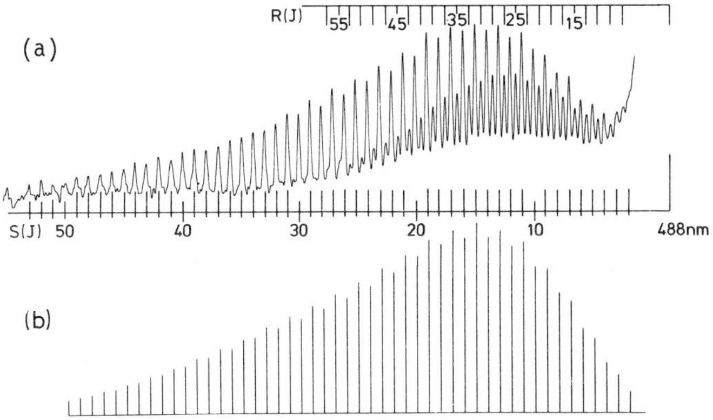

Fig. 3.12a and b. Pure rotation spectrum of $^{11}BF_3$. (a) Observed spectrum 488 nm excitation, gas pressure 600 torr. (b) Calculated intensities of S-branch transitions. (From [3.152] with permission)

In the traditional analysis of the observed spectra the assumption is made that the D_{JK}-term of (3.12,13) is ignorable on account of its smallness. Except for the cases of BF_3, C_3H_4 (allene), and C_3H_6 (cyclopropane) the results listed in Table 3.5 were obtained in this manner from the S-branch data. These results are also averages over the ground and excited vibrational states since no corrections were applied for the presence of hot bands. The results obtained from the analysis of the R-branch data are, however, different from those given in Table 3.5. While small and acceptable for B_0, the discrepancy is especially severe for the centrifugal distortion constant D_J. Omitting the D_{JK}-terms in (3.12,13) is equivalent to the assumption that the S[R] line maxima occur at $K = 0$ [1]. For prolate symmetric top molecules the R- and S-branch line maxima do occur for small values of K, but not $K = 0$ or 1. For oblate symmetric tops the line maxima occur for large values of K. However, as shown by WANG [3.141] for a given value of J, the "center of gravity" of an R-branch line occurs at a larger K-value than for an S-branch line. For oblate symmetric tops the omission of the D_{JK}-term will therefore more severely afflict the results gotten

Table 3.5. Molecular constants of symmetric top molecules derived from pure rotational Raman spectra

Molecule		B_0 [cm^{-1}]	$D_J \times 10^8$ [cm^{-1}]	Ref.
Prolate Symmetric Tops				
Allene	C_3H_4	0.296438 ±0.000006[a] 0.296484 ±0.000007[b]	13.2 ±0.3[a] 12.3 ±0.3[b]	[3.146] [3.146]
	H_2CCCD_2[c]	[0.26190 ±0.0001][c]	[~7.0]	[3.147]
	C_3D_4	[0.23230 ±0.0001]	[7.5]	[3.147]
Disilane	Si_2H_6	0.16998 ±0.00003	5.3 ±1.1	[3.148]
	Si_2D_6	0.13796 ±0.00004	3.1 ±1.0	[3.148]
Ethane	C_2H_6	[0.66215 ±0.0005]	[94]	[3.149]
	CH_3CD_3	[0.54862 ±0.0005]	[42]	[3.149]
	C_2D_6	[0.45915 ±0.0005]	[34]	[3.149]
Dimethylacetylene	C_4H_6	0.112183 ±0.000013	3.70 ±0.51	[3.150]
	C_4D_6	0.088879 ±0.000021	1.69 ±0.61	[3.150]
Dimethyldiacetylene	C_6H_6	0.037270 ±0.000021	0.86 ±0.06	[3.150]
Dimethylzinc	$Zn(CH_3)_2$	[0.13478 ±0.0001]	[5±1]	[3.151]
	$Zn(CD_3)_2$	[0.10485 ±0.0001]	[4±1]	[3.151]
Dimethylcadmium	$Cd(CH_3)_2$	[0.11405 ±0.0001]	[3±1]	[3.151]
	$Cd(CD_3)_2$	[0.08938 ±0.0001]	[2±1]	[3.151]
Dimethylmercury	$Hg(CH_3)_2$	[0.11620 ±0.0001]	[2±1]	[3.151]
	$Hg(CD_3)_2$	[0.09120 ±0.0002]	[2±1]	[3.151]
Oblate Symmetric Tops				
Boron Trifluoride	$^{11}BF_3$	0.34502 ±0.00003	43.8 ±1.0	[3.152]
Sulphur Trioxide	SO_3	0.34858 ±0.00002	26.8 ±0.7	[3.137]
Selenium Trioxide	SeO_3	0.24545 ±0.00008	0.2	[3.137]
Borazine	$^{11}B_3N_3H_6$	0.175581 ±0.000018	2.36 ±0.08	[3.153]
	$^{10}B^{11}B_2N_3H_6$	0.177565 ±0.000018	4.39 ±0.19	[3.153]
sym-Triazine	$C_3N_3H_3$	0.214721 ±0.000002	2.51 ±0.05	[3.150]
	$C_3N_3D_3$	[0.19358 ±0.00015]	~2.5	[3.154]
Cyanuric Fluoride	$C_3N_3F_3$	0.0655954±0.0000014	0.252±0.017	[3.155]
Cyclopropane	C_3H_6	0.670144 ±0.000012[a] 0.67028 ±0.00015[b]	79.9 ±0.9[a] 100 ±9[b]	[3.124a] [3.124b]
	C_3D_6	[0.46079 ±0.00015]	[10 ±10]	[3.156]
Cyclobutane	C_4H_8	[0.35480 ±0.0002]	[~7]	[3.157]
	C_4D_8	[0.25595 ±0.0002]	[~4]	[3.157]

Table 3.5 (continued)

Molecule		B_0 [cm^{-1}]	$D_J \times 10^8$ [cm^{-1}]	Ref.
Oblate Symmetric Tops				
Cyclopentane	C_5H_{10}	0.2161299±0.0000025	3.02 ±0.07	[3.124a]
	C_5D_{10}	[0.16102 ±0.00012]	[4 ±2.2]	[3.158]
Cyclohexane	C_6H_{12}	0.143429 ±0.000002	1.56 ±0.003	[3.159]
	C_6D_{12}	0.109469 ±0.000001	0.61 ±0.01	[3.159]
Cycloheptane	C_7H_{14}	0.0993801±0.0000047	4.08 ±0.17	[3.160]
Cyclooctane	C_8H_{16}	0.073221 ±0.000039	0.784±0.039	[3.160]
Benzene	C_6H_6	0.189675 ±0.000001[a] 0.189665 ±0.000001[b]	2.73 ±0.01[a] 2.28 ±0.04[b]	[3.161] [3.161]
	C_6D_6	0.156894 ±0.000002	1.25 ±0.03	[3.161]
	sym-$C_6H_3D_3$	0.171720 ±0.000002	1.50 ±0.03	[3.162]
1,3,5-Trifluoro-benzene	$C_6H_3F_3$	0.0586517±0.0000034	0.189±0.004	[3.163]
	$C_6D_3F_3$	0.0567955±0.0000004	0.173±0.002	[3.162]
Hexafluorobenzene	C_6F_6	0.0343165±0.0000009	0.070±0.004	[3.163]
1,3,5,7-Cycloocta-tetraene	C_8H_8	0.090385 ±0.000002	1.30 ±0.02	[3.150]
1,5-Cyclooctadiene	C_8H_{12}	0.090345 ±0.000022	0.71 ±0.36	[3.150]

[a] Except where noted constants are based on S-branch Raman shifts, and are uncorrected for unresolved K-structure and presence of rotational hot bands. Results obtained with pre-laser techniques are given in square brackets.

Allene: [a] Determined by assuming D_{JK}=0.

[b] Corrected for K-structure by using D_{JK}=4.6×10^{-6} cm^{-1} determined by A.G. Maki, R.A. Toth: J. Mol. Spectrosc. *17*, 136 (1965) from infrared spectrum.

[c] This molecule is a quasi-symmetric top. The B-value is an average $\bar{B}$=(B+C)/2.

Ethane: Only pre-laser values are available. For other values based on combined infrared and Raman data see [Ref.3.1, Table 8].

Dimethylacetylene: Results represent averages over the ground and low lying vibrational states. The quoted errors are five standard deviations.

Dimethyldiacetylene: See footnote on Dimethylacetylene.

Boron Trifluoride: Constants are based on simultaneous use of R- and S-branch data and consideration of K-structure of lines. From these data D_{JK}=-(9.1±1.0) ×10^{-7}cm^{-1}.

Selenium Trioxide: B-value results from analysis constrained to a calculated D_J=0.2×10^{-8}cm^{-1}.

Borazine: For $^{10}B^{11}B_2N_3H_6$ the quoted B-value is the average $\bar{B}$=1/2(A+B).

Cyanuric Fluoride: H_J=-(1.59±0.59) 10^{-14} cm^{-1}.

Cyclopropane: [a] C_3H_6. Results derived from S-branch data using lines with 19≤J≤30.

[b] C_3H_6. Values determined by including effect of K-structure in analysis; D_{JK}=(1.3±0.4)×10^{-6} cm^{-1}. [3.124b].

Cycloheptane: This molecule is a quasi-symmetric top. B-value represents an average over low lying vibrational states and various conformers present at room temperature.

Cyclooctane: See comments for Cycloheptane.

Benzene: [a] Determined from two constants fit (3.13).

[b] Determined from three constants fit, also giving H_J=-(5.57±0.33)×10^{-13} cm^{-1}.

1,5-Cyclooctadiene: This molecule is a quasi-symmetric top molecule. B-value is an average $\bar{B}$=1/2(A+B).

from the R-branch than those derived from the S-branch. The B_0- and D_J-values derived from the S-branch are therefore judged to be more accurate.

For planar symmetric top molecules the B_0- and D_J-values determined from the S-branch can be used to determine the D_{JK}-constant from (3.12) and the R-branch Raman shifts by substituting for each value of J the appropriate average, $\overline{K^2}$, computed by using (3.14) as weighting function, into (3.12). The D_{JK}-values thus obtained for benzene, benzene-d_6, and cyanuric fluoride [3.141] satisfy the relation

$$D_{JK} \simeq 2D_J[(I_{xx}/I_{zz})^4-1] = -1.875D_J \tag{3.21}$$

derived by ALIEV et al. [3.173] to within 8, 29, and 7%, respectively. The large discrepancy for benzene-d_6 is attributed to the distortion of line positions by hot bands, whose interference with the ground state spectrum is here more pronounced than for benzene and cyanuric fluoride (compare the envelopes for the intensity minima in Figs.3.13 for s-triazine and Fig.3.14 for benzene). This method is clearly of restricted usefulness and relies on the absence of hot band structure. A more generally useful, but still only approximate, approach is to determine B_0, D_J, and D_{JK} directly using both R- and S-branch data and the computed average values of K^2 in a single calculation.

In an elegant analysis of the observed rotation spectrum of $^{11}BF_3$ (Fig.3.12) FREEDMAN and JONES [3.152] accounted for the presence of a hot band due to the population of the lowest vibrational state $\nu_4(e) = 480\ cm^{-1}$, overlap of the R- and S-branch lines, and possible dependence of the computed molecular constants on the spectral line shape, and determined B_0, D_J, and D_{JK} by an iterative least squares fit from the measured Raman shifts. This is the only case so far known where a D_{JK} constant has been reliably determined from the pure rotational Raman spectrum of a symmetric top molecule. Most of the molecules investigated so far are more complex than BF_3 and little information is available on the α-values for their excited vibrational states which are needed to account for the hot bands. The method used by FREEDMAN and JONES should, however, give good results for cyclopropane for which hot bands may be ignored since the lowest vibrational state, $\nu_{14}(e'')$, lies 740 cm^{-1} above the ground state, having a room temperature population of 5%, while 86% of all molecules are in the ground state (see [3.124b]).

The observed spectrum of $^{11}BF_3$ (Fig.3.12) also exhibits an interesting intensity alternation in the S-branch lines, this being quite noticeable in the region S(J) = 7-31. This alternation is due to the nuclear spin statistical weight g_{ns} which has the values 4,2,2,4 for $K = 3p$, $3p+1$, $3p+2$, $3p+3$ (p=integer$\neq$0), while for $K = 0$, $g_{ns} = 0$ or 4 for J = even or odd (see Table 3.4). The observed alternation of s:w for J = odd : J = even is *prima facie* evidence for the planarity and D_{3h} symmetry of the BF_3 molecule.

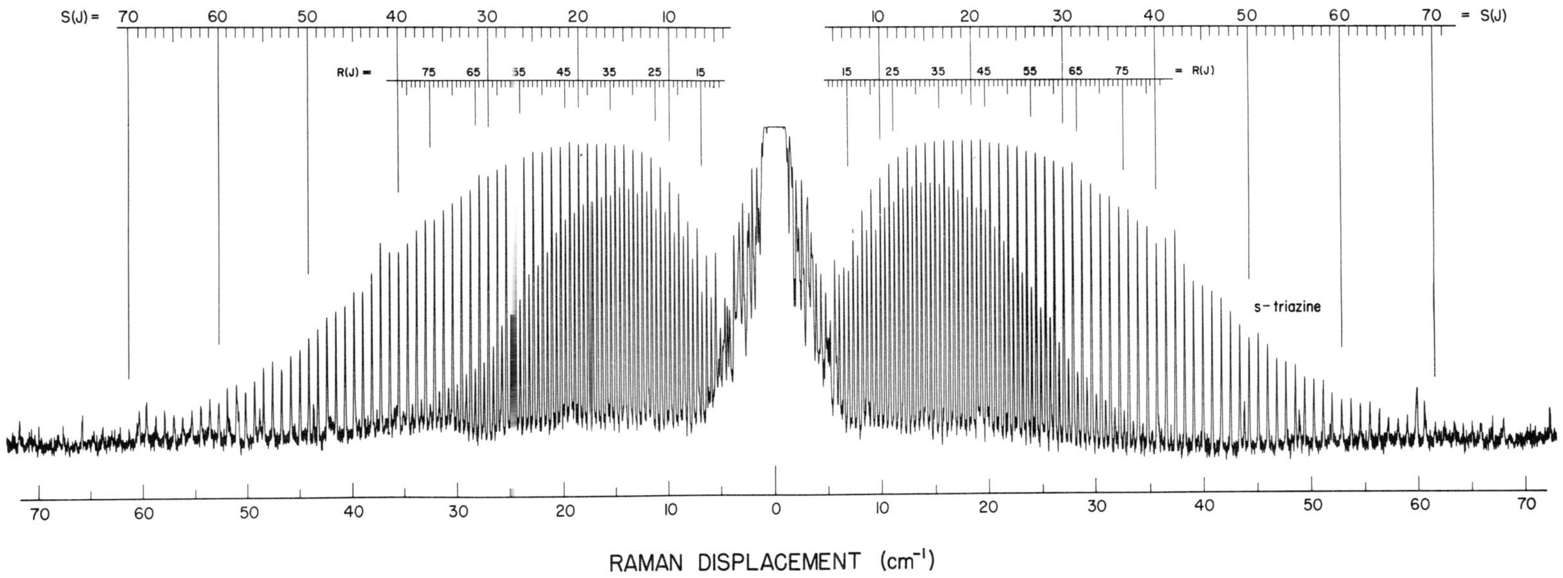

Fig. 3.13. Pure rotation spectrum of s-triazine, $C_3H_3N_3$. $\lambda = 4880$ Å, pressure 4 torr, exposure time 6 h [3.150] The Stokes side lies to the right

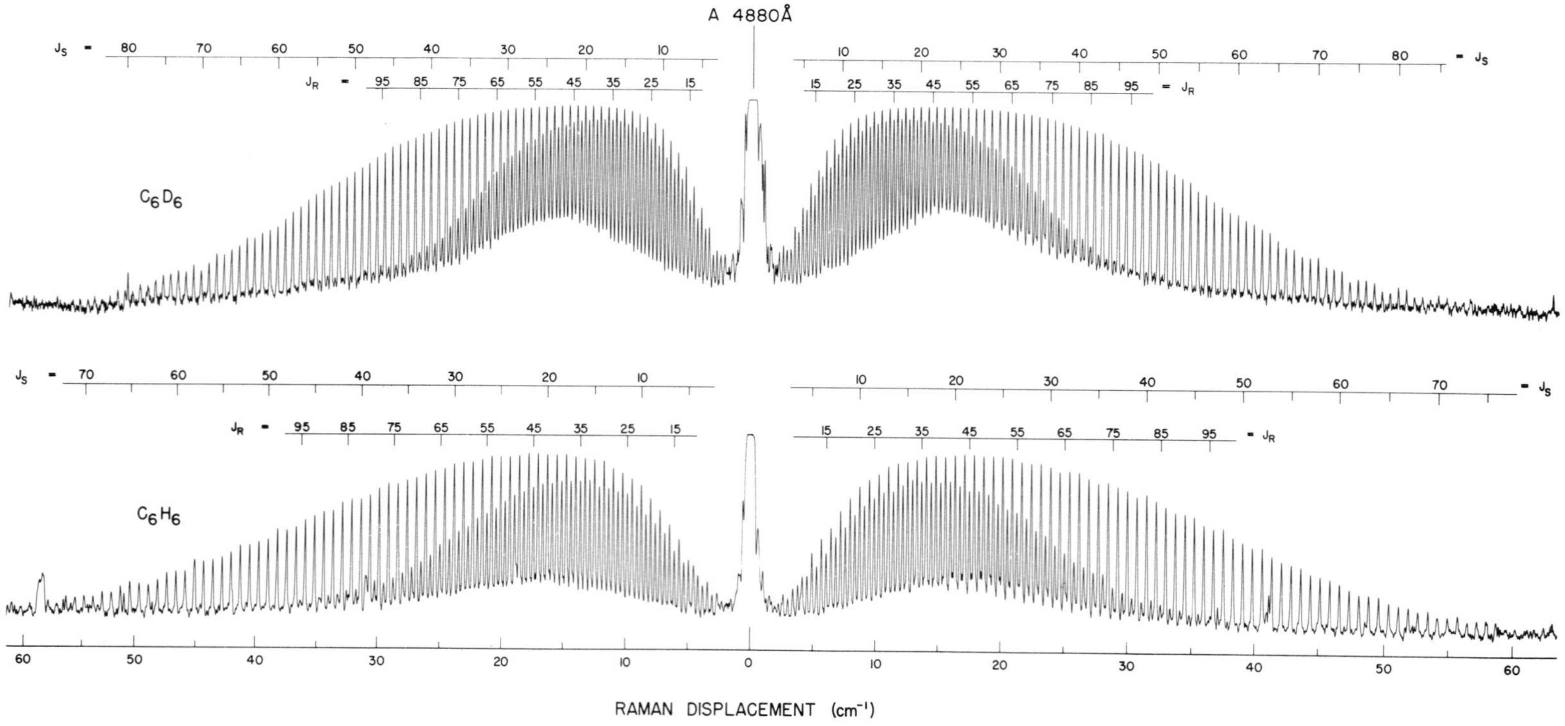

Fig. 3.14. Pure rotation spectra of benzene-d_0 and -d_6 [3.161]. The Stokes side lies to the right

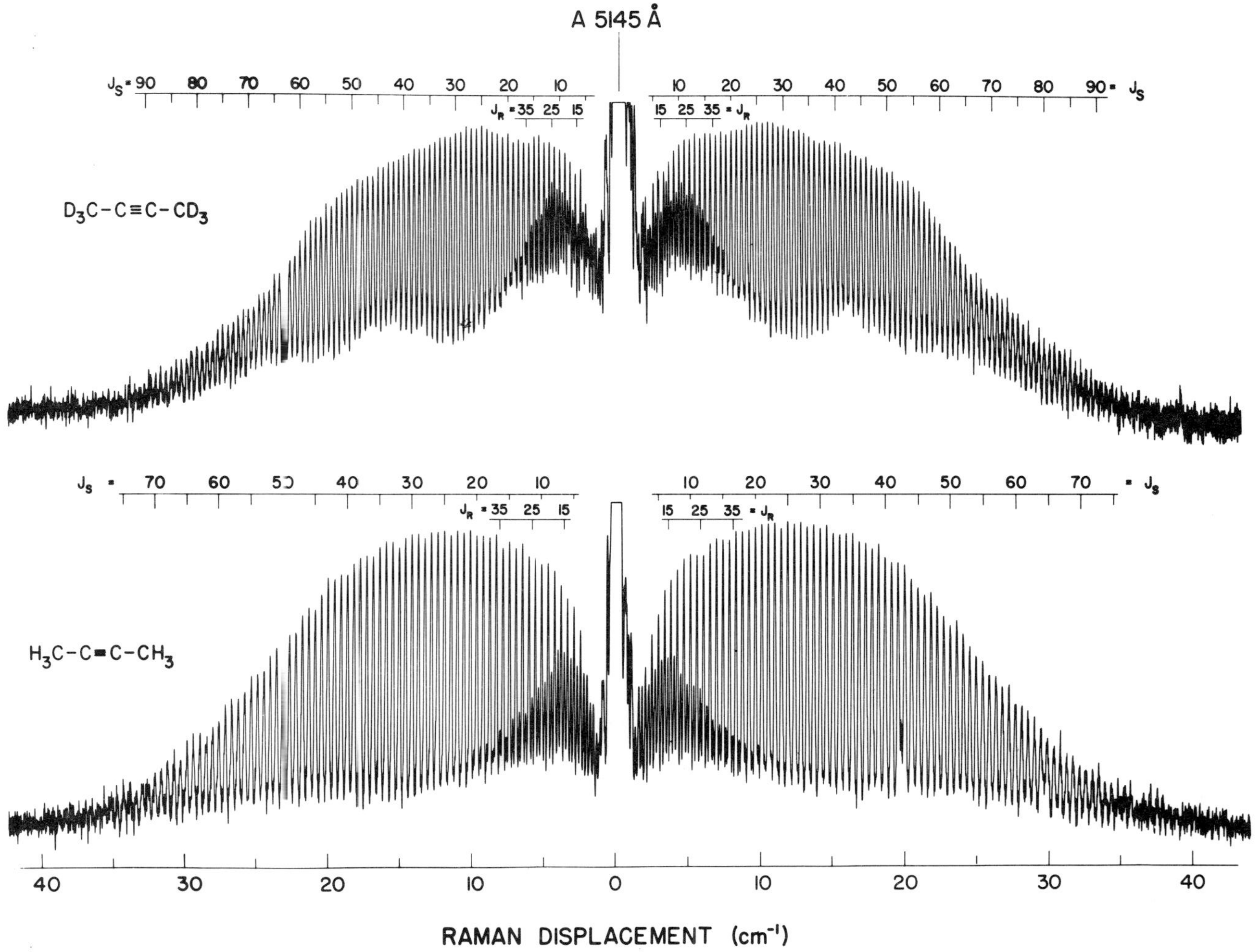

Fig. 3.15. Rotation spectra of dimethylacetylene-d_0 and -d_6; (top), pressure: 40 torr, exposure: 10 h; (bottom), pressure: 50 torr, exposure: 2 h [3.150]. The Stokes side lies to the right

The presence of hot bands is very strikingly shown in Figs.3.15 and 3.16. Not only are the Raman lines superposed on a continuum due to unresolved hot band structure, but there is also a very regular modulation of the envelopes of the intensity maxima and minima, expecially for C_4D_6. Comparison of Figs.3.12-16 also demonstrates the dependence of the relative intensities of the R- and S-branches on the ratio of the moments of inertia I_A/I_B [3.168]. This ratio is very small for C_6H_6 (Fig.3.16) and an R-branch is no longer discernable in this spectrum. In the limit as $I_A/I_B \rightarrow 0$ (linear molecule) only the K = 0 levels have a significant population and the R-branch intensity vanishes.

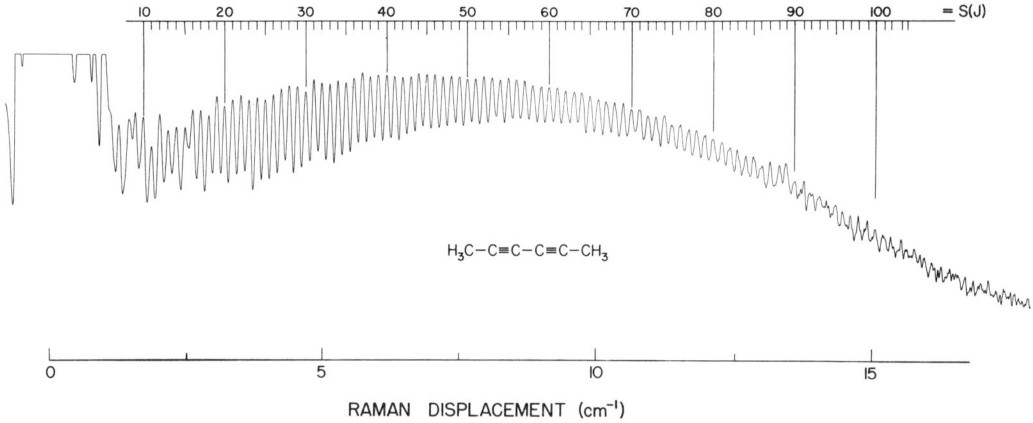

Fig. 3.16. Rotation spectrum of dimethyldiacetylene (Stokes side); λ = 4880 Å, pressure: 2 torr, exposure: 11 h [3.150]

3.2.4 Asymmetric Top Molecules

Since the publication of the previous review article [3.1] there have been significant advances in the art of dealing with the rotational Raman spectra of asymmetric top molecules. Only these will be considered here and the still useful results of the older studies are available in [3.1].

The theory of the rotational energies of an asymmetric top molecule is well developed and details concerning it are readily available elsewhere [3.166,167,174, 175]. Briefly, the energy levels of the general rigid asymmetric top rotor are given by

$$E(J_\tau) = \frac{1}{2}(A+B)\,J(J+1) + \frac{1}{2}(A-C)\,E_\tau^J(\kappa) \tag{3.22}$$

where

$$\kappa = \frac{2B-A-C}{A-C} \tag{3.23}$$

is Ray's asymmetry parameter, A, B, and C are the rotation constants, J is the total rotational angular momentum of the molecule and τ is a pseudoquantum number which replaces the "good" quantum number K of the symmetric top theory. The 2J+1 sublevels of each J-state are labeled by the values of τ: $\tau = -J, -J+1, \ldots, J-1, J$. The quantity $E_\tau^J(\kappa)$ is the solution of a secular determinant for given value of J and κ. The limiting values $\kappa = -1$ and $\kappa = +1$ describe the prolate and oblate symmetric top rotators respectively. Modification of the energy due to centrifugal distortion have been derived in general form by WATSON [3.176] and require in lowest order of approximation five quartic coefficients. In the limiting case of the symmetric top these reduce to the well known D_J, D_{JK}, and D_K coefficients used in (3.11).

The general Raman selection rules are $\Delta J = 0, \pm1, \pm2$. For molecules which possess at least one two-fold axis of symmetry (point groups C_{2v}, D_2, V_h) transitions are further restricted by the rule that only levels with the same ro-vibrational (rotation-vibrational) species may combine with one another. For quasi-symmetric tops it is convenient to label the sublevels of each J-state by the quantum numbers K_P and K_0 for the prolate and oblate symmetric top limiting case. For such molecules the selection rule on K is that $\Delta K = 0, \pm1, \pm2$. Not all transitions have the same intensity, however, these being determined by the non-vanishing components of the polarizability tensor. The $\Delta K = 0$, $\Delta J = +1, +2$ pure rotational transitions [labeled ${}^QR_K(J)$ and ${}^QS_K(J)$, respectively] are most intense and correspond to the symmetric top transitions. The $\Delta K = +1$, $\Delta J = 0, +1, +2$ [labeled, respectively, ${}^RQ_K(J)$, ${}^RR_K(J)$, and ${}^RS_K(J)$] and $\Delta K = +2$, $\Delta J = 0,+1,+2$ [labeled, respectively, ${}^SQ_K(J)$, ${}^SR_K(J)$, and ${}^SS_K(J)$] transitions are weak and owe their intensity to the asymmetry of the molecule. In the symmetric top case the quantities that are associated with the quantum number K exclusively, i.e., (C-B) [or (A-B)] and D_K in (3.11) cannot be determined from the pure rotation spectrum since $\Delta K = 0$. The same holds true for the near symmetric top case where these quantities are determined by the $\Delta K \neq 0$ transitions.

The study of the Raman spectra of asymmetric tops has been greatly aided by the work of HILLS and JONES [3.177,179,180], see also [3.90]. The advances in the art of dealing with the rotational Raman spectra of asymmetric top molecules are four in kind. The first and obvious one is the improvement in the quality of the spectra and accuracy of results that can now be obtained (cf. ethylene [3.177] and hydrazoic acid [3.178]). The second relates to the ability to observe and analyze resolved hot band structure (cf. butadiene [3.179]). The ability to extract useful information from partially resolved spectra which exhibit an intensity beat pattern [3.3,180] constitutes a third advance while the fourth one is the ability to determine the components of the polarizability tensor [3.177,180,181]. In all of these, use of a computer and computer generated spectra, initially introduced into Raman spectroscopy by H.L. WELSH and co-workers (see [3.182]), play a decisive role.

Table 3.6. Molecular constants of ethylene determined from Raman and infrared spectra

	Raman[a]	Infrared[b]
A_0 [cm^{-1}]	4.8666 ± 0.0040	4.86596 ± 0.00018
B_0 [cm^{-1}]	1.00076 ± 0.00013	1.001329 ± 0.000061
C_0 [cm^{-1}]	0.82848 ± 0.00016	0.828824 ± 0.000062
$\Delta_J/D_J \times 10^6$[cm^{-1}]	(1.59 ± 0.03)	(1.447 ± 0.070)
$\Delta_{JK}/D_{JK} \times 10^5$[$cm^{-1}$]	[1.02][c]	(1.468 ± 0.063)
$\Delta_K/D_K \times 10^5$[cm^{-1}]	[8.87][c]	(9.17 ± 0.16)
$\Delta^0 \times 10^{20}$[amu-m^2]	+0.039 ± 0.006	+0.049 ± 0.002

[a] From [3.177]. Centrifugal distortion constants are Δ_J, Δ_{JK} and Δ_K.
[b] D. van Leberghe, I.J. Wright, J.L. Duncan: J. Mol. Spectrosc. *42*, 251 (1972) Centrifugal distortion constants are D_J, D_{JK}, and D_K.
[c] Computed from values given by van Leberghe et al. (see footnote b) and constrained in fit of Raman data.

a) *Ethylene*

The rotational constants for ethylene, C_2H_4, were determined from the observed spectrum by standard procedures [3.177]. Only unblended lines were used to derive the constants (see Table 3.6) these lines being identified by a computer simulation of the whole spectrum. Although there still remains a slight discrepancy between the Raman and infrared determined B_0-values the overall agreement with the results from the infrared work is satisfactory. For this molecule the tensor components α_{xx}, α_{yy}, and α_{zz} were also determined (see below).

b) *Hydrazoic Acid*

Hydrazoic acid, HN_3, is a near prolate symmetrical top molecule with $\kappa = -0.992$. The pure rotational Raman spectra of this as well as of the DN_3 (κ=-0.9977) molecule were studied by BENDTSEN [3.178]. Figure 3.17 shows the Stokes side of the observed DN_3 spectrum. The analysis of both HN_3 and DN_3 spectra is based on the approximate energy expression for a prolate slightly asymmetric top

$$E = \frac{1}{2}(B+C)J(J+1) + [A - \frac{1}{2}(B+C)]K^2 - D_J J^2(J+1)^2 - D_{JK}(J+1)K^2 - D_K K^4 + H_K K^6 \quad (3.24)$$

where the sextic centrifugal distortion term $H_K K^6$ was found to be necessary for the analysis. The results are given in Table 3.7. The contribution of the Raman work focuses on the determination of the D_K and H_K constants and the values for these constants were then used to derive improved values for A, $\overline{B}$, and D_J. The determination of the A_0-value of DN_3 from the Raman spectrum was of significant aid in assigning the b-type microwave spectrum of DN_3. Without this Raman A_0-value there would

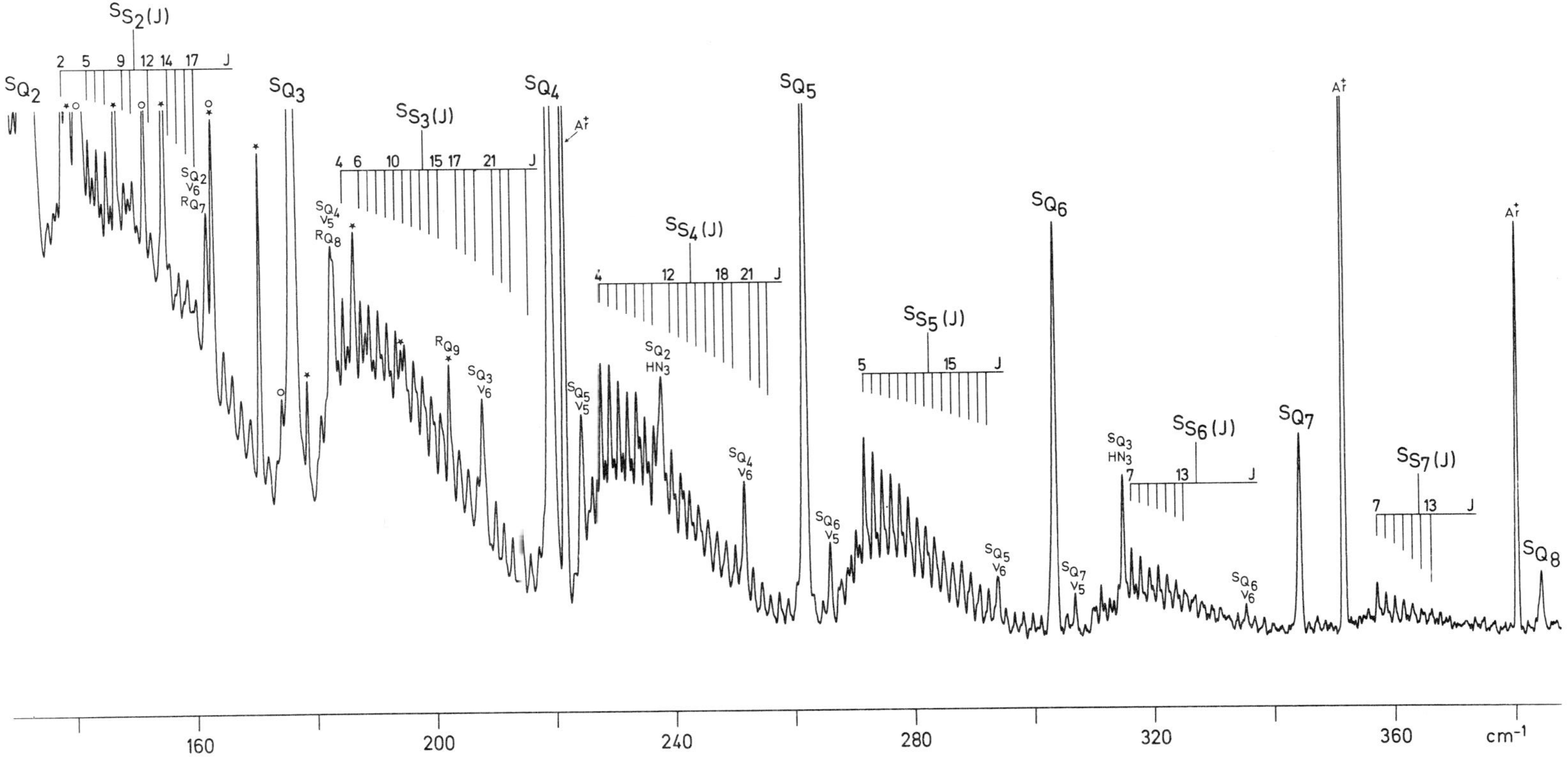

Fig. 3.17. High frequency part of the pure rotational Raman spectrum of DN_3. The spectrum was excited by the argon 4880 Å line using a gas pressure of 350 torr and was photographed with the Aarhus instrument (see Sect.2.2.2) on a Kodak IIa-D plate in 104 h with a plate factor of 2.8 cm^{-1} mm^{-1}. The slit width is 0.25 cm^{-1}. The stars and O indicate impurities due to N_2 and O_2, respectively. $^SQ_K^{\nu_n}$ denotes the ^SQ_K transition in the vibrational state ν_n. (From [3.178])

probably have been no microwave A_0-value for DN_3. This work demonstrates the significant interplay between the Raman, infrared, and microwave techniques in deriving highly accurate molecular constants.

Table 3.7. Ground state rotational constants of hydrazoic acid [3.178][a]

Constant [cm^{-1}]	HN_3	DN_3
A_0	20.38158(5)[b]	11.499456(40)[b]
$\overline{B}$	0.3972014(3)[c]	0.37218633(35)[c]
$D_J \times 10^6$	0.1548(18)[c]	0.13667(37)[c]
$D_{JK} \times 10^5$	2.6478(43)[c]	1.5097(40)[c]
$D_K \times 10^3$	8.60(5)[d]	2.931(36)[e]
$H_K \times 10^6$	7.38(93)[e]	2.00(32)[e]

[a] The errors are standard errors quoted in units of the last significant digit.
[b] Combined Raman and microwave result.
[c] Microwave result.
[d] Combined Raman and near infrared result.
[e] Raman value.

c) *Trans-Butadiene*

Trans-Butadiene with $\kappa = -0.977$ is also a prolate near symmetric top. The $\Delta K = 0$, $\Delta J = 0, +1, +2$ transitions were previously observed by MARAIS et al. [3.183] who reported the A_0 and $\overline{B}_0 = (B_0+C_0)/2$ ground state constants. These transitions form the regular, typical pattern of a symmetric top molecule. The clean spectrum obtained with laser excitation [3.179,180] also reveals, however, an intensity anomaly in the form of intensity beats not noticed in the mercury arc lamp excited spectrum [3.183] but observed previously in the partially resolved laser excited spectra of other slightly asymmetric top molecules [3.3]. Of greater interest here, however, is the observation by HILLS and JONES [3.179] of pure rotational $\Delta K = +2$, $\Delta J = 0$ [$^SQ_K(J)$] transitions in the ground and excited vibrational states ν_{13}, ν_{29}, and $2\nu_{13}$.

The frequencies of the $^SQ_K(J)$ lines are given by

$$\Delta\nu = 4\left(A_v-\overline{B}_v-2D_K^v\right)(K+1) - 8D_K^v(K+1)^3 \tag{3.25}$$

if the D_{JK}^v term is ignored. While good values of $(A_0-\overline{B}_0)$ and D_K^0 were obtained for the ground state constants (see Table 3.8) it was found necessary to take account of the dominant Coriolis interaction between ν_{13} and ν_{24} in order to get acceptable excited state constants. These were obtained by an iterative least squares procedure, using for the Coriolis coupling constant the value $|\zeta_{13,24}| = 0.64$ determined from infrared spectra [3.184]. Although allowed, the $\Delta K = 1$ transitions were not observed. Table 3.8 summarizes the results. A reproduction of the photographically recorded spectrum is given in [3.179].

Table 3.8. Molecular constants of butadiene [3.179]

State	$(A-\bar{B})$ [cm^{-1}]	$D_K \times 10^6$ [cm^{-1}]
Ground	1.24782 ± 0.00009	3.61 ± 0.18
ν_{13} = 162 cm^{-1}	1.22670 ± 0.00024	2.6 ± 0.5
ν_{24} = 301 cm^{-1}	1.2703 ± 0.0010	30 ± 3
$2\nu_{13}$	1.167 ± 0.002	---

The resolved irregular structure of the asymmetric top pure rotation spectrum offers the possibility of obtaining information about the components of the polarizability tensor. For molecules having orthorhombic symmetry (point groups C_{2v}, V_h, V) the intensities of the pure rotational transitions in the (totally symmetric) ground state are determined by the non-vanishing irreducible spherical tensor components $\alpha_2^2 + \alpha_{-2}^2 = \alpha_{xx} - \alpha_{yy}$ and $\alpha_0^2 = 6^{-1/2}(2\alpha_{zz}-\alpha_{xx}-\alpha_{yy})$, see (2.44), whose ratio

$$R = \frac{\alpha_2^2+\alpha_{-2}^2}{\alpha_0^2} = \frac{6^{1/2}(\alpha_{xx}-\alpha_{yy})}{(2\alpha_{zz}-\alpha_{xx}-\alpha_{yy})} \tag{3.26}$$

determines the relative intensity distribution within the rotational spectrum. For quasi-symmetric tops for which K is approximately a "good" quantum number, $(\alpha_{xx}-\alpha_{yy})$ determines the intensities of the $\Delta K = 2$ transitions whereas those of the $\Delta K = 0$ transitions are determined by $6^{-1/2}(2\alpha_{zz}-\alpha_{xx}-\alpha_{yy})$. [HILLS and JONES [3.177] as well as HERRANZ and STOICHEFF [3.185] included a factor $2^{-1/2}$ in their combination $(\alpha_{xx}-\alpha_{yy})$. HILLS and JONES' ratio for R therefore differs from that given in (3.26) by a factor $2^{-1/2}$]. For molecules belonging to the monoclinic groups C_s, C_2, and C_{2h} (e.g., butadiene with symmetry C_{2h}) the component α_{xy} also belongs to the totally symmetric species (see [Ref.3.1, Table 2]) and is responsible for the intensities of the $\Delta K = 1$ transitions. Since the Cartesian axes xyz are the principal axes of the moment of inertia tensor (cf. the diagonalization of the rotational energy matrix) the existence of a nonvanishing component α_{xy} demonstrates that for molecules belonging to the monoclinic point groups the principal axes of the polarizability tensor are not coincident with those of the moment of inertia tensor, i.e., the distribution of mass in such molecules does not also describe the distribution of electric charge. The absence of $\Delta K = 1$ transitions from the observed spectrum of butadiene indicates that for this molecule α_{xy} and thus the mismatch between the mass and charge distributions is small.

Different assumed values of R are used to generate theoretical spectra which are then compared to the observed one. The value of R that furnishes the best match between the observed and simulated spectra is then taken as the true value of the polarizability ratio R. Additional data then allow the determination of the individual components of the polarizability tensor. These are the mean value of the polarizability

$$\bar{\alpha} = \frac{1}{3}(\alpha_{xx}+\alpha_{yy}+\alpha_{zz}) \quad , \tag{3.27}$$

the depolarization factor for the Rayleigh line (for linearly polarized incident light and a 90° scattering configuration)

$$\rho = \frac{3\gamma^2}{45\bar{\alpha}^2+4\gamma^2} \quad , \tag{3.28}$$

where γ is the anisotropy of the polarizability defined by

$$\gamma^2 = \frac{1}{2}\left[(\alpha_{xx}-\alpha_{yy})^2+(\alpha_{yy}-\alpha_{zz})^2+(\alpha_{zz}-\alpha_{xx})^2+6(\alpha_{xy}^2+\alpha_{yz}^2+\alpha_{zx}^2)\right] \quad , \tag{3.29}$$

and also the Kerr constant.

The distinction between the depolarization factor of the "Rayleigh line" and the "Rayleigh band", which includes contribution from the pure rotation spectrum, must be emphasized [Ref.3.1, pp.721,722]. Figure 3.18 shows the comparison between

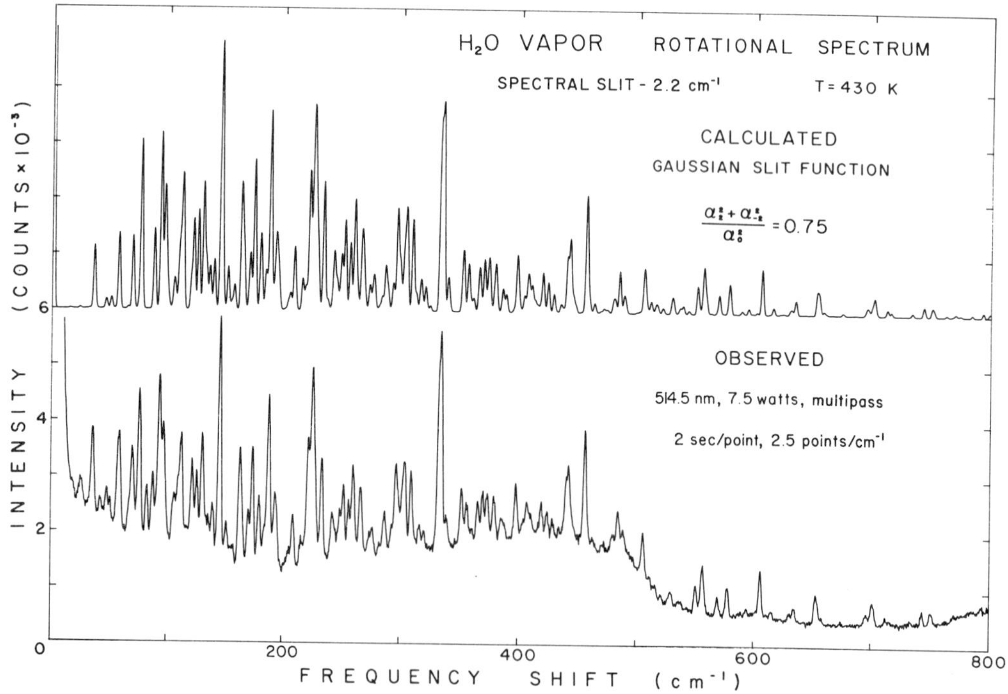

Fig. 3.18. Pure rotational Raman spectrum of water vapor. Top: best calculated spectrum for R = 0.75; Bottom: observed spectrum (photoelectric recording). (From [3.181] with permission)

the best simulated and the photoelectrically recorded spectrum of water vapor performed by MURPHY [3.181]. The match is excellent, giving $R = 0.75$, the base line continuum in the observed spectrum being attributed to parasitic light. With a value $\bar{\alpha} = (1.4703 \pm 0.0025) \times 10^{-24}$ cm^3 determined from refractive index data, and the measured depolarization ration for the Rayleigh line, $\rho = (2.99 \pm 0.45) \times 10^{-4}$, the values for the polarizability components listed in Table 3.9 were obtained by MURPHY [3.181]. A similar procedure was employed by HILLS and JONES [3.177] in their study of ethylene.

Table 3.9. Polarizability tensor components of ethylene and water (in 10^{-24} cm^3)

	Water[a]	Ethylene[b]
α_{xx}	1.4679 ± 0.0031	3.86
α_{yy}	1.4146 ± 0.0128	3.40
α_{zz}	1.5284 ± 0.0134	5.40
$\alpha_{\parallel}$	0.759 ± 0.0301	
$\alpha_{\perp}$	0.7232 ± 0.0155	
Definition of Axes	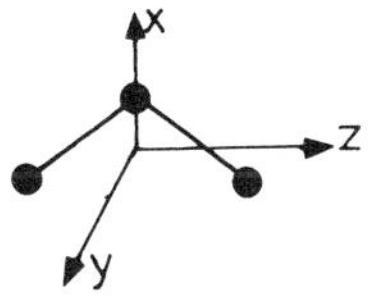	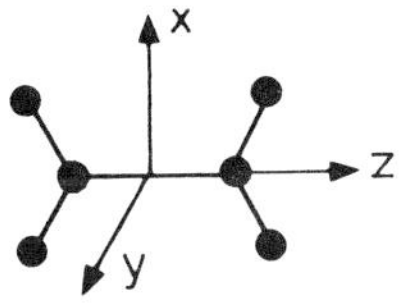

[a] $\alpha_{\parallel}$ and $\alpha_{\perp}$ are the parallel and perpendicular components of the O-H bond polarizabilites [3.181].
[b] Three different sets of bond polarizabilities are presented by HILLS and JONES [3.177].

For partially resolved spectra of near symmetric tops the same procedure for the determination of the polarizability components is in principle applicable, provided that good data on the rotational energy levels are available. The work on butadiene by HILLS and JONES [3.180] illustrates the limitations of the technique. Indeed, the current problem is more often to extract rotational constants from a partially resolved spectrum. Fig.3.19 shows the Stokes side of the pure rotation spectrum of n-pentane. Besides the fact that there are many rotational hot bands as well as the spectra of several conformers present, the unique feature is the modulation of the inner and outer intensity envelopes, called "intensity beats" in analogy to the beat phenomenon that is the result of the coherent superposition of wavetrains. These intensity beats have also been observed in trans-difluoroethylene, trans-dichloroethylene [3.3], butadiene [3.179,180] and other molecules. HILLS and JONES have analyzed this phenomenon and have shown how to utilize it in the determination of molecular constants. Details of the procedure are given in the original publication [3.180].

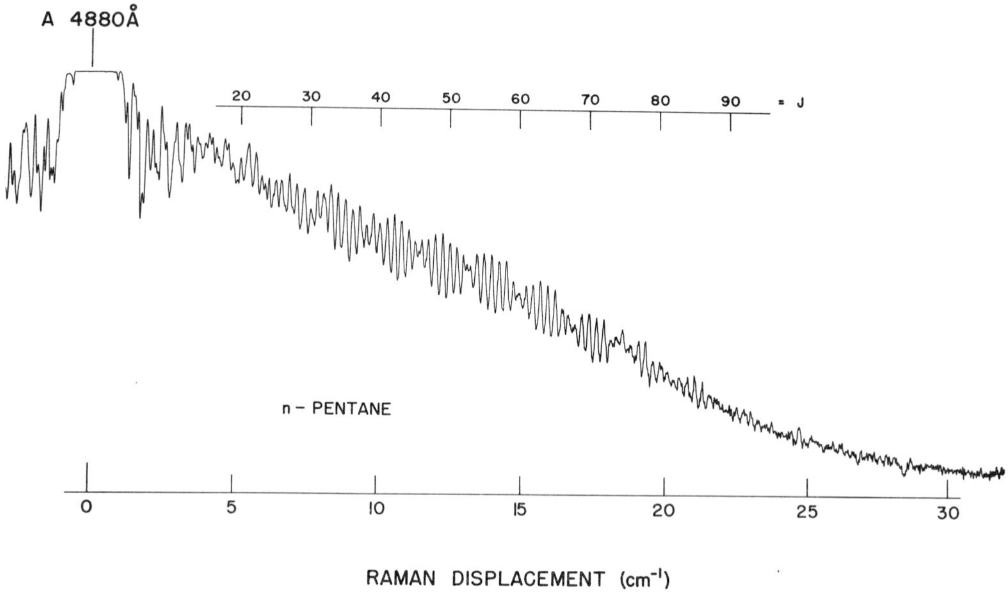

Fig. 3.19. Pure rotation spectrum of n-pentane (Stokes side) [3.162]

3.2.5 "Forbidden" Pure Rotational Spectra

The previous discussion addressed itself exclusively to those pure rotational Raman transitions that are "allowed" in the rigid rotor approximation. The centrifugal distortion effects that were accounted for only modified the rigid rotor energies but did not affect the transition probabilities. In the rigid rotor model the pure rotational Raman spectra of spherical top molecules (e.g., CH_4, point group T_d; SF_6, point group O_h) are strictly forbidden, while those of symmetric top molecules, for example, are allowed with a selection rule $\Delta J = J' - J'' = 0, +1, +2$, $J'+J'' \geqq 2$, $\Delta K = K' - K'' = 0$. For the non-rigid rotator the presence of centrifugal distortion and rotation-vibration interaction modifies the transition probabilities and pure rotational Raman transitions, forbidden in the rigid rotor model, are now allowed. The theory of the forbidden pure rotational Raman spectra has been developed recently by several authors [3.186-188] but no experimental results have been reported so far. A comprehensive review of "forbidden" infrared and microwave rotational transitions is available in the article by OKA [3.189].

a) *Spherical Top Molecules* - T_d *Symmetry*

When the methane molecule, for example, rotates about one of its four C_3 axes (any C-H bond) the centrifugal forces acting on the H-atoms not located on the C_3 axis cause the molecule to become non-spherical, with C_{3v} symmetry, exhibiting a centrifugally induced dipole moment and anisotropy of polarizability. In consequence of this effect both infrared and Raman pure rotational transitions are allowed.

Rotations about any of the three C_2 axes will induce an anisotropy in the polarizability only and a pure rotational Raman spectrum, but no infrared or microwave spectrum, will be allowed. RAO [3.188] has developed the theory for the centrifugal distortion induced Raman spectra of molecules with T_d symmetry while ROSENBERG and CHEN [3.187] also include the vibration-rotation interaction with the doubly and triply degenerate normal modes of vibration. The selection rules for these "forbidden" pure rotational spectra are

$$\Delta J = J' - J'' = 0, +1, +2, \quad J' + J'' \geqq 4$$
$$\Delta J = K' - K'' = 0, \pm 4 \quad . \qquad (3.30)$$

These spectra have not yet been observed. They would, indeed, be very weak, the strongest of their rotational lines in methane (J=9 to J=12 in the $\Delta J=+1$ band at 300 K) having an intensity estimated to be only 10^{-5} that of the Q-branch intensity of the Raman active $\nu_3(f_2)$ band [3.187].

b) *Symmetric Top Molecules* - C_{3v} *Symmetry*

Pure rotational spectra with the selection rule $\Delta J = 0, +1, +2, \Delta K = 0$ are already allowed in the rigid rotor approximation. RAO [3.188] has shown that for a centrifugally distorted rotor Raman transitions with selection rule

$$\Delta J = 0, +1, +2, \quad J' + J'' \geqq 3$$
$$\Delta K = +3, +6 \qquad (3.31)$$

are also present in addition to the ordinary rotation spectrum, while the rigid rotor transitions with $\Delta K = 0$ have their intensities modified. For NH_3 the effective rotation induced polarizability responsible for the occurrence of the "forbidden" transitions is estimated to be only of the order of 10^{-5} Å^3. The presence of the much stronger ordinary Raman spectrum makes the detection of these "forbidden" transitions extremely difficult.

3.3 Summary

This chapter contains a detailed exposition of the experimental techniques used in and the results that have been obtained so far from the study of high-resolution pure rotational Raman spectra of gases. The substance of the chapter is limited to the spontaneous Raman effect and excludes rotational spectra based on resonance and nonlinear phenomena (see, however, Chaps.6,7). The attention of the work described here is focused on non-polar molecules since these are particulary amenable to study by means of the Raman effect. These range from the simplest stable diatomic to rather complex polyatomic molecules. Except for the simplest molecules the spectra are not fully resolved even though the presently attained resolution is near 0.06 cm^{-1}. While generally of lesser precision than the comparable values derived from microwave and infrared spectra, the rotation and centrifugal distortion constants obtained from pure rotational Raman spectra play a useful and important role in the overall schema of molecular structure studies.

Most of the work done so far has been devoted to diatomic, linear, and symmetric top polyatomic molecules. Very few studies have been made of asymmetric tops. The development of the theory of the rotational Raman spectra of these molecules together with the experimental and computational techniques that are now available will enable a greater effort to be expended on them. The question of the practically attainable limit of resolution has not yet been satisfactorily answered. Nearly all of the work done so far used the 90° scattering configuration and no advantage has been taken of the narrow line widths that can be realized by forward scattering [3.190]. The problems that have to be dealt with are, of course, formidable but they can be overcome with currently available techniques.

Several topics are not treated in this chapter. Among these are the determination of the bond lengths and interbond angles, and the Doppler broadening of the Raman lines. A discussion of these may be found in the earlier review by WEBER [3.1] while other topics are treated in Chap.4 (intensities, Raman scattering cross sections) and Chap.5 (pressure broadening effects). One topic of current interest is the rotational Raman scattering from dimer molecules. The little work that has been done here [3.191] indicates that the rotation spectrum is masked by collision-induced scattering, at least at the experimental conditions that prevailed. Observation of resolved rotational Raman spectra of simple van der Waals-bound dimer molecules would indeed constitute a major advance over the sensitivity and resolution capabilities of present day Raman spectroscopy. It is hoped that this review article contributes to the attainment of this goal.

Acknowledgments. The financial support through grants awarded by the National Science Foundation is gratefully acknowledged. Thanks are also due to Drs. J. Bendtsen, S. Brodersen, W.J. Jones, D.W. Lepard, T. King McCubbin, W.F. Murphy, and H.L. Welsh for furnishing figures and for private communication of results prior to publication.

References

3.1 Weber, A.: "High Resolution Raman Studies of Gases", in *The Raman Effect*, ed. by A. Anderson, Vol.2 (Marcel Decker, New York 1973) pp.543-757
3.2 Weber, A.: "Progress in Raman Spectroscopy of Gases", in *Proc. 5th Intern. Conf. on Raman Spectroscopy*, ed. by Schmid,E.D., Brandmüller,J., Kiefer,W., Schrader,B., Schrötter,H.W. (H.F. Schulz Verlag, Freiburg/Br. 1976) pp.391-400
3.3 Weber, A.: In *Developments in Applied Spectroscopy*, ed. by Perkins,A.J., Grove,E.L., Kaelble,E.M., Westermeyer,J.E., Vol.10 (Plenum Press, New York 1972) pp.137-160
3.4 Edwards, H.G.M.: In *Essays in Structural Chemistry*, ed. by Downs,A.J., Long, D.A., Staveley,L.A.K. (Plenum Press, New York 1971) Chap.6
3.5 Edwards, H.G.M., Long, D.A.: In *Molecular Spectroscopy*, ed. by Barrow,R.F., Long,D.A., Millen,D.J., Vol.1 (Chemical Society, London 1973) Chap.6
3.6 Edwards, H.G.M.: In *Molecular Spectroscopy*, ed. by Barrow,R.F., Long,D.A., Millen,D.J., Vol.3 (Chemical Society, London 1975) Chap.5
3.7 Weber, A., Porto, S.P.S.: J. Opt. Soc. Am. *55*, 1033 (1965)
3.8 Weber, A., Porto, S.P.S., Cheesman, L.E., Barrett, J.J.: J. Opt. Soc. Am. *57*, 19 (1967)
3.9 Barrett, J.J., Weber, A.: J. Opt. Soc. Am. *60*, 70 (1970)
3.10 Walker, W.J., Weber, A.: J. Mol. Spectrosc. *39*, 57 (1971)
3.11 Claassen, H.H., Selig, H., Shamir, J.: Appl. Spectrosc. *23*, 8 (1969)
3.12a Norlén, G: Phys. Scripta *8*, 249 (1973)
3.12b Beck, R., English, W., Gurs, K.: *Table of Laser Lines in Gases and Vapors*, Springer Series in Optical Sciences, Vol.2, 2nd ed. (Springer, Berlin, Heidelberg, New York 1978)
3.13a Pittak, U.J.: Laser Focus *5* (9), 47 (1969)
3.13b Hercher, M.: Appl. Opt. *8*, 1103 (1969)
3.13c Pilz, W., Kriegsmann, H.: J. Raman Spectrosc. *5*, 153 (1976)
3.14 Rich, N., Welsh, H.L.: J. Opt. Soc. Am. *61*, 977 (1971); Ind. J. Pure and Appl. Phys. *9*, 944 (1971)
3.15 Brodersen, S., Bendtsen, J.: J. Raman Spectrosc. *1*, 97 (1973)
3.16 Hendra, P.J., Vear, C.J.: Spectrochim. Acta *28*A, 1949 (1972)
3.17a Butcher, R.J., Willetts, D.V., Jones, W.J.: Proc. R. Soc. A*324*, 231 (1971)
3.17b Butcher, R.J., Jones, W.J.: J. Chem. Soc., Faraday Trans. II *70*, 560 (1974)
3.18a Kiefer, W., Bernstein, H.J., Wieser, H., Danyluk, M.: J. Mol. Spectrosc. *43*, 393 (1972)
3.18b Jessen, P.F., Gaydon, A.G.: Combustion and Flame *11*, 11 (1967); *12th Symp. (International) on Combustion* (Combustion Institute, Pittsburgh, Pa. 1969) p.481
3.19 Freedman, P.A., Jones, W.J., Rogstad, A.: J. Chem. Soc., Faraday Trans. II *71*, 286 (1975)
3.20 Hartley, D.L., Hill, R.A.: J. Appl. Phys. *43*, 4134 (1972)
3.21 Hill, R.A., Hartley, D.A., Appl. Opt. *13*, 186 (1974)
3.22 Hill, R.A., Peterson, C.W., Mulac, A.J., Smith, D.R.: J. Quant. Spectrosc. Radiat. Transfer *16*, 953 (1976)
3.23 Weber, A., Schlupf, J.: J. Opt. Soc. Am. *62*, 428 (1972)
3.24 Welsh, H.L., Stansbury, E.J., Romanko, J., Feldman, T.: J. Opt. Soc. Am. *45*, 338 (1955)
3.25 Gordon, J.P., Zeiger, H.J., Townes, C.H.: Phys. Rev. *99*, 1264 (1955)
3.26 Schawlow, A.L., Twones, C.H.: Phys. Rev. *112*, 1940 (1958)
3.27 Devlin, G.E., Davis, J.L., Chase, L., Geschwind, S.: Appl. Phys. Lett. *19*, 138 (1971)
3.28 Schoen, P.E., Jackson, D.A.: J. Phys. E. (Sci. Instr.) *5*, 519 (1972)
3.29 Simmons, J.D., Hougen, J.T.: J. Research Natl. Bur. Stands. *81*A, 25 (1977)
3.30 Gerstenkorn, S., Luc, P.: "Atlas du Spectre d'Absorption de la Molécule d'Iode", parts I and II, Laboratoire Aimé Cotton, CNRS, Orsay (France 1977)
3.31 Douglas, A.E., Herzberg, G.: J. Opt. Soc. Am. *47*, 625 (1957)
3.32 Reader, J., Marquet, L.C., Davis, S.P.: Appl. Opt. *2*, 963 (1963)
3.33 Reader, J.: Appl. Opt. *12*, 1405 (1973)

3.34 Blackwell, H.E., Shipp, G.S., Ogawa, M., Weissler, G.: J. Opt. Soc. Am. *56*, 665 (1966)
3.35 Lagerquist, A., Lundh, L.E., Neuhaus, H.: Phys. Scripta *1*, 261 (1970)
3.36 van Cittert, P.H.: Rev. d'Opt. *2*, 57 (1923); *5*, 393 (1926)
3.37 Leclercq, M., Wallart, F.: J. Raman Spectrosc. *1*, 587 (1973)
3.38 Rank, D.H., Sheppard, N., Szasz, G.J.: J. Chem. Phys. *16*, 698 (1948)
3.39a Shaw, D.E., Welsh, H.L.: Canad. J. Phys. *45*, 3823 (1967)
3.39b Lepard, D.W.: Dissertation, University of Toronto (1964)
3.40 Burch, D.E.: Appl. Opt. *8*, 649 (1969)
3.41 Fellman, J., Lindblom, P.: Appl. Opt. *16*, 1085 (1977)
3.42 Rank, D.H., Eastman, D.F., Birtley, W.B., Skorinko, G., Wiggins, T.A.: J. Opt. Soc. Am. *50*, 821 (1960)
3.43 Edwards, H.G.M., Good, E.A.M., Long, D.A.: J. Chem. Soc., Faraday Trans. II *72*, 865 (1976)
3.44 Willetts, D.V., Freedman, P.A., Jones, W.J.: J. Raman Spectrosc. *2*, 249 (1974)
3.45 Berger, H., Faivre, M., Champion, J.P., Morêt-Bailly, J.: J. Mol. Spectrosc. *45*, 298 (1973)
3.46 Champion, J.P., Berger, H.: J. Mol. Spectrosc. *55*, 15 (1975)
3.47 Fletcher, W.H., Rayside, J.S.: J. Raman Spectrosc. *2*, 3 (1974)
Fletcher, W.H., Rayside, J.S., Wiley, W.J.: Appl. Opt. *13*, 734 (1974)
3.48 Fast, H., Welsh, H.L.: J. Mol. Spectrosc. *41*, 203 (1972)
3.49 Deroche, J., Graner, G., Bendtsen, J., Brodersen, S.: J. Mol. Spectrosc. *62*, 68 (1976)
3.50 See, for example, Sawyer, R.A.: *Experimental Spectroscopy*, 3rd ed. (Dover, New York 1963)
3.51 Edlén, B.: Metrologia *2*, 71 (1966)
3.52 Owens, J.C.: Appl. Opt. *6*, 51 (1967)
3.53 Barostatting the spectrograph room (see [3.15]) is in principle possible but difficult to carry out and is not recommended
3.54 The displacement of the spectrum due to changes in the barometric pressure and the attendant wavelength changes have been treated by a number of authors. See, for example
Guidarelli, S.: Atti Fondaz. Ronchi *12*, 423 (1957)
3.55 Beutler, H., Fred, M.: Phys. Rev. *57*, 69 (1940)
3.56 [Ref.3.50, pp.187-189]
3.57 Bacis, R., Férmelat, B.: Appl. Opt. *12*, 2881 (1973)
3.58 Brodersen, S., Bendtsen, J.: J. Raman Spectrosc. *3*, 207 (1975)
3.59 See, for example, *Advances in Electron Physics*, Vol.40A,B, *Photo-Electronic Image Devices*, ed. by Morgan,B.L., Airey,B.W., McMullan,D. (Academic Press, New York 1976) and earlier volumes in this series
3.60 Detailed information on the properties of Kodak spectroscopic plates is contained in *Kodak Plates and Films for Scientific Photography* (Eastman Kodak Co., 1973) Publication No.P-315
3.61a Hoag, A.A., Miller, W.C.: Appl. Opt. *8*, 2417 (1969)
3.61b Babcock, T.A.: AAS Photo-Bulletin, Issue *13* (3), 3 (1976)
3.61c Allbright, G.S.: J. Photogr. Sci. *24*, 115 (1976)
3.62 Smith, A.G., Schrader, H.W., Richardson, W.W.: Appl. Opt. *10*, 1597 (1971)
3.63 Schlupf, J., Weber, A.: Appl. Opt. *10*, 2373 (1971)
3.64 Miller, W.C.: AAS Photo-Bulletin, Issue *2* (2), 15 (1970)
3.65 Miller, W.C.: AAS Photo-Bulletin, Issue *9* (2), 3 (1975)
3.66 Schoening, W.E.: AAS Photo-Bulletin, Issue *10* (3), 18 (1975)
3.67 Jammu, K.S., Welsh, H.L.: Appl. Opt. *15*, 423 (1976)
3.68 Primik, W., Weiss, W.W.: Astron. Astrophys. *35*, 307 (1974)
3.69 Babcock, T.A., Ferguson, P.M., James, T.H.: Astron. J. *79*, 92 (1974)
3.70 Solomon, S.J.: AAS Photo-Bulletin, Issue *15* (2), 16 (1977)
3.71 Babcock, T.A., Sewell, M.H., Lewis, W.C., James, T.H.: Astron. J. *79*, 1479 (1974)
3.72 Babcock, T.A., Ferguson, P.M., Lewis, W.C., James, T.H.: Photog. Sci. Eng. *19*, 49, 211 (1975)
3.73 Barbieri, C., Cepaccioli, M., Favero, G.: AAS Photo-Bulletin, Issue *2* (2), 3 (1976)

3.74 Scott, R.L., Smith, A.G.: AAS Photo-Bulletin, Issue *12* (2), 6 (1976)
3.75 Latham, D.W.: AAS Photo-Bulletin, Issue *13* (3), 9 (1976)
3.76 Schoening, W.E.: AAS Photo-Bulletin, Issue *14* (4), 3 (1976)
3.77 Sim, M.E.: AAS Photo-Bulletin, Issue *14* (4), 9 (1976)
3.78 Scott, R.L., Smith, A.G., Leacock, R.J.: AAS Photo-Bulletin, Issue *15* (2), 12 (1977)
3.79 Scott, R.L., Smith, A.G.: Astron. J. *79*, 656 (1974)
3.80 Kaye, A.L.: Mon. Not. Roy. Astron. Soc. *180*, 147 (1977); see also the discussion on the gains of the DQE resulting from pre-exposing photographic plates by
Ables, H.D., Hewitt, A.V., James, K.A.: AAS Photo-Bulletin, Issue *3* (1), 18 (1971); as well as
Latham, D.W., Furenlid, I.: AAS Photo-Bulletin, Issue *11* (1), 11 (1976)
3.81 Difley, J.A.: Astron. J. *73*, 762 (1968)
3.82 Gollnow, H., Hagemann, G.: Astron. J. *61*, 399 (1956)
3.83 Altman, J.H., Ball, R.C.: Photog. Sci. Eng. *5*, 278 (1961)
3.84 Levinson, F.H., Ianna, P.A.: Astron. J. *82*, 299 (1977)
3.85 Crosswhite, H.M.: J. Res. Natl. Bur. Stands. *79*A, 17 (1975)
3.86 Giachetti, A., Stanley, R.W., Zalubas, R.: J. Opt. Soc. Am. *60*, 474 (1970)
3.87 Valero, F.P.J.: J. Opt. Soc. Am. *60*, 1675 (1970)
3.88 Weber, A.: Unpublished work
3.89 Bendtsen, J.: J. Raman Spectrosc. *2*, 133 (1974)
3.90 Foster, R.B., Hills, G.W., Jones, W.J.: Mol. Phys. *33*, 1589 (1977)
3.91 Bendtsen, J., Edwards, H.G.M.: J. Raman Spectrosc. *2*, 407 (1974)
3.92 Gerstenkorn, S., Luc, P.: Nouv. Rev. Optique *7*, 149 (1976);
Gerstenkorn, S., Luc, P., Perrin, A., Chauville, J.: Astron. Astrophys. *58*, 255 (1977)
3.93 Schildkraut, E.R., Hirschfeld, T.B.: "Fourier Transform Raman Spectroscopy", in *Laser Raman Gas Diagnostics*, ed. by Lapp,M., Penney,C.M. (Plenum Press, New York 1974) pp.379-384; see also the discussion concerning this paper on pp.384-388
3.94 The suitability of the molecular iodine absorption lines measured by means of the Fourier transform technique to serve as standards for the calibration of spectrographs may also be judged by the results of the rotational analysis of the 5350 Å band reported by
Gerstenkorn, S., Luc, P., Perrin, A.: J. Mol. Spectrosc. *64*, 56 (1977)
3.95 Tomkins, F.S., Fred, M.: J. Opt. Soc. Am. *41*, 641 (1951)
3.96 Dieke, G.H., Dimmock, D., Crosswhite, H.M.: J. Opt. Soc. Am. *46*, 456 (1956)
3.97 Djurle, E., Gran, G.: J. Sci. Instrum. *35*, 304 (1958)
3.98 Steinhaus, D.W.: J. Opt. Soc. Am. *49*, 468 (1959)
3.99 Bennett, J.M., Koehler, W.F.: J. Opt. Soc. Am. *49*, 466 (1959)
3.100 Bennett, J.M.: Appl. Opt. *2*, 1330 (1963)
3.101 Ebbinghausen, E.G.: Pub. Astron. Soc. Pac. *74*, 488 (1962)
3.102 Gunnvald, P.: Ark. Fys. *22*, 333 (1962)
3.103 Gollnow, H.: Mon. Not. Roy. Astron. Soc. *123*, 391 (1962)
3.104 Gollnow, H., Rudge, P., Thomas, D.G.: In *Determination of Radial Velocities and Their Applications*, Proc. I.A.U. Symposium No.30, ed. by Batten,A.H., Heard,J.F. (Academic Press, London 1967) p.23
3.105 Bennett, J.M.: J. Opt. Soc. Am. *51*, 1133 (1961)
3.106 Moore, F.L. Jr., Yarosewick, S.J.: Appl. Opt. *5*, 1465 (1966)
3.107 See, for example, part 3 of *Image Processing Techniques in Astronomy*, ed. by deJager,C., Nieuwenhuijzen,H. (D. Reidel Publ., Dordrecht 1975)
3.108 Steinhaus, D.W., Fisher, K.J., Engleman, J. Jr.: Chem. Instrum. *3*, 141 (1971)
3.109 Brodersen, S., Rasmussen, F., Bendtsen, J.: J. Raman Spectrosc. *6*, 67 (1977)
3.110 Tomkins, F.S., Fred, M.: Appl. Opt. *2*, 715 (1963)
3.111 Spieweck, F.: Metrologia *12*, 43 (1976)
3.112 Spieweck, F.: In *Atomic Masses and Fundamental Constants 5*, ed. by Sanders, J.H., Wapstra,A.H. (Plenum Press, New York 1976) p.372
3.113 Hackel, L.A., Hackel, R.P., Ezekiel, S.: Metrologia *13*, 141 (1977)
3.114 Minkowski, R.: Ap. J. *96*, 306 (1942)
3.115 von Keussler, V.: Z. Astrophys. *24*, 230, 247, 252 (1948)
3.116 Tull, R.G.: Appl. Opt. *8*, 1635 (1969)

3.117 See [3.17] and
Jones, W.J.: Contemp. Phys. *13*, 419 (1972); also see [3.3]
3.118 Barrett, J.J., Myers, S.A.: J. Opt. Soc. Am. *61*, 1246 (1971)
Barrett, J.J., Harvey, A.B.: J. Opt. Soc. Am. *65*, 392 (1975)
Barrett, J.J.: Opt. Engineering *16*, 85 (1977)
Barrett, J.J.: In *Laser Raman Gas Diagnostics*, ed. by Lapp,M., Penney,C.M. (Plenum Press, New York 1974) pp.63-85
Barrett, J.J., Steinberg, G.N.: Appl. Opt. *11*, 2100 (1972)
3.119 Armstrong, R.L.: J. Opt. Soc. Am. *64*, 871 (1974)
3.120 Hargis, P.J., Hill, R.A.: J. Opt. Soc. Am. *65*, 219 (1975)
3.121 Smith, W.H.: In *Advances in Raman Spectroscopy*, ed. by J.P. Mathieu, Vol.1 (Heyden and Sons, London 1973) pp.102-108
3.122 King, R.A., Smith, W.H.: J. Raman Spectrosc. *5*, 339 (1976)
3.123 Herzberg, G.: *Molecular Spectra and Molecular Structure, I. Spectra of Diatomic Molecules*, 2nd ed. (Van Nostrand, New York 1950)
3.124a Kainnady, J.V.: Dissertation, Fordham University (1976)
3.124b Rubin, B., Steiner, D.A., McCubbin, T.K. Jr., Polo, S.R.: J. Mol. Spectrosc. *72*, 57 (1978)
McCubbin, T.K. Jr.: Private communication
3.125 Lofthus, A., Krupenie, P.H.: J. Phys. Chem. Ref. Data *6*, 113 (1977)
3.126 Jammu, K.S., St.John, G.E., Welsh, H.L.: Canad. J. Phys. *44*, 797 (1966)
3.127 Renschler, D.L., Hunt, J.L., McCubbin, T.K. Jr., Polo, S.R.: J. Mol. Spectrosc. *31*, 173 (1969)
3.128 Rich, N., Lepard, D.W.: J. Mol. Spectrosc. *38*, 549 (1971)
3.129 Altmann, K., Strey, G., Hochenbleicher, J.G., Brandmüller, J.: Z. Naturforsch. *27*A, 56 (1972)
3.130 Lepard, D.: Canad. J. Phys. *48*, 1664 (1970)
3.131a Steinbach, W., Gordy, W.: Phys. Rev. A*8*, 1753 (1973); A*11*, 729 (1975)
3.131b Tomuta, L., Mizushima, M., Howard, C.J., Evenson, K.M.: Phys. Rev. A*12*, 974 (1975)
3.132 Albritton, D.L., Harrop, W.J., Schmeltekopf, A.L., Zare, R.: J. Mol. Spectrosc *46*, 103 (1973)
3.133 Shotton, K.C., Jones, W.J.: Canad. J. Phys. *48*, 632 (1970)
3.134 Renschler, D.L., Hunt, J.L., McCubbin, T.K. Jr., Polo, S.R.: J. Mol. Spectrosc *32*, 347 (1969)
3.135 Fast, H., Welsh, H.L., Lepard, D.W.: Canad. J. Phys. *47*, 2879 (1969)
3.136 Stoicheff, B.P.: Canad. J. Phys. *35*, 730 (1957)
3.137 Long, D.A., Edwards, H.G.M., Brassington, N.J., Farwell, D.W., Mansour, H.R., Najm, K.A.B., Skinner, M.: In 5th Colloquium on High Resolution Molecular Spectroscopy, Tours, France (1977)
3.138 Edwards, H.G.M., Good, E.A.M., Long, D.A.: J. Chem. Soc., Faraday Trans. II *72*, 984 (1976)
3.139 Edwards, H.G.M., Good, E.A.M., Long, D.A.: J. Chem. Soc., Faraday Trans. II *72*, 927 (1976)
3.140 Baierl, P., Hochenbleicher, J.G., Kiefer, W.: Appl. Spectrosc. *29*, 356 (1975)
3.141 Wang, Chiu-Hua W.: Dissertation, Fordham University (1975)
3.142 Wang, I-Yan, Weber, A.: Ind. J. Pure and Appl. Phys. *16*, 358 (1978)
3.143 Jones, W.J., Stoicheff, B.P., Tyler, J.K.: Canad. J. Phys. *41*, 2098 (1963)
3.144 Callomon, J.H., Stoicheff, B.P.: Canad. J. Phys. *35*, 373 (1957)
3.145 Wang, I-Yan, Weber, A.: J. Chem. Phys. *67*, 3084 (1977)
3.146 Butcher, R.J., Jones, W.J.: J. Raman Spectrosc. *1*, 393 (1973)
3.147 Stoicheff, B.P.: Canad. J. Phys. *33*, 811 (1955)
3.148 Shotton, K.C., Lee, A.G., Jones, W.J.: J. Raman Spectrosc. *1*, 243 (1973)
3.149 Stoicheff, B.P.: Canad. J. Phys. *40*, 358 (1962)
3.150 Thomas, P.M., Weber, A.: Unpublished work
3.151 Rao, K. Suryanarayana, Stoicheff, B.P., Turner, R.: Canad. J. Phys. *38*, 1516 (1960)
3.152 Freedman, P.A., Jones, W.J.: J. Mol. Spectrosc. *54*, 182 (1975)
3.153 Kainnady, J.V., Weber, A.: J. Raman Spectrosc. *5*, 35 (1976)
3.154 Lancaster, J.E., Stoicheff, B.P.: Canad. J. Phys. *34*, 1016 (1956)
3.155 Schlupf, J., Weber, A.: J. Mol. Spectrosc. *54*, 10 (1975)
3.156 Jones, W.J., Stoicheff, B.P.: Canad. J. Phys. *42*, 2259 (1964)

3.157 Lord, R.C., Stoicheff, B.P.: Canad. J. Phys. *40*, 725 (1962)
3.158 Tanner, K., Weber, A.: J. Mol. Spectrosc. *10*, 381 (1963)
3.159 Peters, R.A., Walker, W.J., Weber, A.: J. Raman Spectrosc. *1*, 159 (1973); *2*, 443 (1974)
3.160 Kainnady, J.V., Weber, A.: J. Raman Spectrosc. *4*, 275 (1976)
3.161 Kainnady, J.V., Weber, A.: To be published
3.162 Weber, A.: Unpublished work
3.163 Schlupf, J., Weber, A.: J. Raman Spectrosc. *1*, 3 (1973)
3.164 See [3.155]. In this reference the coefficient of (J+3/2) should read $[4B-6D_J+27H_J/4-4(D_{JK}-3H_{JK}/2)K^2+H_{KJ}K^4]$
3.165 Herzberg, G.: *Molecular Spectra and Molecular Structure III. Electronic Spectra and Electronic Structure of Polyatomic Molecules* (Van Nostrand Reinhold, New York 1966) pp.90-94
3.166 Gordy, W., Cook, R.L.: *Microwave Molecular Spectra* (Wiley and Sons, New York 1970) pp.61-70
3.167 Townes, C.H., Schawlow, A.L.: *Microwave Spectroscopy* (McGraw-Hill, New York 1955)
3.168 Placzek, G., Teller, E.: Z. Phys. *81*, 209 (1933)
3.169 Wilson, E.Br. Jr.: J. Chem. Phys. *3*, 276 (1935); *6*, 740 (1938)
3.170 Schäfer, K.: Z. Phys. Chem. B*40*, 357 (1938)
3.171 Maue, A.W.: Ann. Phys. (Leipzig) *30*, 555 (1937)
3.172 Mizushima, M.: J. Chem. Phys. *21*, 1222 (1953)
3.173 Aliev, M.R., Subbotin, S.I., Tyulin, V.I.: Opt. Spectrosc. *24*, 47 (1968)
3.174 Wollrab, J.E.: *Rotational Spectra and Molecular Structure* (Academic Press, New York 1967)
3.175 Kroto, H.W.: *Molecular Rotation Spectra* (Wiley and Sons, New York 1975)
3.176 Watson, J.K.G.: J. Chem. Phys. *46*, 1935 (1967)
3.177 Hills, G.W., Jones, W.J.: J. Chem. Soc., Faraday Trans. II *71*, 812 (1975)
3.178 Bendtsen, J.: Private communication; and J. Raman Spectrosc. *6*, 306 (1977)
3.179 Hills, G.W., Jones, W.J.: J. Chem. Soc., Faraday Trans. II *71*, 827 (1975)
3.180 Hills, G.W., Jones, W.J.: J. Chem. Soc., Faraday Trans. II *71*, 835 (1975)
3.181 Murphy, W.F.: Private communication and papers presented at 4th Intern. Conf. on Raman Spectroscopy, New Brunswick, Maine (1974) Abstract 1.2.13; and 30th Annual Symposium on Molecular Spectroscopy, Ohio State University (1975) Abstract ME3; J. Chem. Phys. *67*, 5877 (1977)
3.182 See Ref. No. 82, 94, 95, 95a, 96 cited in [3.1]
3.183 Marais, D.J., Sheppard, N., Stoicheff, B.P.: Tetrahedron *17*, 163 (1962)
3.184 Cole, A.R.H., Green, A.A., Osborne, G.A.: J. Mol. Spectrosc. *48*, 212 (1973)
3.185 Herranz, J., Stoicheff, B.P.: J. Mol. Spectrosc. *10*, 448 (1963)
3.186 Verlan, E.M.: Opt. Spectrosc. *35*, 627 (1973)
3.187 Rosenberg, A., Chen, K.-M.: J. Chem. Phys. *64*, 5304 (1976)
3.188 Rao, Ch.V.S. Ramachandra: J. Mol. Spectrosc. *65*, 24 (1977); *72*, 165 (1978)
3.189 Oka, T.: "'Forbidden' Rotational Transitions", in *Molecular Spectroscopy: Modern Research*, ed. by K. Narahari Rao, Vol.2 (Academic Press, New York 1976) pp.229-253
3.190 See the discussion of linewidth problems in [Ref.3.1, pp.729-742]
3.191 See, for example,
Medina, F.D., Daniels, W.B.: J. Chem. Phys. *66*, 2229 (1977) and references to earlier word cited therein

4. Raman Scattering Cross Sections in Gases and Liquids

H. W. Schrötter and H. W. Klöckner

With 14 Figures

The photographic technique of recording high-resolution Raman spectra of gases allows the precise measurement of wavenumber shifts and the accurate determination of molecular constants derived therefrom. The corresponding methods and results are treated in Chap.3 for pure rotational and in Chap.2 for rotational-vibrational Raman spectra.

On the other hand, the measurement of Raman intensities and scattering cross sections requires the use of photoelectric recording techniques which were introduced by RANK et al. [4.1-3] into Raman spectroscopy over 30 years ago. Another decade passed, before the many corrections required for careful intensity measurements were considered [4.4] and the method was applied to the determination of absolute Raman scattering cross sections (then called "absolute Raman intensities") in liquid carbon tetrachloride [4.5] and in a number of gases [4.6-8]. It has to be mentioned that in one case rather reliable data were obtained also by the photographic technique [4.9]. The "pre-laser" data on gas phase Raman intensities were summarized in a review by MURPHY et al. [4.10], with exception of some further results also obtained with mercury arc excitation [4.11,12].

The application of lasers as excitation sources for Raman spectra of liquids [4.13-16] and gases [4.17] opened also new possibilities for an improvement of the methods to measure Raman intensities. A direct cross section measurement was performed for liquid benzene [4.18] and served as reference value, until KATO and TAKUMA [4.19,20] determined absolute cross sections for a few liquids by comparison with a blackbody source. For gaseous nitrogen similar attempts [4.21] led to erroneous results. It soon turned out that for gases the most reliable values [4.22] are obtained by comparison with the scattering cross section of a rotational Raman line of hydrogen, a method which was first applied by GOLDEN et al. [4.23]. Once the absolute scattering cross section for nitrogen is firmly established, for all other gases only measurements relative to N_2 as standard are necessary.

In this chapter a critical review of Raman scattering cross section measurements in gases and liquids is given including new results from our laboratory. It has been known for a long time that Raman cross sections in liquids are considerably higher than in the gaseous state [4.5,24-26], but a direct measurement of this ratio was lacking. In Sect.4.3 we present such measurements for a number of substances and compare them with previous more indirect results.

4.1 Raman Scattering Cross Sections in Gases

4.1.1 Theoretical Considerations

The intensity of a Raman line is determined by the scattering cross section $\sigma(m \leftarrow \ell)$ of the transition between the energy levels E_ℓ and E_m. These levels are characterized by the quantum numbers J, J' of the total angular momentum and M, M' of the projection of the total angular momentum onto a space fixed z-axis, the vibrational quantum numbers v, v', and the remaining quantum numbers summarized as n, n'. For a system of freely oriented molecules the Raman scattering intensity further depends on the number of scattering molecules N, the sum of the population probabilities of degenerate initial levels, the statistical weight factor g_s of the initial level due to the nuclear spin, and the exciting power density I_0 [W/cm^2]. At thermal equilibrium the population probability of the initial level is described by Boltzmann statistics. The power scattered into the full solid angle 4π is given by

$$I(\nu)_{m\leftarrow\ell} = \sigma(\nu)_{m\leftarrow\ell} \cdot g_s \cdot N \cdot (2J+1) \cdot \frac{1}{Z} \cdot \exp(-E_\ell/kT) \cdot I_0 \qquad (4.1)$$

and depends on the wavenumber of the exciting line ν_0 as does the scattering cross section σ. E_ℓ denotes the energy of the initial state, k Boltzmann's constant, and T the absolute temperature. The state sum Z is calculated as

$$Z = \sum_{v,J,n} g_s \cdot (2J+1) \cdot \exp[-E(v,J,n)/kT] \quad .$$

The total scattering cross section of a transition $m \leftarrow \ell$ can be expressed through the matrix elements of the components of the polarizability tensor α_{rs}, when the conditions for the applicability of PLACZEK's polarizability theory [4.27] are fulfilled. In the absence of external fields initial and final states are degenerate with respect to the magnetic quantum number M, M'. Therefore one has to sum over all possible states with quantum numbers M, M' and to divide by the degree of degeneracy of the initial state (2J+1)

$$\sigma(\nu)_{m\leftarrow\ell} = \frac{2^7\pi^5}{3^2} \cdot (\nu_0+\nu_{m\leftarrow\ell})^4 \cdot \frac{1}{2J+1} \cdot \sum_{M,M'} \sum_{r,s} \left|(\alpha_{rs})_{vJnM}^{v'J'n'M'}\right|^2 \qquad (4.2)$$

with r,s = x,y,z. $\nu_{m\leftarrow\ell} = (E_\ell - E_m)/hc$ is the wavenumber of the Raman transition $m \leftarrow \ell$. If the polarizability tensor is real and symmetric, it can be separated into an isotropic and an anisotropic part.

In order to carry out the summation over M, M' in (4.2) it is practical to transform the cartesian components of the polarizability tensor into irreducible spherical tensor components f_{tq}.

Following PLACZEK [4.27,28] and ALTMANN [4.29] one obtains for the dependence of the total scattering cross section on the isotropic and anisotropic parts of the irreducible spherical polarizability tensor, respectively, which are denoted as line strengths $L^{(0)}$ and $L^{(2)}$ here,

$$\sigma(\nu)_{m\leftarrow\ell} = \frac{2^7\pi^5}{3^2} \cdot (\nu_0+\nu_{m\leftarrow\ell})^4 \cdot \frac{1}{2J+1} \cdot \left(L^{(0)v'J'n'}_{vJn} + L^{(2)v'J'n'}_{vJn}\right) \quad . \tag{4.3}$$

The line strengths are independent of the magnetic quantum numbers and are determined by the components of the irreducible spherical polarizability tensor in a space-fixed coordinate system, i.e.,

$$L^{(t)v'J'n'}_{vJn} = \sum_{qMM'} \left|(f_{tq})^{v'J'n'M'}_{vJnM}\right|^2 \quad , \quad q = -t,-t+1,\dots,t-1,t \quad , \tag{4.4}$$

and the index $t=0$ or 2 distinguishes trace and anisotropic scattering, respectively.

Equation (4.3) gives the scattering cross section of a Raman line assigned to a single transition, i.e., a rotational-vibrational line or a pure rotational line. As ALTMANN and STREY [4.29,30] have shown, the line strengths for a symmetric top molecule (including linear molecules) can be expressed explicitly using Wigner's 3-j-symbols (see [Ref. 4.31, §106])

$$L^{(t)v'J'k'}_{vJk} = (2J'+1)(2J+1)\begin{pmatrix} J' & t & J \\ -k' & \overline{q} & k \end{pmatrix}^2 \left|(\overline{f}_{t\overline{q}})^{v'}_{v}\right|^2 \quad , \quad \overline{q} = k' - k \quad , \tag{4.5}$$

where $\overline{f}_{t\overline{q}}$ are the irreducible spherical components of the polarizability tensor in the molecular coordinate system and n, n' has been replaced by k, k', the quantum number of the projection of the angular momentum on the axis of the symmetric top.

In order to calculate the total scattering cross section of a rotational-vibrational band one has to sum over all rotational transitions belonging to the vibrational band in question.

The result of this calculation (see [4.27,30]) is, that the scattering cross section of a rotational-vibrational Raman band is independent of the rotational excitation and equal to the average cross section of space-fixed, non-rotating molecules with all possible orientations

$$\sigma(\nu)^{v'}_{v} = \frac{2^7\pi^5}{3^2} \cdot (\nu_0+\nu^{v'}_{v})^4 \cdot \sum_{r,s} \left|(\overline{\alpha}_{rs})^{v'}_{v}\right|^2 \quad . \tag{4.6}$$

$\overline{\alpha}_{rs}$ are the components of the polarizability tensor in the molecular coordinate system. The dependence of the wavenumber $\nu^{v'}_{v}$ on the rotational quantum number is neglected.

When the conditions of PLACZEK's polarizability theory are fulfilled, the sum in (4.6) can be replaced by the invariants of the polarizability tensor derived with respect to the normal coordinates q_j of the normal vibration ν_j, namely, the linear average a_j and the anisotropy γ_j^2. Restricting the calculation to a transition where only the vibrational quantum number v_j of one vibration ν_j is changed, $v_j \to v_j + 1$, one obtains for the total scattering cross section (see [Ref.4.32, p.62])

$$\sigma(\nu_j) = \frac{2^7\pi^5}{3^2} \cdot \frac{(\nu_0-\nu_j)^4 \cdot b_j^2}{1-\exp(-hc\nu_j/kT)} \cdot g_j \cdot \left(3a_j^2 + \frac{2}{3}\gamma_j^2\right) \quad . \tag{4.7}$$

Here, g_j denotes the degree of degeneracy of the normal vibration ν_j, $b_j^2 = h/8\pi^2 c\nu_j$ the square of the zero point amplitude, h Planck's constant, c the velocity of light in vacuum,

$$a_j = \frac{1}{3}\left[\left(\frac{\partial\alpha_{xx}}{\partial q_j}\right)_0 + \left(\frac{\partial\alpha_{yy}}{\partial q_j}\right)_0 + \left(\frac{\partial\alpha_{zz}}{\partial q_j}\right)_0\right] \quad ,$$

and

$$\gamma_j^2 = \frac{1}{2}\left\{\left[\left(\frac{\partial\alpha_{xx}}{\partial q_j}\right)_0 - \left(\frac{\partial\alpha_{yy}}{\partial q_j}\right)_0\right]^2 + \left[\left(\frac{\partial\alpha_{yy}}{\partial q_j}\right)_0 - \left(\frac{\partial\alpha_{zz}}{\partial q_j}\right)_0\right]^2 + \left[\left(\frac{\partial\alpha_{zz}}{\partial q_j}\right)_0 - \left(\frac{\partial\alpha_{xx}}{\partial q_j}\right)_0\right]^2 + 6\left[\left(\frac{\partial\alpha_{xy}}{\partial q_j}\right)_0^2 + \left(\frac{\partial\alpha_{yz}}{\partial q_j}\right)_0^2 + \left(\frac{\partial\alpha_{zx}}{\partial q_j}\right)_0^2\right]\right\}$$

the invariants of the derived polarizability tensor.

Equation (4.7) gives the total average scattering cross section of a molecule into the full solid angle 4π. However, usually the observation is made with a limited acceptance angle into a certain direction. Therefore, the differential scattering cross section into a certain direction is the more useful quantity. It is given by [4.33]

$$\frac{d\sigma_j}{d\Omega}(\nu_0,\Theta) = \frac{2^4\pi^4}{45} \cdot \frac{(\nu_0-\nu_j)^4 \cdot b_j^2}{1-\exp(-hc\nu_j/kT)} \cdot g_j \cdot \left(45a_j^2 + 7\gamma_j^2\right) \cdot \frac{2\rho_j + (1-\rho_j)\sin^2\Theta}{1+\rho_j} \quad , \tag{4.8}$$

where Θ is the angle between the direction of observation (without polarization analyzer) and the polarization vector of the linearly polarized exciting radiation, and

$$\rho_j = \frac{3\gamma_j^2}{45a_j^2 + 4\gamma_j^2}$$

the depolarization ratio of the Raman band ν_j. Integration of (4.8) over the solid angle Ω leads back to the total scattering cross section in (4.7). HIRSCHFELD [4.34] has tabulated the evaluation of (4.8) for all practically important scattering geometries. For the most important case $\Theta = 90^\circ$ (linearly polarized incident light, observation at 90° to both the polarization vector and the direction of incidence) one obtains

$$\frac{d\sigma_j}{d\Omega}(\nu_0) = \frac{2^4\pi^4}{45} \cdot \frac{(\nu_0-\nu_j)^4 \cdot b_j^2}{1-\exp(-hc\nu_j/kT)} \cdot g_j \cdot \left(45a_j^2+7\gamma_j^2\right) \quad . \tag{4.9}$$

This quantity is usually designated as the absolute differential Raman scattering cross section of a Stokes shifted vibrational Raman band ν_j. It depends on the temperature through the exponential, which is negligible (<1%) for $\nu_j < 1000\ \mathrm{cm}^{-1}$.

Far from resonance of the exciting frequency with electronic absorption frequencies (or vibrational-rotational frequencies) of the molecule the polarizability invariants a_j and γ_j^2 are constant. Therefore the scattering cross section depends on the wavenumber of the exciting radiation only through the factor $(\nu_0-\nu_j)^4$. On the other hand, a deviation of the frequency dependence of the scattering cross section from the $(\nu_0-\nu_j)^4$-law indicates a resonance Raman effect where a_j and γ_j^2 are no longer constant. For rigorous resonance the derived polarizability tensor may become asymmetric and a third invariant δ_j^2 is needed [4.27].

4.1.2 Determination of the Absolute Scattering Cross Section of Nitrogen

Relative Raman scattering cross sections of gases measured with laser excitation [4.21,35-40] were determined in relation to that of the Q-branch of the vibrational Raman band of nitrogen, because N_2 is non-reactive and can easily be mixed with other gases to serve as an internal standard. Moreover, N_2 is also used as standard in air pollution measurements with the Raman LIDAR method [4.41,42], because its partial pressure is nearly constant.

Therefore it is very important to determine the absolute scattering cross section for nitrogen as accurately as possible. Pre-laser values were measured by STANSBURY et al. [4.9] and YOSHINO and BERNSTEIN [4.6]. LEONARD [4.35] calculated the absolute cross section from the old photographic value erroneously to be ten times too high and many optimistic estimates of the detection limits of the Raman LIDAR method were based upon it, until REISS [4.43] pointed out the error.

With laser excitation at 514.5 nm the nitrogen cross section was determined by FOUCHE and CHANG [4.36,37] and by PENNEY et al. [4.38] by measuring the Raman/Rayleigh intensity ratio and also by utilization of an absolute spectrometer calibration. FENNER et al. [4.21] obtained with 488 nm excitation too low a value by the latter method and a somewhat better, but still low result by comparison with liquid benzene. In the same year HYATT et al. [4.22] reported a new measurement using the pure rotational Raman line (J=1→3) of H_2 at 587 cm^{-1} as reference [4.23] and the result was in good agreement with the earlier data [4.6,9,37,38]. They also established that the nitrogen cross section closely follows the $(\nu_0-\nu_j)^4$ dependence of (4.9) for the visible exciting lines of the argon ion laser from 458 to 514.5 nm.

Because for nitrogen no resonance enhancement is expected except in the far uv, the cross section of N_2 determined in the visible was also used for measurements with nitrogen laser excitation at 337 nm [4.39]. However, in 1974 PENNEY [4.44] reported a deviation by a factor of 2 from the $(\nu_0-\nu_j)^4$ dependence for excitation with a frequency doubled dye laser at 300 nm.

In order to bridge the gap between 458 and 300 nm and to establish a better basis for measurements in the near uv we decided to remeasure the absolute scattering cross section of N_2 using the ultraviolet lines of an argon ion laser at 351 and 364 nm as well as the visible lines [4.45]. Hydrogen was used both as internal and as external standard in order to avoid systematic errors, such as imperfect mixing of the gases or de-adjustment of the optical alignment upon exchange of sample cells. Only those measurements were retained where the internal and external standard gave the same result.

The scattering cross section of the rotational lines of H_2 was chosen as reference, because the optical anisotropy and its wavelength dependence are well known both experimentally and theoretically. To check the results a measurement of the cross section of the Q-branch of the vibrational band relatively to the strongest pure rotational line ($J=6\rightarrow8$) of nitrogen was also made.

The differential scattering cross section of a Stokes shifted pure rotational Raman line in the S-branch ($\Delta J=+2$) of a linear molecule follows from (4.3) and (4.5) with $t=2$

$$\frac{d\sigma_{rot}}{d\Omega}(\nu_0,J) = \frac{2^4\pi^4}{45}(\nu_0-\nu_{rot})^4 \cdot \frac{3(J+1)(J+2)}{2(2J+3)(2J+1)} \cdot 7\gamma_0^2 \quad , \tag{4.10}$$

where ν_{rot} is the wavenumber shift of the rotational line and γ_0^2 the anisotropy of the polarizability tensor. The scattering geometry is the same as that for (4.9).

Taking into account the thermal population of the initial level and the statistical weight factor g_s due to the nuclear spin one obtains from (4.1) and using (4.10) for the intensity of a rotational line, again for 90° scattering,

$$\begin{aligned} I_{rot} &= N' \cdot (d\sigma_{rot}/d\Omega) \cdot I_0 \\ &= \frac{2^4\pi^4}{45} \cdot g_s \cdot N \cdot (\nu_0-\nu_{rot})^4 \cdot \exp[-J(J+1)hcB/kT] \\ &\quad \cdot \frac{3(J+1)(J+2)(2J+1)}{2(2J+3)(2J+1)\cdot Z} \cdot 7\gamma_0^2 \cdot I_0 \quad . \end{aligned} \tag{4.11}$$

B denotes the rotational constant. The spectroscopic temperature can be determined from the intensity ratio of a Stokes and an anti-Stokes rotational Raman line (see [4.46])

$$\frac{I(J)_{St}}{I(J)_{aSt}} = \frac{(\nu_0-\nu_{rot})^4}{(\nu_0+\nu_{rot})^4} \cdot \exp\left[4\left(J+\frac{3}{2}\right) hcB/kT\right] \tag{4.12}$$

and was found to be (299±1) K for hydrogen at a room temperature of (296±2) K.

As usual the cross section measurements were carried out against the strongest rotational line of H_2 at 587 cm^{-1}. Moreover, the intensities of the other rotational lines of hydrogen at 354 cm^{-1} ($J=0\to2$), 814 cm^{-1} ($J=2\to4$), and 1034 cm^{-1} ($J=3\to5$) were used to check the accuracy of the measurements and of the calibration of the spectral sensitivity of the spectrometer.

The Raman spectrometer consists of a Spectra-Physics Model 170-03 argon ion laser, a Jarrell-Ash 1 m double monochromator equipped with holographic gratings from Jobin-Yvon with 2400 grooves/mm, an RCA C-31034-A-02 photomultiplier cooled to -30°C, and a dual channel photon-counting system PC-07 from Photonic. Parallel to the recorder (Rikadenki Model B-361) the spectra are stored in the memory of a multichannel analyzer (Canberra Model 8100) which allows the determination of integrated intensities through the electronic summation over all channels containing the signals from one band. The spectral sensitivity was measured by recording the emission of a calibrated tungsten band lamp.

The calculated values of the optical anisotropy of hydrogen and its dependence on the wavelength [4.47-49] are in good agreement with the values derived from depolarization measurements [4.50,51]; from a plot of γ_0^2 as a function of the wavenumber the values for the different exciting lines were obtained through interpolation and inserted into (4.11). In this way the differential scattering cross sections for the 4 rotational Raman lines were calculated for 5 exciting wavelengths from 351 to 514.5 nm. The measured ratios of the intensities of the $J=0\to2$, $J=2\to4$, and $J=3\to5$ lines to that of the $J=1\to3$ line differed from the calculated ratio by less than 2.6% in the visible and less than 10% in the uv. This gives an estimate of the uncertainty of the spectral sensitivity calibration, which is greater in the uv because of the reduced emissivity of the tungsten lamp in this region.

In Table 4.1 the results of the measurements of the absolute differential Raman scattering cross section of the Q-branch of N_2 at 2331 cm^{-1} are presented and compared with the results of other groups. In order to facilitate the comparison and the detection of a possible deviation from the $(\nu_0-\nu_j)^4$ dependence, also the normalized cross sections obtained by division by the $(\nu_0-\nu_j)^4$ factor are given in the third column. These normalized absolute differential scattering cross sections are also plotted in Fig.4.1 as a function of $\nu_s = \nu_0 - 2331$ cm^{-1}. It is obvious that the value is constant within experimental accuracy, except for PENNEY's value [4.44]. Therefore it is justified to calculate a weighted average from all data except the one at 300 nm. The result is

Table 4.1. Absolute differential Raman scattering cross section of the vibrational-rotational band of N_2 (Q-branch)

Exciting line λ_0 [nm]	$\frac{d\sigma}{d\Omega}$ $\left[10^{-32}\ \frac{cm^2}{sr}\right]$	$\frac{d\sigma}{d\Omega}\cdot(\nu_0-2331\ cm^{-1})^{-4}$ $\left[10^{-48}\ \frac{cm^6}{sr}\right]$	Ref.
632.8	21 ± 3	6.4 ± 1	[4.52]
514.5	44 ± 17	5.1 ± 2	[4.37]
	43 ± 2	5.0 ± 0.3	[4.38]
	42 ± 2	4.9 ± 0.3	[4.22]
	43.2 ± 0.8	5.05 ± 0.11	[4.45]
488.0	33 ± 11	3.0 ± 1	[4.21]
	43	4.0 ± 1	[4.21]
	54 ± 3	5.0 ± 0.3	[4.22]
	55.8 ± 2	5.13 ± 0.2	[4.45]
457.9	76 ± 5	5.2 ± 0.4	[4.22]
	73.7 ± 3	5.09 ± 0.25	[4.45]
435.8	92 ± 10	5.1 ± 0.5	[4.10]
363.8	204 ± 25	5.1 ± 0.6	[4.45]
351.1	243 ± 30	5.2 ± 0.7	[4.45]
300	970	(10.5)	[4.44]
Weighted average (except value in parentheses)		5.05 ± 0.08	

$$\left(\frac{d\sigma}{d\Omega}\right)_{Q,N_2} = (5.05\pm0.1)\cdot10^{-48}\cdot(\nu_0-2331\ cm^{-1})^4\quad cm^6\ sr^{-1}$$

and can be used throughout the visible and in the uv to about 330 nm. Further extrapolation into the uv remains uncertain without additional measurements in this region.

Further support is given to this value by a measurement against the pure rotational Raman line $J=6\rightarrow8$ of N_2 at the exciting lines 488 and 514.5 nm [4.45]. The result

$$\left(\frac{d\sigma}{d\Omega}\right)_{Q,N_2} = (4.97\pm0.2)\cdot10^{-48}\cdot(\nu_0-2331\ cm^{-1})^4\quad cm^6\ sr^{-1}$$

agrees well with the average given above.

4.1.3 Measurement of Relative Scattering Cross Sections

Most of the differential Raman scattering cross sections in the literature are given relative to the cross section of the Q-branch of N_2 measured at the same exciting wavelength and can therefore be readily converted into absolute values with the results given in Sect.4.1.2. The relative differential scattering cross section slightly depends on the exciting wavelength used,

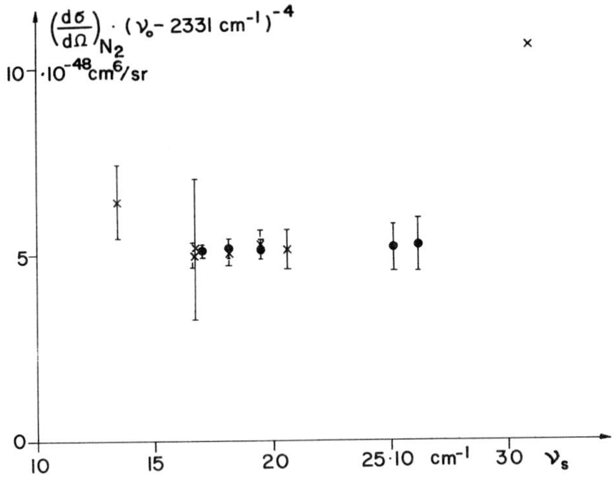

Fig. 4.1. Absolute normalized differential Raman scattering cross section of the Q-branch of nitrogen at different exciting lines. ● our measurement, x data from the literature

$$\frac{d\sigma_j/d\Omega}{(d\sigma/d\Omega)_{Q,N_2}} = \frac{(\nu_0-\nu_j)^4}{(\nu_0-2331\ cm^{-1})^4} \cdot \frac{2331\ cm^{-1}}{\nu_j[1-\exp(-hc\nu_j/kT)]} \cdot \frac{g_j\cdot\left(45a_j^2+7\chi_j\gamma_j^2\right)}{\left(45a_{N_2}^2+7\chi_{N_2}\gamma_{N_2}^2\right)} \quad , (4.13)$$

especially when the Raman shift ν_j is very different from 2331 cm^{-1}. In (4.13) the exponential Boltzmann factor for N_2 has been omitted, because it is negligible at ambient temperature, and χ_j denotes the fraction of anisotropic Raman scattering which is concentrated in the Q-branch. In order to compare results measured at different wavelengths they are usually recalculated for one particular wavelength, see for instance the tables given by CHERLOW and PORTO [4.53] and INABA [4.42]. This procedure can be avoided by defining the "relative normalized differential Raman scattering cross section"

$$\Sigma_j \equiv \frac{d\sigma_j/d\Omega}{(d\sigma/d\Omega)_{Q,N_2}} \cdot \frac{(\nu_0-\nu_j)^{-4}}{(\nu_0-2331\ cm^{-1})^{-4}} \cdot \left[1-\exp\left(-\frac{hc\nu_j}{kT}\right)\right]$$

$$= \frac{2331\ cm^{-1}}{\nu_j} \cdot \frac{g_j\cdot\left(45a_j^2+7\chi_j\gamma_j^2\right)}{\left(45a_{N_2}^2+7\chi_{N_2}\gamma_{N_2}^2\right)} \quad , \qquad (4.14)$$

which is now independent of ν_0, as long as a_j^2 and γ_j^2 can be regarded as constant, and of the temperature.

It is obvious that the quantity Σ_i is closely related to the relative scattering coefficient or "standard intensity" S_j defined for liquids by BERNSTEIN and ALLEN [4.4] and the "scattering activity" $g_j\cdot\left[45a_j^2+7\gamma_j^2\right]$ [4.10].

a) *Experimental Procedure*

Before we present the results for relative Raman cross section measurements from our laboratory [4.54,45] and compare them with those of other groups, we briefly describe the experimental procedure using as example hydrogen sulfide H_2S. Figure 4.2 shows a survey spectrum of H_2S at low gain. Apart from the strong Q-branch of the ν_1 band at 2611 cm^{-1} only weak rotational wings are discernible. At right one sees the unresolved pure rotational spectrum near the exciting line. For analytical purposes obviously only the ν_1 Q-branch is of significance. At 16-fold gain (Fig.4.3) the ν_2 band with its origin at 1183 cm^{-1} appears and the structure of the rotational wings around 2600 cm^{-1} becomes clearer. It contains not only the S-, R-, P-, and O-branches of ν_1, but also the complete structure of the antisymmetric ν_3 band at 2627 cm^{-1}.

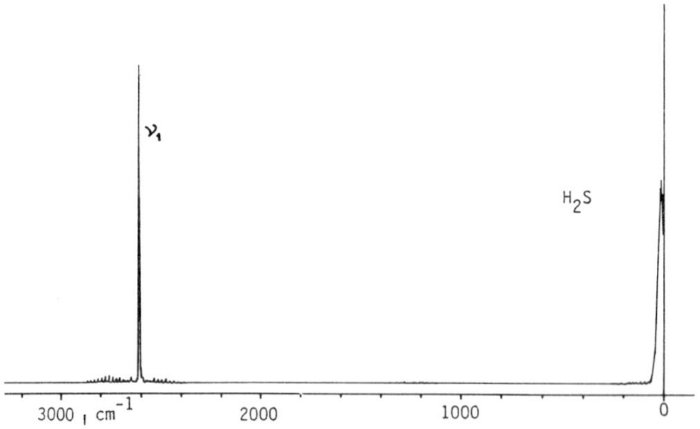

Fig. 4.2

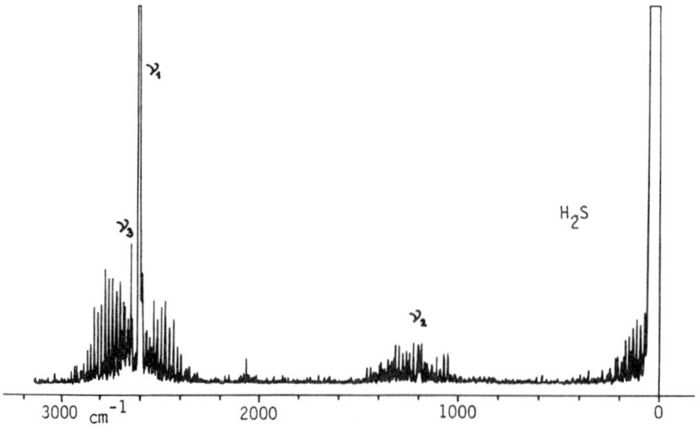

Fig. 4.3

Fig. 4.2. Raman spectrum of H_2S at low gain. (Laser: 1 W at 514.5 nm, slit: 2 cm^{-1}, scanning speed: 50 cm^{-1}/min., time constant: 1 s, 4096 pps for full scale)

Fig. 4.3. Raman spectrum of H_2S at higher gain. Conditions as in Fig.4.2, except 256 pps for full scale

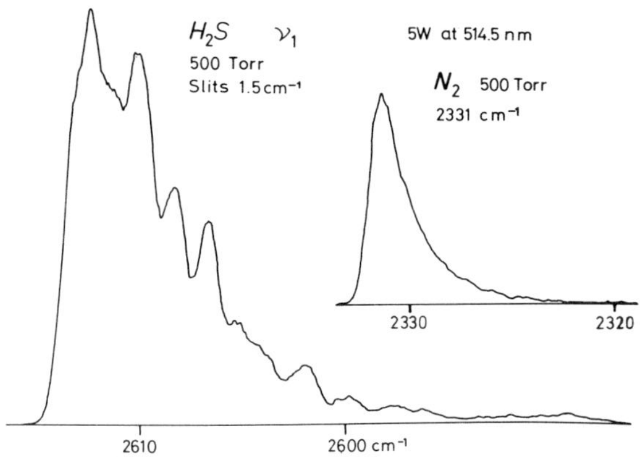

Fig. 4.4. Q-branches of ν_1 band of H_2S and of N_2 at equal pressures

The cross section was measured only for the Q-branch of ν_1. For this purpose the sample cell was filled with a mixture of H_2S and N_2 at partial pressures of 0.667 bar (500 Torr). The laser beam was focused with a lens of 3 cm focal length into the sample cell and the focus projected with two quartz lenses onto the entrance slit of the double monochromator. Only a single pass of the laser beam was used and the stabilized laser power was monitored with a power meter. Using the exciting lines 514.5, 488, and 457,9 nm the Q-branches of the H_2S band and the N_2 reference band were recorded several times. Figure 4.4 shows the recorder traces for the exciting line 514.5 nm. The area between the recorder trace and the baseline was determined either with a planimeter or by electronic integration in the multichannel analyzer. For the calculation of the relative cross section the spectral sensitivity was taken into account. In front of the slit we used an analyzer transmitting only horizontally polarized light, so that the measured signal was

$$I_{j\parallel} \sim \left(45a_j^2+4\chi_j\gamma_j^2\right) \quad . \tag{4.15}$$

Therefore the result had to be multiplied by $1+\rho_j$, with ρ_j the depolarization ratio, in order to obtain the desired quantity

$$I_j = I_{j\parallel} + I_{j\perp} = I_{j\parallel}(1+\rho_j) \sim \left(45a_j^2+7\chi_j\gamma_j^2\right) \quad . \tag{4.16}$$

Without analyzer the instrument polarization would require complicated corrections. Alternately a scrambler can be used.

After exchange of the laser mirrors the same procedure was repeated for the UV exciting lines 363.8 and 351.1 nm. To avoid systematic errors (as incomplete mixing of the gases) the measurement was repeated also with the pure gases in the sample cell at the same pressure as the partial pressure used before. Only when all results agreed the measurement series was retained for further evaluation.

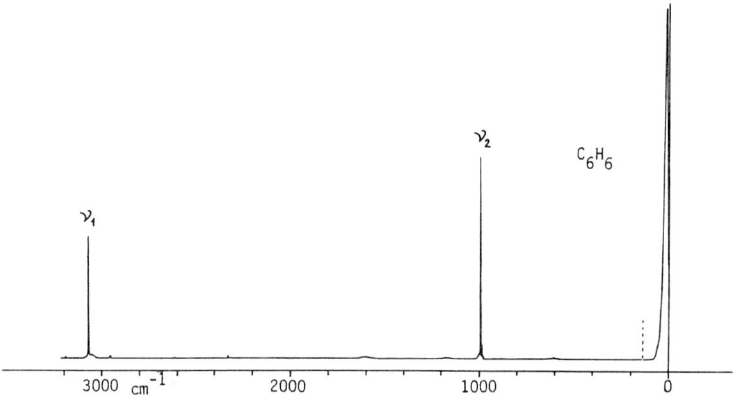

Fig. 4.5

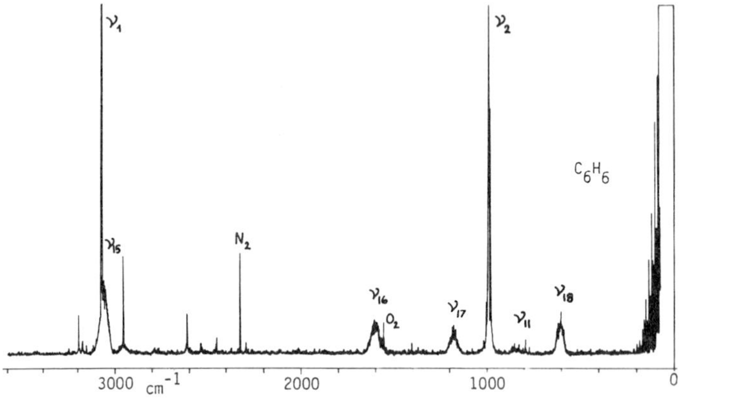

Fig. 4.6

Fig. 4.5. Raman spectrum of benzene vapor at 0.1 bar at low gain. (Laser: 1 W at 514.5 nm, slit: 2 cm^{-1}, scanning speed: 20 cm^{-1}/min, time constant: 1 s, 4096 pps for full scale)

Fig. 4.6. Raman spectrum of benzene vapor at higher gain. Conditions as in Fig.4.5, except 256 pps for full scale

As a further example we show the survey spectrum of benzene C_6H_6. At low gain in Fig.4.5 only the Q-branches of the totally symmetric vibrations ν_1 (3070 cm^{-1}) and ν_2 (992 cm^{-1}) appear, while at 16-fold gain in Fig.4.6 also the degenerate bands at 3049, 1600, 1180, and 606 cm^{-1} are clearly discernible. Even some sharp overtones and combination bands are observed under these conditions in the region 2200-3200 cm^{-1}. The appearance of the N_2 and O_2 Q-branches indicates air in the sample cell. Figure 4.7 shows the profile of the ν_2 band together with the ^{13}C isotope line at 984 cm^{-1}. The integration limits are indicated. Difficulties arise for overlapping bands as ν_1 and ν_{15} in Fig.4.8. The depolarized ν_{15} band including the depolarized rotational wings of ν_1 was graphically separated from the Q-branch of ν_1, the two lines at 3068 and 3065 cm^{-1} being considered as hot bands belonging to the Q-branch.

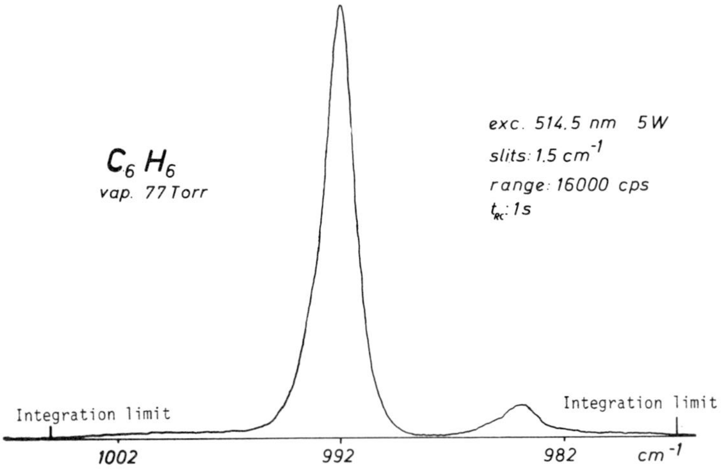

Fig. 4.7. Profile of Q-branch of ν_2 of benzene vapor at 0.1 bar

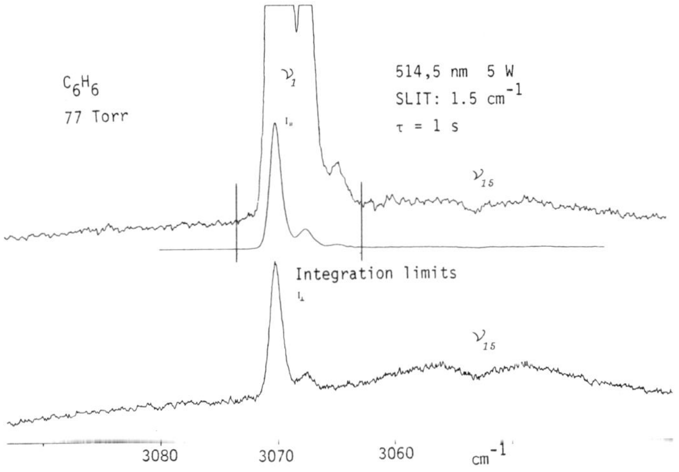

Fig. 4.8. $I_{\parallel}$ and $I_{\perp}$ components of ν_1 and ν_{15} Raman bands of benzene vapor. (Upper trace: range 1024 pps, middle trace: range 16384 pps, lower trace: range 512 pps)

A substance which is important in environmental analysis is vinyl chloride H_2C=CHCl, and we show its Raman spectrum in Fig.4.9. The cross section was measured for the C=C stretching band at 1632 cm^{-1}.

b) *Results*

The results for the relative normalized differential scattering cross sections Σ_j are presented in Table 4.2 and compared with the results of other groups. The data are arranged in columns for the different exciting lines in order to facilitate

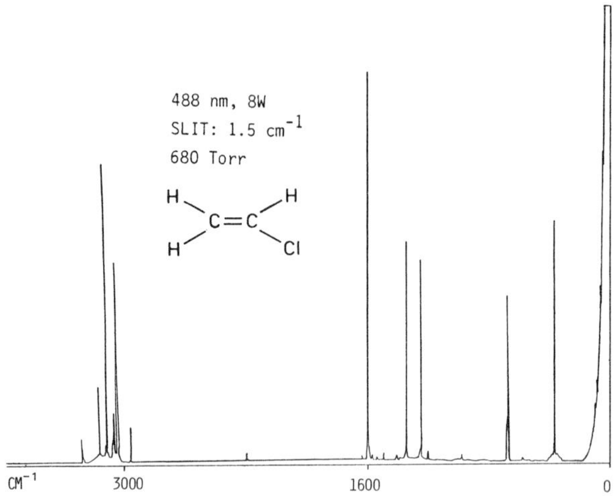

Fig. 4.9. Raman spectrum of vinyl chloride H_2CCHCl at 0.9 bar. (Laser: 8 W at 488 nm, slit: 1.5 cm^{-1})

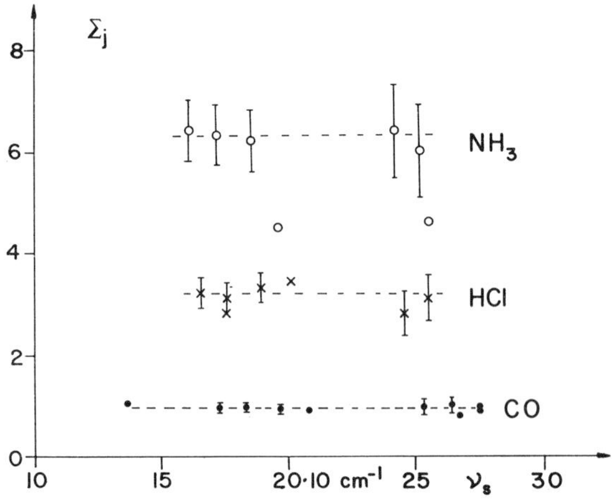

Fig. 4.10. Relative normalized differential Raman scattering cross sections of Q-branches of CO, HCl, and ν_1 of NH_3 at different exciting lines. Our measurements with error bars

the comparison and to make the detection of a resonance Raman effect easier. The latter cases will be discussed in the following subsection.

The accuracy of the data from our laboratory and also from most other groups is ±10% for the visible exciting lines and ±15% in the near uv (LEVATTER et al. [4.40] claim ±5% at 337 nm), and within these limits most of the values agree quite well. Some of the values measured by FENNER et al. [4.21] appear to be too low, e.g., for NO, CH_4, and C_6H_6, and also some of the other results for benzene show greater discrepancies.

Figure 4.10 shows three examples for practically constant values of $\sum_j$. Whereas in these cases no significant resonance enhancement is observed, it is possible to take the average of all available data for one vibration of a molecule. In this way one gets an estimate of the relative detectability of a molecule by Raman scattering.

Table 4.2. Relative normalized differential Raman scattering cross sections Σ_j for Q-branches of vibrational bands measured at the exciting lines λ_0

		Exciting wavelength λ_0 [nm]									
Molecule	$\nu_j[cm^{-1}]$	287	337	347	351	364	436	458	488	515	633
N_2	2331	1	1	1	1	1	1	1	1	1	1
O_2	1555		$1.2_5{}^e$ $1.0_9{}^f$	$1.0_8{}^h$	1.0^p	1.0^p	$1.0_5{}^a$	1.0^p	$1.0_9{}^d$ 1.0^p $1.0_3{}^{dd}$	$1.0_2{}^b$ $1.0_2{}^c$ 1.0^p	$0.9_1{}^o$
H_2	4156		4.2^e				$4.1_5{}^a$		$3.6_4{}^d$	$3.4_5{}^b$	
D_2	2986						2.6^a				
F_2	893								$0.3_2{}^g$		
Cl_2	554			2.5^h	2.0^p	2.2^p		2.4^p	2.1^p	2.3^p	
Br_2	318				$13^{p,v}$	$18^{p,v}$		$5^{p,v}$	$4^{p,v}$		
I_2	212				$30^{p,v}$	$50^{p,v}$		$180^{p,v}$	$200^{p,v}$		
HF	3962								$1.3_0{}^x$	$1.3_7{}^x$	
HCl	2886				3.1^p	2.8^p	3.4^a	3.3^p	$2.8_2{}^m$ 3.1^p	3.2^p	
HBr	2558								$4.4_2{}^m$		
HI	2230							$5.8_7{}^m$	$5.8_7{}^m$	$5.6_6{}^m$	
CO	2143		$0.8_7{}^e$ $0.9_6{}^f$	$0.7_9{}^h$	$0.9_8{}^p$	$0.9_6{}^p$	$0.9_0{}^a$	$0.9_2{}^p$	$0.9_6{}^d$ $0.9_3{}^p$ $0.7_2{}^{dd}$	$0.8_7{}^b$ $0.9_6{}^c$ $0.9_5{}^p$	$1.0_4{}^o$

Table 4.2 (continued)

		Exciting wavelength λ_0 [nm]									
Molecule	ν_j[cm^{-1}]	287	337	347	351	364	436	458	488	515	633
NO	1877		0.4_3[e] 0.4_4[f]	0.07_5[h?]	0.3_4[p]	0.3_4[p]		0.3_4[p]	0.2_5[d] 0.4_9[g] 0.4_0[k] 0.3_2[p]	0.4_8[b] 0.4_2[c] 0.3_5[p]	
CO_2	1285		0.7_2[e] 0.8_0[f]				0.7_3[a]		0.7_1[d] 0.9_2[dd]	0.7_8[b]	0.7_8[o]
	1388		1.0_4[e] 1.1_9[f]	1.0_3[h]			1.0_7[a]		1.1_4[d] 1.3_5[dd]	1.2_1[b] 1.2_1[c]	1.1_3[o]
N_2O	1285			2.5 [h]	2.0 [p]	2.0 [p]		1.8_5[p]	1.7_6[d] 1.8 [p]	2.1_2[b] 1.8 [p]	
	2224								0.5_0[d]	0.5_1[b]	
ONF	1844								0.6_6[g]		
H_2O	3652		3.1_2[e] 7.7_4[bb]	3.9_8[h]					2.5_6[k] 3.8_4[dd]	3.4_6[c] 6.8_9[bb]	
	1595									0.008[z]	
H_2S	2611			5.7 [h]	7.1 [p]	7.4 [p]	7.0 [a]	6.8 [p]	6.8_1[d] 6.6_5[p]	7.0_4[b] 6.7 [p]	
O_3	1103	816 [n]								3 [b]	
SO_2	1151		4.8 [e] 4.6 [f]		6 [p]	5 [p]		3.8 [p]	4.0_4[d] 3.7 [p] 4.3 [k]	4.2 [b] 4.1_2[c] 3.8 [p]	

Table 4.2 (continued)

		Exciting wavelength λ_0 [nm]									
Molecule	ν_j[cm^{-1}]	287	337	347	351	364	436	458	488	515	633
SO_2	519								0.08[d]		
NO_2	1320 750		13 [e] 5.8 [e]								
C_2H_2	1973		11.8 [u]				6.6 [a]			5.5 [u]	
C_2N_2	2330 845								11.9[cc] 0.8[cc]		
NH_3	3334			4.6 [h]	6.0 [p]	6.4 [p]	4.5 [a]	6.2 [p]	6.2_7[d] 6.3 [p]	6.4 [p]	
ND_3	2420								3.0_6[d]		
C_2H_4	3020 1623 1342		5.9 [e] 1.8 [e] 3.5 [i] 2.8 [e]				6.9 [a] 1.7 [a] 2.8 [a]			 1.5 [i]	
cis-$C_2H_2Cl_2$	1591		9.6 [i]							2.5 [i]	
C_2H_3Cl	1632				2.6 [q]	2.4 [q]		1.7_8[q]	1.5_5[q]	1.6_1[q]	
CH_3OH	2955 2846		2.3 [e] 12.7 [bb,w] 4.2 [e] 8.2 [bb]				2.8 [a] 5.2 [a]			10.9 [bb,w] 6.9 [bb]	
C_2H_6	993						2.2 [a]		1.2_4[d]		

Table 4.2 (continued)

Molecule	ν_j[cm^{-1}]	Exciting wavelength λ_0 [nm] 287	337	347	351	364	436	458	488	515	633
C_2H_5I	1216								1.9 [l]		
	512								7.1 [l]		
	263								1.7 [l]		
C_2H_5OH	2943		6.0 [e] 27.8 [bb,w]							25.6 [bb,w]	
C_3H_6	1297		2.2_7 [f]								
	920		0.9_5 [f]								
CH_4	2917		8.2 [e]	7.8 [h]	8.4 [p]	8.3 [p]	8.0 [a] 9.5 [j]	9.2 [p]	6.8 [d] 9.2 [p] 8.7 [r]	9.1 [b] 8.7 [c] 9.3 [p]	
CD_4	2108						4.7 [a] (6.6) [j,w]		3.9 [r] (6.2) [r,w]		
CCl_4	459		6.5 [e]		8.5 [q]	7.2 [q]	6.0 [a]	6.1 [q]	6.1 [q] 6.6 [t]	6.2 [q] 7.7 [t]	
CH_2Cl_2	713		5.6 [u]		4.5 [q]	4.4 [q]		3.1 [q]	4.0 [l] 3.2 [q]	3.5 [u] 3.2 [q]	
$CHCl_3$	773		3.2 [u]		1.8 [q]	1.9 [q]	1.9 [a]	1.7 [q]	1.8 [q]	1.6 [q] 1.4 [s] 1.7 [u]	
	674		4.2 [u]		2.8 [q]	2.7 [q]	2.4 [a]	2.8 [q]	2.8 [q]	2.7 [q] 2.4 [s] 3.1 [u]	

Table 4.2 (continued)

Molecule	ν_j[cm^{-1}]	Exciting wavelength λ_0 [nm] 287	337	347	351	364	436	458	488	515	633
SF_6	775		3.3 [e]				3.4 [a]		3.7_6 [aa]	3.9_6 [aa]	
UF_6	665								12.6 [y]		
C_3H_8	2890						32 [a,w]			6.6 [b]	
	1451		2.8 [f]								
	867						1.8 [a]			1.7 [b]	
C_4H_6	1643		88 [i]				13.5 [a]			18.8 [i]	
$(CH_3)_2CO$	782								2.4 [l]		
C_6H_6	3070		11.3 [e]		19.7 [q]	16.6 [q]	11.8 [a]	15.6 [q]	8.3 [d]	15.3 [q]	
									15.7 [l]		
									16.0 [q]		
	992		9.7_5 [e]			10.9 [q]	12.5 [a]	10.8 [q]	6.8_5 [d]	12.7 [i]	
			19.8 [i]						10.8 [l]	10.8 [q]	
			23.1 [u]						10.5 [q]	12.8 [u]	
C_6D_6	945		23.5 [u]							12.4 [u]	
C_6H_{12}	802		4.7 [u]				3.7 [a]		3.6 [l]	4.1 [u]	

[a][4.10], [b][4.36,37], [c][4.38], [d][4.21], [e][4.39], [f][4.40], [g][4.55],
[h][4.56], [i][4.57] total bands, [j][4.11], [k][4.58], [l][4.59], [m][4.53,60], [n][4.44]
[o][4.52], [p][4.54], [q][4.45], [r][4.61], [s][4.62], [t][4.63], [u][4.64] total bands,
[v] not normalized with temperature dependent factor, [w] including all Fermi-resonance components.
[x][4.75], [y][4.92], [z][4.97], [aa][4.100], [bb][4.101] total bands, [cc][4.99], [dd][4.102].

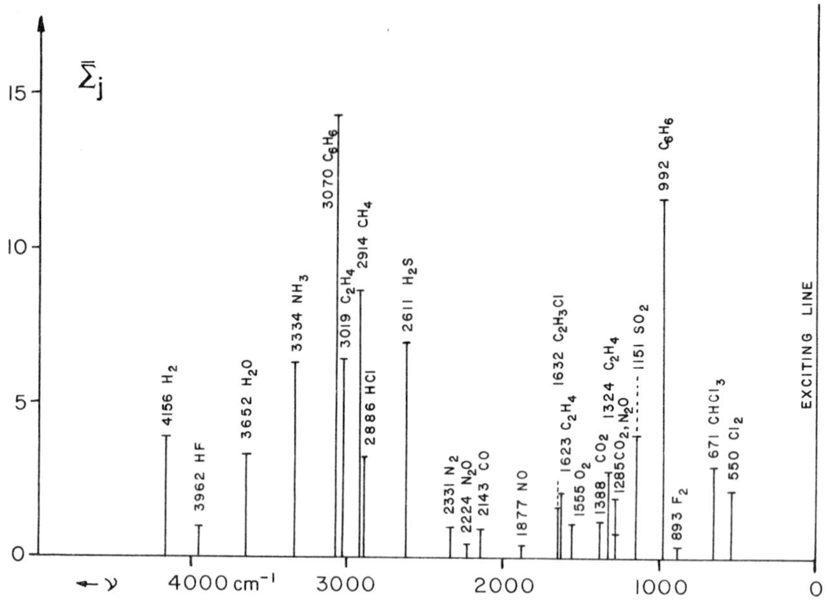

Fig. 4.11. Averaged relative normalized differential Raman scattering cross sections for important gases as function of Raman wavenumber shift. Dotted lines indicate resonance increase in the near UV

In Fig.4.11 we have plotted the averaged relative normalized cross sections $\bar{\Sigma}_j$ against the Raman wavenumber shift ν_j for the more important molecules listed in Table 4.2. For SO_2 and C_2H_3Cl the average was only taken over the values measured at visible exciting lines, the resonance increase in the near uv is indicated by dotted lines. It can be seen that at a resolution of 5 cm^{-1} or better all the gases included in Fig.4.11 can be detected separately, however, the coincidence with rotational-vibrational wings of the bands of more abundant gases may severely limit the detection limit, e.g. for CO in N_2 or air.

Quite a number of cross sections of gases which were investigated by mercury excitation [4.10,12] are not yet remeasured by laser excitation. For the sake of completeness we have converted also these values to the scale used in Table 4.2 and present the data in Table 4.3. Because all these older data were originally given on the scale of scattering activities $g_j \cdot \left(45a_j^2 + 7\gamma_j^2\right)$, the normalized absolute cross sections $(d\sigma_j/d\Omega)\cdot(\nu_0-\nu_j)^{-4}$ are also included in Table 4.3.

In some publications it was tried to derive bond polarizabilities from the scattering activities and to test the possibility to predict Raman intensities by a transfer of these bond polarizabilities to other molecules, but the results remained inconclusive.

Three recent investigations treated the same problem starting from more accurate experimental data. BERMEJO et al. [4.61] studied methane and its deuterated derivatives, BERJEMO [4.62] and one of us [4.45] chlorine–substituted methanes, and CLARK and MITCHELL [4.63] the group IV tetrachlorides. For this type of work it is necessary to determine the absolute scattering cross sections for all normal vibrations of a molecule. Therefore a rather complete set of data exists for methane derivatives and tetrahalides. In Table 4.4 we have collected wavenumbers, depolarization ratios, and relative and absolute normalized Raman scattering cross sections for these tetrahedral and quasi-tetrahedral molecules. Also included here instead in Table 4.3 are the data on fluorine–substituted methanes measured by HOLZER [4.65-67]. Wherever the measurements were performed at more than one wavelength, the average was taken for all exciting lines in the visible region.

HILLS and JONES [4.93] have determined single polarizability components from the pure rotational Raman spectrum of ethylene based upon the mean value and the anisotropy of the polarizability tensor as given by BRIDGE and BUCKINGHAM [4.94]. Their result is

$$\alpha_{xx} = 3.86 \cdot 10^{-24}\ \mathrm{cm}^3 \quad ,$$
$$\alpha_{yy} = 3.40 \cdot 10^{-24}\ \mathrm{cm}^3 \quad , \text{ and}$$
$$\alpha_{zz} = 5.40 \cdot 10^{-24}\ \mathrm{cm}^3 \quad .$$

They also have deduced sign and magnitude of the derived polarizability component ratio [4.95,97]

$$R_{20}^{(j)} = \frac{\left(\alpha_{xx}^{(j)} - \alpha_{yy}^{(j)}\right) \cdot 6^{1/2}}{2\alpha_{zz}^{(j)} - \alpha_{xx}^{(j)} - \alpha_{yy}^{(j)}} \quad ,$$

where the index j denotes the derivation of the polarizability component with respect to the normal coordinate q_j, for the vibrations ν_1 and ν_3 of ethylene from the intensity distribution in these rotational-vibrational bands [4.96]. These results show that the polarizability ellipsoid of ethylene changes considerably when the molecule vibrates. However, for the simple reason that there exists no good value for the anisotropy of the derived polarizability tensor for these vibrations to our knowledge, it is not yet possible to deduce absolute values for the components $\alpha_{rr}^{(j)}$.

For the ν_2 band of H_2O on the other hand MURPHY [4.97] has measured a complete set of data. From the absolute scattering cross section $d\sigma_2/d\Omega = (3.3 \pm 0.8) \cdot 10^{-32}$ $\mathrm{cm}^2\ \mathrm{sr}^{-1}$ at 515 nm, the depolarization ratio $\rho_2 = 0.736 \pm 0.002$, and the polarizability ratio $R_{20}^{(2)} = -(1.1 \pm 0.05)$ he derived four sets of components of the derived polarizability tensor which are compatible with the measured data because of sign ambiguities. Through considerations of the resulting bond polarizability parameters two of the sets could be eliminated, however, no clear choice between the remaining two is possible at the present stage [4.97].

Table 4.3. Relative and absolute normalized differential Raman scattering cross sections measured with mercury arc excitation[c]

ν_j [cm^{-1}]	Σ_j	$\frac{d\sigma_j}{d\Omega}(\nu_0-\nu_j)^{-4}$
CS_2 [a]		
658, 802	15.2	80.0
SeF_6 [a]		
708	5.0	26.0
TeF_6 [a]		
701	4.5	23.7
CH_3OH [a]		
3697, 3683	1.5	7.8
2999	0.3	1.7
2955	2.8	14.0
2927	0.6	2.9
2846	5.2	26.2
1071	0.2	1.1
1033	0.3	1.5
CH_3CN [a]		
2942	6.3	31.8
2249	3.8	19.1
918	0.9	4.7
C_2H_2 [a]		
3372	2.0	9.9
1973	6.9	34.9
CHF_2Cl [a]		
3024	2.4	12.3
1350	0.3	1.4
1312	0.1[Q]	0.7[Q]
824	1.9	9.9
598	0.6	3.3
417	0.5	2.8
400	1.1	6.4
$CHFCl_2$ [a]		
3023	2.5	12.7
1313	0.4	2.2
1079	0.7	3.4
806	0.8	4.0
741	2.0	10.4
455	1.4	7.8
368	0.9	5.4
270	1.3	9.1
CHF_2Br [a]		
3020	2.9	14.5
1328	0.2	1.1
1278	0.4	2.1
1123	0.5	2.3
716	2.6	13.4
577	0.4	2.0
323	2.4	15.2
C_2F_6 [a]		
1417	0.2	1.0
1250	0.2	1.2
807	1.7	8.9
619	0.3	1.9
372, 348	0.5	3.1
CH_3CH_2Cl [a]		
2971	3.7	18.7
2942	3.4	17.2
2894	0.9	4.4
2883	0.5	2.6
2744	0.3	1.6
1460	1.3	6.7
1075	0.4	1.9
675	2.2[Q]	11.5[Q]
335	0.3[Q]	1.8[Q]

Table 4.3 (continued)

ν_j [cm^{-1}]	$\sum_j$	$\frac{d\sigma_j}{d\Omega}(\nu_0-\nu_j)^{-4}$
CH_3CH_2Br [a]		
2980	3.4	17.0
2936	2.9	14.9
2881	1.1	5.6
1001	0.3	1.7
573	4.4	23.5
285	0.6[Q]	3.8 [Q]
CH_3CHF_2 [a]		
2976	2.4	12.0
2955	4.0	20.4
2882	0.3	1.4
2812	0.3	1.3
868	0.7	3.7
570	0.1	0.5
CH_3CF_2Cl [a]		
3033	1.5	7.4
2962	4.2	21.3
1446	0.8	3.9
898	0.3	1.3
683	2.7	13.9
435	1.8	10.0
CF_3CF_2Cl [a]		
1348	0.4	2.0
1236	0.3	1.7
1187	0.2	1.1
1131	0.2	1.1
763	1.7	8.7
648	0.3	1.5
561	0.2	1.2
440	1.5	8.4
366	0.6	3.8
331	0.4	2.4
315	0.6	4.2
n-C_4H_{10} [a]		
$\sum \nu_{CH}$	36.5	184
n-C_5H_{12} [a]		
$\sum \nu_{CH}$	39.2	198
$C(CH_3)_4$ [a]		
$\sum \nu_{CH}$	40.4	204
1425	<0.8[Q]	<4[Q]
734	2.9[Q]	15.3[Q]
n-C_6H_{14} [a]		
$\sum \nu_{CH}$	47	238
C_4H_6(1,3-butadiene)[a]		
3100	3.1	15.7
3014	7.8	39.2
1643	24.4	123
1442	4.2	21.1
1303, 1279	6.8	34.3
1205	6.2	31.5
900	1.6	8.4
513	1.5[Q]	8.3[Q]
C_4H_4(vinyl acetylene)[a]		
3029	15.6	78.7
2118	5.6	28.5
1617, 1601	10.8	54.6
$(CH_3)_3COH$ [a]		
3640	0.9	4.6
2978, 2917	15	74
1445	4.2	21.1
1205	2.6	13.4
1017	0.3	1.8
910	1.7	8.9
748	1.7	8.7

Table 4.3 (continued)

ν_j [cm^{-1}]	$\sum_j$	$\frac{d\sigma_j}{d\Omega}(\nu_0-\nu_j)^{-4}$
C_4F_8 [a]		
1431	0.2	1.0
1285	0.1	0.7
1220	0.4	1.9
1008	0.6	2.9
699	2.5	13.2
439	1.2	6.7
358	1.8	11.3
273, 258	0.8	5.4
trans-$C_2H_2Cl_2$ [b]		
3155	0.6	2.9
3081	2.7	13.8
1584	2.2	11.3
1275	2.2	11.2
852	0.6	3.1
350	2.3	14.0
1,1-$C_2H_2Cl_2$ [b]		
3138	2.0	10.2
3045	3.3	16.9
2993	0.3	1.4
1622	2.5	12.6
604	3.2	16.9
372	0.9	5.3
296	1.3	8.4
CH_2BrCl [b]		
3002	2.1	10.4
1424	0.6	2.9
728	0.7	3.7
612	3.6	19.1
225	1.7	12.6
$H_2C{=}CHCH_2Cl$ [b]		
3095	1.9	9.7
3031, 2999, 2966	12.1	61.0
1645	3.3	16.8
1418	1.2	6.0
749	3.3	17.0
594	0.9	4.9
412	1.0	5.6
1,1-$C_2H_4Cl_2$ [b]		
3015, 2996	4.9	24.8
2946	4.2	21.2
705	1.3	6.7
651	3.0	16.0
404	0.7	4.3
273	1.0	6.9
1,1,1-$C_2H_3Cl_3$ [b]		
3013	2.7	13.8
2951	3.8	19.2
726	2.1	10.8
525	4.1	22.2
339	2.1	13.4
236	1.4	10.3
$CF_2ClCHCl_2$ [b]		
3012, 2996	2.4	12.0
994, 976	0.7	3.5
837	1.0	4.9
752	1.3	6.7
642	1.0	5.0
438	2.1	12.3
374	0.9	5.1
344	0.6	3.8
302	0.7	4.7

Table 4.3 (continued)

ν_j [cm^{-1}]	Σ_j	$\frac{d\sigma_j}{d\Omega}(\nu_0-\nu_j)^{-4}$
$CF_2ClCFCl_2$ [b]		
1050	0.6	3.3
907	0.7	3.5
654	1.3	6.8
633	0.3	1.4
528	0.4	2.0
504	0.5	2.9
457	1.4	7.7
442	1.3	7.2
315, 305	0.6	3.8
289	0.5	3.4
241	0.7	4.8
$CF_3CHClBr$ [b]		
3004, 2982	2.3	11.4
967	1.3	6.8
797, 789	1.3	6.9
634	0.5	2.5
610	0.2	1.2
590	0.6	3.1
471	0.3	1.8
423	0.6	3.7
378	0.2	1.4
332	1.9	11.8
316, 303	0.7	4.8
230	0.7	5.2
$n\text{-}C_3H_7Br$ [b]		
2971, 2946, 2910, 2890, 2866	20.8	105
1448	1.6	7.9
1029	0.8	4.0
777	0.4	1.9
660	1.5	7.7
571	2.9	15.5
310	0.8	5.0
$n\text{-}C_3H_7Cl$ [b]		
2966, 2945, 2919, 2892	20.4	103
1449	1.6	8.0
789	0.4	2.0
742	0.9	4.5
661	1.7	9.0
362	0.6	3.6
$(CH_3)_2CHBr$ [b]		
2999, 2963, 2942, 2931, 2874	24.1	122
1455	1.4	7.1
879	0.9	4.8
549	5.8	31.6
405	0.6	3.3
295	2.4	15.6
$(CH_3)_2CHCl$ [b]		
2996, 2945, 2927, 2878	22.0	111
1449	1.3	6.7
1066	0.4	2.0
887	0.8	4.2
631	3.5	18.4
422	0.5	2.7
339	1.0	6.1

Table 4.3 (continued)

ν_j [cm^{-1}]	Σ_j	$\frac{d\sigma_j}{d\Omega}(\nu_0-\nu_j)^{-4}$
$(CH_3)_2CHCH_2Br$ [b]		
3001 2970 2964 2916 2884	24.7	125
1453	1.5	7.8
945	1.2	6.1
802	0.9	4.5
663	2.9	15.5
628	1.1	5.8
473	0.6	3.2
418	0.4	2.1
302	1.0	6.4
$(CH_3)_2CHCH_2Cl$ [b]		
2978 2956 2916 2885	23.0	116
1457	1.5	7.7
815 801	1.0	5.0
745	2.1	10.8
696	0.8	4.0
425 404	0.7	4.3
338	0.5	2.9
$CH_3CH_2CHBrCH_3$ [b]		
2975 2933 2893 2855	25.6	129
1454	1.3	6.8
626	0.8	4.4
587	0.4	2.2
541	2.4	13.0
489	0.5	2.8
458	0.6	3.5
317	0.6	3.9
289	0.9	6.0
$CH_3CH_2CHClCH_3$ [b]		
2976 2949 2918 2892	25.3	128
1455	1.8	9.1
841 789	1.0	5.0
683	0.6	3.0
640	0.4	2.0
622	1.8	9.4
463	0.6	3.2
377	0.6	3.5
$(CH_3)_3CBr$ [b]		
2978 2934 2900	25.4	128
1454	2.1	10.4
805	1.7	8.7
523	5.7	31.1
303	3.8	25.0
$(CH_3)_3CCl$ [b]		
2980 2936 2909	24.7	125
1451	2.0	9.9
817	1.6	8.5
570	4.4	23.8
374	1.0	6.2
301	0.9	5.7

[Q] Q-branch only

[a] [4.10]

[b] [4.12]

[c] The absolute normalized cross sections are given in units of [10^{-48} cm^6 sr^{-1}]

Table 4.4. Depolarization ratios, relative and absolute normalized differential Raman scattering cross sections of tetrahedral and quasi-tetrahedral molecules

Molecule and vibration	ν_j [cm^{-1}]	ρ_j	Σ_j	$\frac{d\sigma_j}{d\Omega} \cdot (\nu_0-\nu_j)^{-4}$ $\left[10^{-48} \frac{cm^6}{sr}\right]$	Ref.
CH_4					
$\nu_1(a_1)$	2916.7	0	8.56 [a]	43.2	
$\nu_2(e)$	1534.8	0.75	0.52	2.6	[4.61]
			0.66	3.3	[4.11]
$\nu_3(f_2)$	3020	0.75	4.82	24.3	[4.61]
			4.0	20.3	[4.11]
$\nu_4(f_2)$	1306	0.75	<0.01	<0.04	[4.61]
CD_4					
$\nu_1(a_1)$	2108.6	0	6.23 [b]	31.4	[4.61]
			6.7 [b]	34.0	[4.11]
$\nu_2(e)$	1092.7	0.75	0.37	1.9	[4.61]
$\nu_3(f_2)$	2258.5	0.75	3.55	17.9	[4.61]
$\nu_4(f_2)$	996	0.75	<0.01	<0.04	[4.61]
CF_4					
$\nu_1(a_1)$	908	0	1.2	5.9	[4.10]
$\nu_2(e)$	435	0.75	0.26	1.3	[4.10]
$\nu_3(f_2)$	1283	0.75	0.62	3.2	[4.10]
$\nu_4(f_2)$	631	0.75	0.64	3.2	[4.10]
CCl_4					
$\nu_1(a_1)$	459	0	6.0	34.0	[4.10]
			6.13	34.6	[4.45]
			7.15	40.6	[4.63]
$\nu_2(e)$	221	0.75	1.7	13.1	[4.10]
			2.3	17.7	[4.45]
			2.9	22.8	[4.63]
$\nu_3(f_2)$	794	0.75	2.7	14.1	[4.10]
			3.3	16.9	[4.45]
			3.6	18.8	[4.63]
$\nu_4(f_2)$	318	0.75	2.7	17.3	[4.10]
			3.0	19.5	[4.45]
			5.3	34	[4.63]
SiF_4					
$\nu_1(a_1)$	800	0	1.1	5.4	[4.10]
$SiCl_4$					
$\nu_1(a_1)$	423	0	9.8	56.6	[4.63]
$\nu_2(e)$	145	0.75	4.5	44	[4.63]
$\nu_3(f_2)$	616	0.75	2.1	11.2	[4.63]

Table 4.4 (continued)

Molecule and vibration	ν_j [cm^{-1}]	ρ_j	$\sum_j$	$\frac{d\sigma_j}{d\Omega} \cdot (\nu_0-\nu_j)^{-4}$ $\left[10^{-48}\ \frac{cm^6}{sr}\right]$	Ref.
$SiCl_4$ (cont.)					
$\nu_4(f_2)$	220	0.75	4.3	32.8	[4.63]
$GeCl_4$					
$\nu_1(a_1)$	397	0	16.3	96	[4.63]
$\nu_2(e)$	125	0.75	6.4	71	[4.63]
$\nu_3(f_2)$	459	0.75	6.4	36	[4.63]
$\nu_4(f_2)$	171	0.75	5.5	49	[4.63]
$SnCl_4$					
$\nu_1(a_1)$	369	0	22.2	134	[4.63]
$\nu_2(e)$	95	0.75	7.7	105	[4.63]
$\nu_3(f_2)$	408	0.75	9.8	57	[4.63]
$\nu_4(f_2)$	126	0.75	9.5	105	[4.63]
$TiCl_4$					
$\nu_1(a_1)$	389	0	32.6	194	[4.63]
$\nu_2(e)$	114	0.75	19.8	235	[4.63]
$\nu_3(f_2)$	498	0.75	9.8	54	[4.63]
$\nu_4(f_2)$	136	0.75	6.7	70	[4.63]
CH_3D					
$\nu_1(a_1)$	2967	0.04	6.45 [b] 7.4 [b]	32.6 37.3	[4.61] [4.11]
$\nu_2(a_1)$	2200	0.13	2.94 2.78	14.8 14.1	[4.61] [4.11]
$\nu_3(a_1)$	1300		$\simeq$0.005	$\simeq$0.02	[4.61]
$\nu_4(e)$	3017	0.75	3.46	17.5	[4.61]
$\nu_5(e)$	1473	0.75	0.54	2.7	[4.61]
$\nu_6(e)$	1158	0.75	<0.004	<0.02	[4.61]
CHD_3					
$\nu_1(a_1)$	2992	0.21	2.80 3.4	14.1 16.9	[4.61] [4.11]
$\nu_2(a_1)$	2142	0.02	5.42 5.1	27.4 25.9	[4.61] [4.11]
$\nu_3(a_1)$	1003		<0.01	<0.05	[4.61]
$\nu_4(e)$	2252	0.75	2.99	15.1	[4.61]
$\nu_5(e)$	1295	0.75	0.28	1.4	[4.61]
$\nu_6(e)$	1037	0.75	0.12	0.6	[4.61]

Table 4.4 (continued)

Molecule and vibration	ν_j [cm^{-1}]	ρ_j	Σ_j	$\frac{d\sigma_j}{d\Omega} \cdot (\nu_0-\nu_j)^{-4}$ $\left[10^{-48} \frac{cm^6}{sr}\right]$	Ref.
CHF_3					
$\nu_1(a_1)$	3033	0.12	2.7	13.8	[4.10]
$\nu_2(a_1)$	1138	0.04^Q	0.5^Q	2.3^Q	[4.10]
$\nu_3(a_1)$	702	0.33	0.8	4.1	[4.10]
$\nu_4(e)$	1376	0.8	0.7	3.3	[4.10]
$\nu_5(e)$	1160	0.75	1	5	[4.10]
$\nu_6(e)$	505	0.55	0.7	4.0	[4.10]
CH_3Cl					
$\nu_1(a_1)$	2965	0.02	7.0 [b]	35.1 [b]	[4.45]
		0.02	6.1 [b]	30.6 [b]	[4.10]
$\nu_2(a_1)$	?				
$\nu_3(a_1)$	725	0.24	2.4	12.6	[4.45]
		0.25	3.0	15.7	[4.10]
$\nu_4(e)$	3043	0.75	3.7	18.8	[4.45]
		0.8	5	27	[4.10]
$CHCl_3$					
$\nu_1(a_1)$	3032	0.18	1.9	9.8	[4.45]
		0.05	1.2	5.9	[4.10]
		0.19	1.8	9.0	[4.62]
$\nu_2(a_1)$	674	0.01	2.7	14.3	[4.45]
		0.09	2.4	12.5	[4.10]
		0.02	2.4	12.7	[4.62]
			3.1	16.0	[4.64]
$\nu_3(a_1)$	365	0.14	2.2	13.5	[4.45]
		0.4	2.0	12.2	[4.10]
		0.12	2.0	12.0	[4.62]
$\nu_4(e)$	1221	0.75	0.4	1.8	[4.45]
			0.3	1.7	[4.62]
$\nu_5(e)$	773	0.75	1.7	8.6	[4.45]
			1.9	10.0	[4.10]
			1.4	7.2	[4.62]
			1.7	9.0	[4.64]
$\nu_6(e)$	261	0.75	2.1	14.6	[4.45]
		0.8	1.7	11.6	[4.10]
			2.0	13.7	[4.62]
$CDCl_3$					
$\nu_1(a_1)$	2265	0.18	1.1	5.6	[4.45]
		0.21	1.0	5.3	[4.62]
$\nu_2(a_1)$	655	0.02	2.8	14.5	[4.45]
		0.02	2.5	12.9	[4.62]

Table 4.4 (continued)

Molecule and vibration	ν_j [cm^{-1}]	ρ_j	Σ_j	$\frac{d\sigma_j}{d\Omega} \cdot (\nu_0-\nu_j)^{-4}$ $\left[10^{-48} \frac{cm^6}{sr}\right]$	Ref.
$CDCl_3$ (cont.)					
$\nu_3(a_1)$	365	0.11 0.13	2.1 1.9	12.9 11.5	[4.45] [4.62]
$\nu_4(e)$	914	0.75	0.5 0.5	2.7 2.7	[4.45] [4.62]
$\nu_5(e)$	745	0.75	1.3 1.1	6.7 5.6	[4.45] [4.62]
$\nu_6(e)$	259	0.75	2.0 1.8	14.5 12.7	[4.45] [4.62]
CH_3Br					
$\nu_1(a_1)$	2971	0.0	5.2	26.3	[4.10]
$\nu_2(a_1)$	1304	0.33^Q	0.2^Q	1^Q	[4.10]
$\nu_3(a_1)$	609	0.23	5.2	27.8	[4.10]
$\nu_4(e)$	3050	0.75	5.0	25.2	[4.10]
$\nu_5(e)$	1450	0.75	1.9	9.5	[4.10]
CF_3Cl					
$\nu_1(a_1)$	1106	0.33	0.4	1.8	[4.10]
$\nu_2(a_1)$	782	0.03	1.4 1.7	7.2 8.9	[4.10] [4.37]
$\nu_3(a_1)$	476	0.37	1.4 1.2	7.9 6.7	[4.10] [4.37]
$\nu_4(e)$	1217	0.8	0.8	3.8	[4.10]
$\nu_5(e)$	560	0.7	0.5	2.4	[4.10]
$\nu_6(e)$	350	0.6	0.8	4.6	[4.10]
$CFCl_3$					
$\nu_1(a_1)$	1090		0.6	3.2	[4.10]
$\nu_2(a_1)$	535	0.13	3.6	19.7	[4.10]
$\nu_3(a_1)$	350	0.18	1.5	9.0	[4.10]
$\nu_4(e)$	847	0.73	1.6	8.3	[4.10]
$\nu_5(e)$	394	0.8	1.4	8.2	[4.10]
$\nu_6(e)$	241	0.77	1.6	11.7	[4.10]
CF_3Br					
$\nu_1(a_1)$	1082	0.27	0.8	4.2	[4.10]
$\nu_2(a_1)$	761	0.03	1.4	7.0	[4.10]
$\nu_3(a_1)$	349	0.2	2.4	14.9	[4.10]
$\nu_4(e)$	1207	0.48	0.7	3.6	[4.10]

Table 4.4 (continued)

Molecule and vibration	ν_j [cm^{-1}]	ρ_j	$\sum_j$	$\frac{d\sigma_j}{d\Omega} \cdot (\nu_0-\nu_j)^{-4}$ [$10^{-48}\ \frac{cm^6}{sr}$]	Ref.
CF_3Br (cont.)					
$\nu_5(e)$	541	0.8	0.5	2.5	[4.10]
$\nu_6(e)$	306	0.43	0.9	5.5	[4.10]
CH_2D_2					
$\nu_1(a_1)$	2975	0.08	4.7 4.9	23.7 24.8	[4.61] [4.11]
$\nu_2(a_1)$	2202	0.08	4.5 4.2	22.7 21.0	[4.61] [4.11]
$\nu_3(a_1)$	1435	0.65	0.2	1.0	[4.61]
$\nu_4(a_1)$	1031		$\simeq 0.003^Q$	$\simeq 0.02^Q$	[4.61]
$\nu_5(a_2)$	1325 1343	0.75	0.2	1.1	[4.61]
$\nu_6(b_1)$	3021	0.75	2.0	9.9	[4.61]
$\nu_7(b_1)$	1090	0.75	$\lesssim 0.002^Q$	$\lesssim 0.01^Q$	[4.61]
$\nu_8(b_2)$	2241	0.75	1.1	5.4	[4.61]
$\nu_9(b_2)$	1234	0.75	$\lesssim 0.002^Q$	$\lesssim 0.004^Q$	[4.61]
CH_2F_2					
$\nu_1(a_1)$	3025 2960 2948 2836	0.04	3.2 [b]	16.2 [b]	[4.10]
$\nu_2(a_1)$	?				
$\nu_3(a_1)$	1112	0.04^Q	0.5^Q	2.6^Q	[4.10]
$\nu_4(a_1)$	527	0.33	0.4	2.2	[4.10]
$\nu_6(b_1)$	2995	0.5	1.8	9.0	[4.10]
$\nu_8(b_2)$?	1508	0.8	0.4	1.8	[4.10]
$\nu_9(b_2)$?	1085	0.4	0.7	3.5	[4.10]
CH_2Cl_2					
$\nu_1(a_1)$	2997	0.05	4.6 4.6	23.4 23.4	[4.45] [4.59]
$\nu_2(a_1)$	1428	0.54	1.1	5.6	[4.45]
$\nu_3(a_1)$	713	0.06	3.2 4.0 3.5	16.7 20.9 18.2	[4.45] [4.59] [4.64]
$\nu_4(a_1)$	282	0.36	1.6 1.7	10.8 11.7	[4.45] [4.59]

Table 4.4 (continued)

Molecule and vibration	ν_j [cm^{-1}]	ρ_j	Σ_j	$\frac{d\sigma_j}{d\Omega} \cdot (\nu_0-\nu_j)^{-4}$ $\left[10^{-48} \frac{cm^6}{sr}\right]$	Ref.
CD_2Cl_2					
$\nu_1(a_1)$	2207	0.06	2.6	13.0	[4.45]
$\nu_2(a_1)$	1063	0.38	0.6	3.0	[4.45]
$\nu_3(a_1)$	687	0.05	3.2	16.8	[4.45]
$\nu_4(a_1)$	281	0.35	1.5	9.9	[4.45]
CF_2Cl_2					
$\nu_1(a_1)$	1098	0.48 [c]	0.8 [c]	4.1 [c]	[4.10]
$\nu_2(a_1)$	667	0.07	1.9	9.8	[4.10]
$\nu_3(a_1)$	458	0.27 [d]	3.1 [d]	17.5 [d]	[4.10]
$\nu_4(a_1)$	261	0.6	0.7	4.7	[4.10]
$\nu_5(a_2)$	322	0.75	0.3	2.0	[4.10]
$\nu_6(b_1)$	923	0.8	0.4	2.2	[4.10]
$\nu_7(b_1)$	433 [d]				
$\nu_8(b_2)$	1167 [c]				
$\nu_9(b_2)$	446 [d]				

[Q] Q-branch only
[a] average from Table 4.2
[b] all Fermi-resonance components
[c] data for ν_1 include also values for ν_8
[d] data for ν_3 include also values for ν_7 and ν_9

From an investigation of the intensity distribution in the rotational-vibrational bands ν_2 and ν_5 of CH_3F, which are coupled through second order Coriolis interaction, ESCRIBANO et al. [4.98] were able to deduce relative magnitudes and relative signs for the spherical components of the polarizability derivatives for these two bands. However, since neither scattering cross sections nor depolarization ratios have been measured for this molecule, no absolute values can be given for these quantities.

c) *Influence of the Resonance Raman Effect*

When Raman scattering is excited with a wavelength which is close to or coincides with an absorption band of a molecule, the Raman scattering cross section is enhanced considerably, see Chap.6. A few such cases occur in Table 4.2. The most pronounced effects were found for Br_2 and I_2 [4.54,68], where the exciting lines of the argon ion laser fall into a region of continuous absorption above the dissociation limit [4.69].

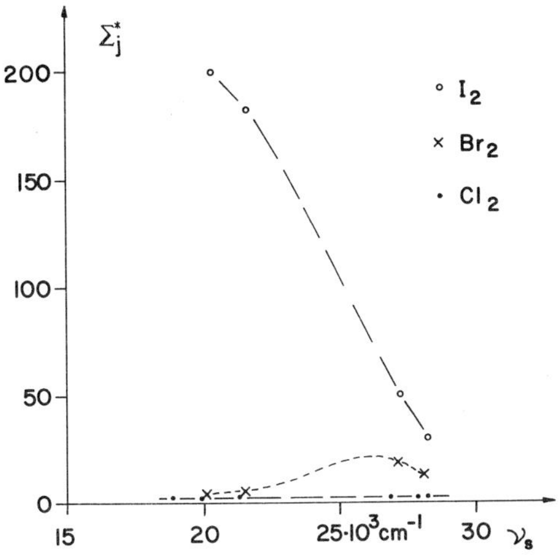

Fig. 4.12. Relative normalized differential Raman scattering cross sections for fundamental bands of the halogens at different exciting lines. Not normalized with temperature dependent factor

Figure 4.12 shows a plot of $\sum_j^*$ against the wavenumber of the scattered light for the fundamental vibrational bands of the halogens Cl_2, Br_2, and I_2. Whereas for Cl_2 the normalized scattering cross section is constant, the values for I_2 rise with increasing extinction coefficient from 30 at 351 nm to 200 at 488 nm. These values are not normalized with the temperature dependent factor, because the temperature, which is different for every exciting line due to the different extinction coefficient, was not determined in these measurements. Therefore the numbers given in Table 4.2 and in Fig.4.12 correspond to $\sum_j^* = \sum_j \cdot [1-\exp(-hc\nu_j/kT)]^{-1}$ and are useful only as rough estimates of the scattering cross section. The exciting line 514.5 nm coincides with discrete energy levels of I_2 and of Br_2 and gives rise to resonance

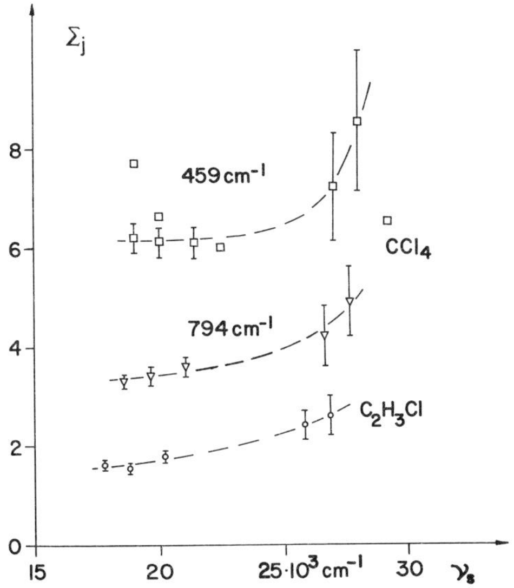

Fig. 4.13. Relative normalized differential Raman scattering cross sections for CCl_4 and vinyl chloride at different exciting lines. Our measurements with error bars

fluorescence [4.69]. In Br_2 the resonance increase from 4 at 488 nm to 18 at 363.8 nm does not follow the wavelength dependence of the extinction coefficient. It may be that interference effects between two electronic states responsible for resonance scattering in Br_2 [4.70] play a role, but further measurements and a quantitative calculation has to be made to clear up this point.

In some other cases only a slight resonance enhancement was observed, a so called "pre-resonance" Raman effect [4.71], mainly for the uv exciting lines. Figure 4.13 shows two examples where the effect is clearly outside the limits of experimental error, namely, carbon tetrachloride and vinyl chloride. It is not attempted at this stage to correlate the data with any of the resonance enhancement factors known from the literature, e.g., from ALBRECHT's theory (see [4.72]), because the range of exciting wavelengths is too limited.

4.2 Raman Scattering Cross Sections in Liquids

4.2.1 Internal Field Effect and Intermolecular Interactions

Equation (4.7) has been derived using the condition that the molecules are rotating freely in space without restriction to their orientation. This condition is valid for gases at low pressure, but it is not fulfilled for liquids. In the liquid phase the influence of the neighboring molecules changes the electric field at the position of the molecule as well as the polarizability of the molecule itself. This internal field effect and the intermolecular interactions cause deviations of the Raman spectra of liquids from those of gases with respect to line positions, line width, temperature dependence, and line intensity. Generally the scattering cross section of a Raman line increases upon transition from the gaseous to the liquid phase, so that—as long as one wants to retain the applicability of the polarizability theory—a correction factor for the internal field has to be incorporated into (4.7,8,9) and a_j^2 and γ_j^2 have to be regarded as dependent upon intermolecular interactions.

In the literature one finds several correction factors for the internal field effect [4.33,73]. Generally they are functions of the refractive index of the liquid of the form [4.33]

$$L = \frac{n_s}{n_0} \cdot \frac{\left(n_s^2+2\right)^2 \cdot \left(n_0^2+2\right)^2}{81} , \qquad (4.17)$$

where n_0 and n_s are the refractive indices of the liquid for the wavelength of the exciting and Raman scattered radiation, respectively. Equation (4.17) was derived from ONSAGER's model [4.74] of dielectric polarization of the molecules in a liquid for a sequential absorption and emission process.

The differential scattering cross section in a liquid is then given as

$$\frac{d\sigma_j}{d\Omega} = \frac{2^4\pi^4}{45} \cdot \frac{b_j^2 \cdot (\nu_0-\nu_j)^4}{1-\exp(-hc\nu_j/kT)} \cdot g_j \cdot \left(45a_j^2+7\gamma_j^2\right) \cdot L \tag{4.18}$$

for 90° scattering and linearly polarized incident light.

In addition to the internal field effect, specific intermolecular interactions can influence the intensity of Raman lines in liquids and solutions. This is proven by the fact that different Raman lines in the same molecule may show a different intensity change upon the phase transition, whereas the internal field effect alone is the same for all lines of a molecule. The influence of the neighboring molecules may change the electron distribution in a molecule and thereby changes the polarizability and possibly also the symmetry of a molecule.

4.2.2 Absolute Scattering Cross Sections

Direct measurements of absolute Raman scattering cross sections in liquids with laser excitation have been performed by MC CLUNG et al. [4.76], DAMEN et al. [4.77], BRET [4.78], SKINNER and NILSEN [4.18], SCHOEN and CUMMINS [4.79] and KATO and TAKUMA [4.19,20]. The latter authors measured against a blackbody source as reference and it seems worthwhile to describe their procedure in a little more detail.

In order to avoid contributions from unwanted reflections to the scattered intensity, they used a sample cell with oblique plane windows. The laser beam passed through a variable attenuator and a pinhole and was focused into the sample cell with a long-focal-length lens to form a 350 μm wide light column as source of scattered light. The laser power reflected at the entrance window was measured by a calibrated thermopile. Two identical optical systems and a rotating sector mirror alternately projected 1 : 1 images of the scattering volume and the 2 × 20 mm² slit-shaped opening of the blackbody radiator onto the entrance slit (500 μm wide) of a double monochromator, so that all the projected Raman scattered light passed through the slit and the whole slit width was illuminated by the blackbody source. The spectral slit width was large enough to measure integrated intensities of isolated Raman lines [4.80,5]. The laser power was attenuated until the phase-sensitive detection system indicated balance between Raman scattered and blackbody radiation. The scattering cross section of the Raman line is then given by [4.19]

$$\frac{d\sigma_j}{d\Omega} = \frac{R_M}{T_M} \cdot \frac{n_S^2 \cdot W \cdot \rho_B(\lambda_S, T) \cdot \Delta\lambda}{2 \cdot (1-R_G) \cdot N_S \cdot \Phi_0} \quad , \tag{4.19}$$

where R_M and T_M are reflectance and transmittance of the rotating sector for the blackbody and the Raman radiation, respectively, n_S the refractive index at the wavelength λ_S of the scattered light giving the solid-angle correction, W the width of

Table 4.5. Absolute normalized Raman scattering cross sections of liquids at different exciting wavelengths λ_0

	$d\sigma_j/d\Omega \cdot (\nu_0-\nu_j)^{-4}$ $[10^{-48}\ cm^6/sr]$					
λ_0 [nm]	337	436	488	515	633	694
C_6H_6 992 cm^{-1}	580 [d]		225 [a] 260 [b] 250 [g]	220 [a] 250 [d]	165 [a] 150 [e]	180 [c] 350 [f]
$C_6H_5CH_3$ 1002 cm^{-1}			127 [a]	120 [a]	73.5[a]	53 [c] 170 [f]
$C_6H_5NO_2$ 1345 cm^{-1}			766 [a]	840 [a]	360 [a]	290 [c] 390 [f]
CS_2 656 cm^{-1}			280 [a] 340 [h]	263 [a]	180 [a]	360 [f]
CCl_4 459 cm^{-1}		125 [i]	140 [a] 135 [h] 128 [j]	137 [a] 126 [j]	113 [a]	
$CHCl_3$ 666 cm^{-1}	119 [d]		54 [h]	64 [d]		
762 cm^{-1}	58 [d]		32 [h]	25 [d]		
C_6H_{12} 802 cm^{-1}	86 [d]		60 [h] 75 [k]	69 [d]		

[a][4.20], [b][4.18], [c][4.76], [d][4.64], [e][4.77], [f][4.78], [g][4.79], [h][4.59], [i][4.5], [j][4.63], [k][4.81].

the entrance slit, $\rho_B(\lambda_s,T)$ the energy density of the blackbody given by Planck's formula, $\Delta\lambda$ the wavelength interval of the blackbody radiation transmitted through the double monochromator, R_G the reflection loss of the scattered light at the cell-to-air interface, N_s the number density of the molecules, and Φ_0 the radiant flux of the exciting laser beam. When a second Raman line falls into the spectral range transmitted an additional correction has to be applied.

The results of KATO and TAKUMA [4.19,20] are presented in Table 4.5 and compared with those of the previous workers [4.5,18,76-79]. We have again chosen to normalize the data by dividing the absolute cross section over the $(\nu_0-\nu_j)^4$ factor. Also included are selected results of NESTOR and LIPPINCOTT [4.59] and CLARK and MITCHELL [4.63], who based the relative cross sections in liquids on the benzene value of KATO and TAKUMA [4.19], of COLLES and GRIFFITHS [4.81], who used the ν_2 band of C_6D_6 at 944 cm^{-1} as reference assuming its cross section to be equal to that of the 992 cm^{-1} band of C_6H_6 as measured by SKINNER and NILSEN [4.18], and of ABE et al. [4.64], who referred their data to the absolute cross section of the 802 cm^{-1} band of cyclohexane vapor corrected by the internal field factor L from (4.17).

Most of the normalized cross sections show a steady increase for decreasing exciting wavelengths indicating a pre-resonance Raman effect. In their second paper

KATO and TAKUMA [4.20] have evaluated their data, which were measured at 488, 514.5, and 632.8 nm, using a formula from the theory of PETICOLAS et al. [4.82]. The agreement is quite good, however, other formulas (e.g. from [4.72]) reproduce the available data equally well.

4.2.3 Relative Scattering Coefficients

Whereas we have tried to give a complete collection of Raman cross section data for gases in Section 4.1, no such attempt is made for liquids. We only want to discuss the proper procedure for the determination of relative scattering coefficients.

BERNSTEIN and ALLEN [4.4] have defined a standard intensity scale which has been adopted by many authors with few modifications. They proposed the ν_1 band of CCl_4 at 459 cm^{-1} as reference line, whereas Russian workers preferred the ν_4 band at 314 cm^{-1} of the same substance [4.83] or the 802 cm^{-1} band of cyclohexane [4.84]. The latter choice is also favored in the recent work of ABE et al. [4.64] because the intensity of this line follows the refractive index dependence in various solvents described by (4.17) in a very satisfactory way.

Some of the corrections discussed by BERNSTEIN and ALLEN [4.4] are no longer needed for laser excitation, so that their formula for the relative scattering coefficient S_j can be modified as follows

$$S_j \equiv \frac{g_j\cdot\left(45a_j^2+7\gamma_j^2\right)}{g_R\cdot\left(45a_R^2+7\gamma_R^2\right)}\cdot\frac{L}{L_R} = \frac{I_{j\ obs}}{I_{R\ obs}}\cdot\frac{\varepsilon(\nu_R)}{\varepsilon(\nu_j)}\cdot\frac{M\cdot d_R}{M_R\cdot d}\cdot\frac{n^2}{n_R^2}\cdot A(\nu_j)$$

$$\cdot\frac{\nu_j\cdot(\nu_0-\nu_R)^4}{\nu_R\cdot(\nu_0-\nu_j)^4}\cdot\frac{1-\exp(-hc\nu_j/kT)}{1-\exp(-hc\nu_R/kT)}\quad . \qquad (4.20)$$

Here the index R denotes the reference band. $I_{j\ obs}$ is the integrated intensity excited with linearly polarized light and observed using a polarization scrambler; when an analyzer is used in front of the slit it must be replaced by $(I_{j\parallel}+I_{j\perp})_{obs} = I_{j\parallel\ obs}\cdot(1+\rho_j)$. $\varepsilon(\nu_j)$ is the spectral sensitivity, M the molecular mass, d the density of the liquid, and $A(\nu_j)$ a correction factor for absorption, when colored substances are investigated. The refractive index dependent solid-angle correction is represented by n^2/n_R^2 [4.85].

In this form (4.20) applies to the external standard method of measurement where sample and reference substance are inserted sequentially in separate sample cells. This method is recommended by NESTOR and LIPPINCOTT [4.59] because it avoids the complications introduced by the influence of the refractive index of a solution [4.86] and specific solute-solvent interactions that occur for the internal standard method.

In contrast to this COLLES and GRIFFITHS [4.81], CLARK and MITCHELL [4.63] and ABE et al. [4.64] have used the internal standard method, because it is easier to handle experimentally. Extensive lists on cross section data in liquids can be found in these publications. It is also to be considered that cross section data in liquids are only valid for the temperature at which the measurement was made, because the temperature influences the refractive index, the population of states, and the intermolecular interactions separately.

4.3 Ratio of the Raman Scattering Cross Section in the Liquid and in the Gaseous State

There is general agreement that the internal field effect on the Raman scattering cross section in liquids is best described by (4.17) as proposed by ECKHARDT and WAGNER [4.33]. All previous quantitative measurements of the ratio of the cross sections in liquids and vapors, however, were made in a somewhat indirect way. SCHRÖTTER and BERNSTEIN [4.26] compared the absolute cross section of gaseous CCl_4 with the liquid value obtained from the Raman/Rayleigh intensity ratio [4.5,24]. NESTOR and LIPPINCOTT [4.59] measured the liquid cross sections using KATO and TAKUMA's [4.19] value for ν_2 of benzene and for the vapor cross sections the value for the Q-branch of nitrogen as reference. CLARK and MITCHELL [4.63] used the ν_1 band of methane as reference for the vapor measurements and the cross section of the 802 cm^{-1} band of cyclohexane as given by NESTOR and LIPPINCOTT [4.59] to put the relative liquid data on an absolute scale. Finally ABE et al. [4.64] relied upon the exact validity of (4.17) for the 802 cm^{-1} band of cyclohexane.

Therefore we decided that a direct measurement of the ratio of the Raman scattering cross section in the liquid and in the gaseous state for a few substances would be useful [4.45]. It can only be performed analogous to the external standard method. The change of the scattering geometry due to the refractive index has to be taken into account. It influences the focus of the laser beam and thereby the scattering volume in addition to the solid angle seen by the collecting optical system. Different reflection losses in the sample cell also must be accounted for.

In order to circumvent some of these difficulties the laser beam was not focused. Apertures were used to define the collection angle of the scattered light. The sample arrangement is shown schematically in Fig.4.14. The unfocused laser beam passes through the square sample cell close to the exit window. An aperture with 2 mm diameter is fixed to the exit window, a second aperture with 10 mm diameter limits the solid angle seen by lens L_1. The distance from the scattering volume to the lens L_1 was equal to its focal length and a second lens L_2 focused the scattered light onto the entrance slit. The sample cell was connected to a liquid reservoir, a mercury manometer, and a vacuum line.

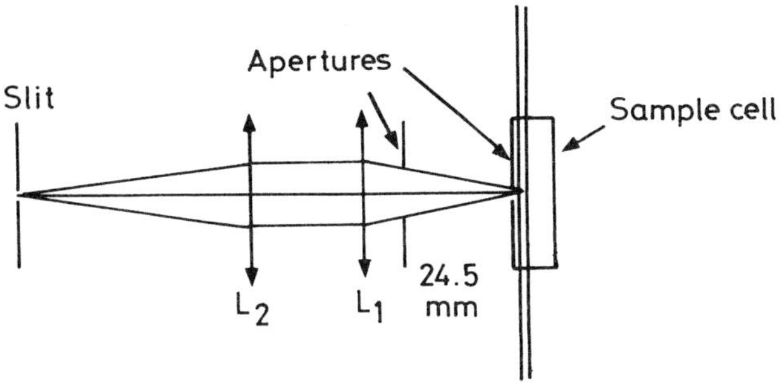

Fig. 4.14. Optical arrangement used for measurement of Raman scattering cross sections in the liquid and gaseous phase

At first the Raman intensities of a substance were measured in the liquid phase, then the sample was transferred to the reservoir by cryogenic pumping. Then a reservoir temperature well below room temperature was chosen to give a suitable vapor pressure which was measured by the manometer and the Raman intensities from the vapor phase were measured. The sample temperatures for liquid and vapor were equal, namely, 295 K for CCl_4 and (299±2)K for all other substances. The laser power before and after the sample cell was carefully monitored.

In order to check the results some measurements for the more intense bands were repeated without the lenses L_1 and L_2. The greatest deviation of the results obtained with and without lenses was ±10%.

From the measured integrated intensities and depolarization ratios the ratio of the scattering cross sections in the liquid and gaseous phase was obtained as

$$\Lambda \equiv \frac{(d\sigma/d\Omega)_\ell}{(d\sigma/d\Omega)_g} = \frac{I_{\parallel\ell}\cdot(1+\rho_\ell)\cdot M\cdot p_0\cdot K(n_\ell^2)}{I_{\parallel g}\cdot(1+\rho_g)\cdot d\cdot 22{,}400\cdot 760} \quad , \tag{4.21}$$

where the subscripts ℓ and g denote the liquid and gaseous phase and p_0 is the gas pressure in Torr reduced to 0°C. The correction factor $K(n_\ell^2)$ for the different solid angles and the different scattering volumes is to a good approximation [4.59] equal to n_ℓ^2. The error of the ratio Λ adds up to about ±15% from the error of the intensity measurement (±5%) and the uncertainty in the geometric correction factor.

In Table 4.6 we present the measured liquid/gas cross section ratios Λ for CCl_4, $CHCl_3$, $CDCl_3$, CH_2Cl_2, CD_2Cl_2, and C_6H_6 benzene and compare them with data from the literature. The considerable scatter of some of the data is most probably due to the different methods of measurement and the systematic errors connected therewith.

In Table 4.6 are also included the data of BURNHAM et al. [4.87] obtained from cross section measurements of the Rayleigh line. In the last column the ratio Λ of the measured cross sections is divided by the internal field correction L calculated from the refractive index according to (4.17).

For the totally symmetric CH- and CD-valence vibrations of the investigated chloromethanes the average $\Lambda/L = 1.39 \pm 0.07$ is significantly higher than the average for

Table 4.6. Ratio of the differential Raman scattering cross sections in the liquid and gaseous phase

Vibration	Λ					Λ/L
CCl_4						
ν_1	3.66[e]	3.15[b]	3.82[c/a]	3.7[d/e,a]	2.6[f]	1.02
ν_2	3.25[e]	2.81[b]	5.23[c/a]			0.90
ν_3	3.12[e]	2.05[b]	4.3 [c/a]			0.87
ν_4	3.31[e]	2.80[b]	4.63[c/a]			0.92
Rayleigh		3.11[k]				
$CHCl_3$						
ν_1	4.91[e]	8.43[c/a]	4.74[g/e]			1.42
ν_2	3.17[e]	4.23[c/a]	4.02[g/e]	3.92[m]		0.92
ν_3	3.38[e]	4.25[c/a]	3.62[g/e]			0.98
ν_5	3.23[e]	3.04[c/a]		2.78[m]		0.93
ν_6	3.67[e]	4.17[c/a]				1.06
Rayleigh		3.17[k]				
$CDCl_3$						
ν_1	5.1 [e]	5.82[g/e]				1.48
ν_2	3.3 [e]					0.96
ν_3	3.26[e]					0.95
ν_5	3.1 [e]					0.9
ν_6	3.58[e]					1.04
CH_2Cl_2						
ν_1	4.28[e]	4.01[c]	4.2 [g/e]			1.32
ν_2	4.1 [e]					1.26
ν_3	3.64[e]	3.59[c]				1.18
ν_4	2.97[e]	2.75[c]				0.91
CD_2Cl_2						
ν_1	4.36[e]					1.34
ν_2	3.2 [e]					0.98
ν_3	3.4 [e]					1.05
ν_4	2.86[e]					0.88
C_6H_6						
ν_1	4.65[e]	4.54[c]	4.47[c/e]	4.73[g/a]		1.15
ν_2	3.57[e]	4.14[c]	3.48[d/a]	2.7 [f]		0.88
Rayleigh		3.54[k]	4.0 [h/i]	3.79[m]		

[a] [4.10], [b] averages from [4.63], [c] [4.59], [d] [4.19], [e] [4.45], [f] [4.26], [g] liquid data from M. WAHL: Dissertation University of Munich (1969), recalculated with absolute value from [4.19], [h] [4.18], [i] [4.57], [k] [4.87], [m] [4.64].

all other vibrations $\Lambda/L = 0.99 \pm 0.1$. A more detailed dependence of Λ/L on specific modes of vibrations cannot be detected in the data of Table 4.6.

The results show with a higher degree of confidence than before, that the increase of the Raman scattering cross section in liquids over that in the gaseous state is described very well by (4.17) and that additional intermolecular interactions are making only minor contributions, even in the case of chloroform, where NESTOR and LIPPINCOTT [4.59] found an Λ/L of 2.4 for the CH-band.

4.4 Conclusion

In this chapter we have tried to give a comprehensive review of the available Raman scattering cross section data in gases and we have discussed the problems associated with the determination of Raman cross sections in liquids.

The absolute Raman scattering cross section of the Q-branch of the vibrational band of N_2 is now known with an accuracy of about 2% throughout the visible and near uv. Relative cross sections for other gases have been determined with an accuracy varying from 5 to 15% and more. A look at Table 4.2 shows that many more measurements are required to clear up discrepancies and to fill the many gaps. An extension of the measurements further into the uv, e.g. with frequency doubled argon ion lasers, also remains a task for the future, especially for those substances where a significant resonance Raman effect is expected.

For liquids it has become clear that the enhancement of the scattering cross section over the value in the gas phase is mainly due to the internal field effect as described by the formula suggested by ECKHARDT and WAGNER [4.33] and that specific molecular interactions play a major role only for systems with hydrogen bonding. The direct measurement of the liquid/gas cross section ratio for a few substances presented in Sect.4.3 has given strong support to this statement.

Finally we want to mention that there are also methods to measure absolute Raman scattering cross sections in crystal powders [4.88-90] and in crystals [4.79,91] which have not been treated in this chapter.

Acknowledgments. We thank J. Brandmüller for his constant encouragement, J.G. Hochenbleicher for many of the measurements, J. Behringer, P. Baierl, and W. Kiefer for useful discussions on the resonance Raman effect, and the "Bundesministerium für Forschung und Technologie" of the Federal Republic Germany for financial support.

References

4.1 Rank, D.H., Pfister, R.J., Coleman, P.D.: J. Opt. Soc. Am. *32*, 390 (1942)
4.2 Rank, D.H., Pfister, R.J., Grimm, H.H.: J. Opt. Soc. Am. *33*, 31 (1943)
4.3 Rank, D.H., Wiegand, R.V.: J. Opt. Soc. Am. *36*, 325 (1946)
4.4 Bernstein, H.J., Allen, G.: J. Opt. Soc. Am. *45*, 237 (1955)
4.5 Brandmüller, J., Schrötter, H.W.: Z. Phys. *149*, 131 (1957)
4.6 Yoshino, T., Bernstein, H.J.: J. Mol. Spectrosc. *2*, 213 (1958)
4.7 Yoshino, T., Bernstein, H.J.: In *Proc. Conf. Mol. Spectrosc.*, London, 1958 (Pergamon Press, London 1959) p.141; Spectrochim. Acta *14*, 127 (1959)
4.8 Yoshino, T., Bernstein, H.J.: J. Mol. Spectrosc. *2*, 241 (1958)
4.9 Stansbury, E.J., Crawford, M.F., Welsh, H.L.: Can. J. Phys. *31*, 954 (1953)
4.10 Murphy, W.F., Holzer, W., Bernstein, H.J.: Appl. Spectrosc. *23*, 211 (1969)
4.11 Kagel, R.O.: Ph. D. Thesis, University of Minnesota (1964)
4.12 Hornischer, R., Moser, H.: Spectrochim. Acta *28*A, 81 (1972)
4.13 Porto, S.P.S., Wood, D.L.: J. Opt. Soc. Am. *52*, 251 (1962)
4.14 Stoicheff, B.P.: In *Proc. Xth Coll. Spectrosc. Intern.*, ed. by Lippincott,E.R., Margoshes,M. (Spartan Books, Washington 1963) p.399
4.15 Leite, R.C.C., Porto, S.P.S.: J. Opt. Soc. Am. *54*, 981 (1964)
4.16 Koningstein, J.A., Smith, R.G.: J. Opt. Soc. Am. *54*, 1061 (1964)
4.17 Weber, A., Porto, S.P.S.: J. Opt. Soc. Am. *55*, 1033 (1965)
4.18 Skinner, J.G., Nilsen, W.G.: J. Opt. Soc. Am. *58*, 113 (1968)
4.19 Kato, Y., Takuma, H.: J. Opt. Soc. Am. *61*, 347 (1971)
4.20 Kato, Y., Takuma, H.: J. Chem. Phys. *54*, 5398 (1971)
4.21 Fenner, W.R., Hyatt, H.A., Kellam, J.M., Porto, S.P.S.: J. Opt. Soc. Am. *63*, 73 (1973)
4.22 Hyatt, H.A., Cherlow, J.M., Fenner, W.R., Porto, S.P.S.: J. Opt. Soc. Am. *63*, 1604 (1973)
4.23 Golden, D.M., Crawford, B. Jr.: J. Chem. Phys. *36*, 1654 (1962)
4.24 Schrötter, H.W.: Z. Elektrochem. Ber. Bunsenges. Phys. Chem. *64*, 853 (1960)
4.25 Schrötter, H.W., Bernstein, H.J.: J. Mol. Spectrosc. *7*, 464 (1961)
4.26 Schrötter, H.W., Bernstein, H.J.: J. Mol. Spectrosc. *12*, 1 (1964)
4.27 Placzek, G.: "Rayleigh-Streuung und Raman-Effekt", in *Handbuch der Radiologie*, ed. by G. Marx, Vol.6, Pt.2 (Akademische Verlagsgesellschaft, Leipzig 1934) p.205
4.28 Placzek, G., Teller, E.: Z. Phys. *81*, 209 (1933)
4.29 Altmann, K.: Dissertation, University of Munich (1975)
4.30 Altmann, K., Strey, G.: J. Mol. Spectrosc. *44*, 571 (1972)
4.31 Landau, L.D., Lifschitz, E.M.: *Lehrbuch der theoretischen Physik*, Bd.III, Quantenmechanik (Akademie Verlag, Berlin 1965)
4.32 Brandmüller, J., Moser, H.: *Einführung in die Raman-Spektroskopie* (Steinkopff, Darmstadt 1962)
4.33 Eckhardt, G., Wagner, W.G.: J. Mol. Spectrosc. *19*, 407 (1966)
4.34 Hirschfeld, T.: Appl. Spectrosc. *27*, 389 (1973)
4.35 Leonard, D.A.: J. Appl. Phys. *41*, 4238 (1970)
4.36 Fouche, D.G., Chang, R.K.: Appl. Phys. Lett. *18*, 579 (1971)
4.37 Fouche, D.G., Chang, R.K.: Appl. Phys. Lett. *20*, 256 (1972)
4.38 Penney, C.M., Goldman, L.M., Lapp, M.: Nature Phys. Sc. *235*, 110 (1972)
4.39 Shimizu, H., Kobayasi, T., Inaba, H.: Ōyō Buturi (Publ. Japan Soc. Appl. Phys.) *42*, 889 (1973)
4.40 Levatter, J.I., Sandstrom, R.L., Lin, Shao-Chi: J. Appl. Phys. *44*, 3273 (1973)
4.41 Inaba, H., Kobayasi, T.: Opto-Electronics *4*, 101 (1972)
4.42 Inaba, H.: "Detection of Atoms and Molecules by Raman Scattering and Resonance Fluorescence", in *Laser Monitoring at the Atmosphere*, ed. by E.D. Hinkley, Topics in Applied Physics, Vol.14 (Springer, Berlin, Heidelberg, New York 1976) Chap.5, pp.153-236
4.43 Reiss, N.M.: J. Appl. Phys. *43*, 739 (1972)
4.44 Penney, C.M.: Paper III-3 at VIth Int. Laser Radar Conf., Sendai, Japan (1974); IEEE J. QE.-*11*, 36 D (1975)
4.45 Klöckner, H.W.: Dissertation, University of Munich (1977)

4.46 Barrett, J.J., Weber, A.: J. Mol. Spectrosc. *50*, 205 (1974)
4.47 Kolos, W., Wolniewicz, L.: J. Chem. Phys. *46*, 1426 (1967)
4.48 Victor, G.A., Dalgarno, A.: J. Chem. Phys. *50*, 2535 (1969)
4.49 Ford, A.L., Browne, J.C.: Phys. Rev. A*7*, 418 (1973)
4.50 Bridge, N.J., Buckingham, A.D.: J. Chem. Phys. *40*, 2733 (1964)
4.51 Rowell, R.L., Aval, G.M., Barrett, J.J.: J. Chem. Phys. *54*, 1960 (1971)
4.52 Kamiyama, H., Okano, S., Maruyama, N.: Paper III-5 at VIth Int. Laser Radar Conf., Sendai, Japan (1974)
4.53 Cherlow, J.M., Porto, S.P.S.: "Laser Raman Spectroscopy of Gases", in *Laser Spectroscopy of Atoms and Molecules*, ed. by H. Walther, Topics in Applied Physics, Vol.2 (Springer, Berlin, Heidelberg, New York 1976) Chap.4, pp.253-282
4.54 Hochenbleicher, J.G.: Dissertation, University of Munich (1977)
4.55 Hoell, J.M., Allario, F., Jarrett, O., Seals, R.K.: J. Chem. Phys. *58*, 2896 (1973)
4.56 Arden, W.: Remote Raman Detection Study Instrument, 6th Quart. Prog. Rpt, Edgewood Arsenal Contract DAAA 15-70-C-0418 (1972), as quoted by DeLong, H.P.: Opt. Engg. *13*, 5 (1974)
4.57 Udagawa, Y., Mikami, N., Kaya, K., Ito, M.: J. Raman Spectrosc. *1*, 341 (1973)
4.58 Schwiesow, R.L.: Am. Inst. Aeron. Astron. J. *11*, 87 (1973)
4.59 Nestor, J.R., Lippincott, E.R.: J. Raman Spectrosc. *1*, 305 (1973)
4.60 Cherlow, J.M., Hyatt, H.A., Porto, S.P.S.: J. Chem. Phys. *63*, 3996 (1975)
4.61 Bermejo, D., Escribano, R., Orza, J.M.: J. Mol. Spectrosc. *65*, 345 (1977)
4.62 Bermejo, D.: Dissertation, University of Madrid (1976)
4.63 Clark, R.J.H., Mitchell, P.D.: J. Chem., Soc., Faraday Trans. II *71*, 515 (1975)
4.64 Abe, N., Wakayama, M., Ito, M.: J. Raman Spectrosc. *6*, 38 (1977)
4.65 Holzer, W., Moser, H.: J. Mol. Spectrosc. *20*, 188 (1966)
4.66 Holzer, W.: J. Mol. Spectrosc. *25*, 123 (1968)
4.67 Holzer, W.: Dissertation, University of Munich (1967)
4.68 Hochenbleicher, J.G., Kiefer, W., Brandmüller, J.: Appl. Spectrosc. *30*, 528 (1976)
4.69 Holzer, W., Murphy, W.F., Bernstein, H.J.: J. Chem. Phys. *52*, 399 (1970)
4.70 Baierl, P., Kiefer, W., Williams, P.F., Rousseau, D.L.: Chem. Phys. Lett. *50*, 57 (1977)
4.71 Behringer, J.: "Observed Resonance Raman Spectra", in *Raman Spectroscopy*, Theory and Practice, Vol.1, ed. by H.A. Szymanski (Plenum Press, New York 1967) Chap.6, pp.168-223
4.72 Albrecht, A.C., Hutley, M.C.: J. Chem. Phys. *55*, 4438 (1971)
4.73 Sidorov, N.K., Stalmakhova, L.S., Bogachyov, N.V.: Opt. Spectrosc. *30*, 375 (1971)
4.74 Onsager, L.: J. Am. Chem. Soc. *58*, 1486 (1936)
4.75 Le Duff, Y., Holzer, W.: J. Chem. Phys. *60*, 2175 (1974)
4.76 Mc Clunq, F.J., Weiner, D.: J. Opt. Soc. Am. *54*, 641 (1964)
Weiner, D., Schwarz, S.E., Mc Clung, F.J.: J. Appl. Phys. *36*, 2395 (1965)
4.77 Damen, T.C., Leite, R.C.C., Porto, S.P.S.: Phys. Rev. Lett. *14*, 9 (1965)
4.78 Bret, G.: Compt. Rend. *260*B, 6323 (1965)
4.79 Schoen, P.E., Cummins, H.J.: "Absolute Cross Sections for Raman and Brillouin Light Scattering in Quartz", in *Light Scattering in Solids*, ed. by M. Balkanski (Flammarion, Paris 1971) pp.460-462
4.80 Sushchinskij, M.M.: Zh. Eksp. Teor. Fiz. *22*, 765 (1952)
4.81 Colles, M.J., Griffiths, J.E.: J. Chem. Phys. *56*, 3384 (1972)
4.82 Peticolas, W.L., Nafie, L., Stein, P., Fanconi, B.: J. Chem. Phys. *52*, 1576 (1970)
4.83 Shorygin, P.P.: Uspechi Chimii *19*, 419 (1950)
4.84 Sterin, Kh.E.: Isv. Akad. Nauk SSSR, Ser. Fiz., *14*, 411 (1950)
4.85 Woodward, L.A., George, J.H.B.: Proc. Phys. Soc. London B*65*, 780 (1951)
4.86 Fini, G., Mirone, P., Patella, P.: J. Mol. Spectrosc. *28*, 144 (1968)
4.87 Burnham, A.K., Alms, G.R., Flygare, W.H.: J. Chem. Phys. *62*, 3289 (1975)
4.88 Blomer, F., Moser, H.: Z. Angew. Phys. *27*, 302 (1969)
4.89 Gorelik, V.S., Sushchinskij, M.M., Novik, A.E.: Sov. Phys.—Solid State *11*, 619 (1969)
4.90 Lopez, M.A., Montero, S.: Appl. Spectrosc. *31*, 469 (1977)

4.91 Gorelik, V.S., Sushchinskij, M.M.: Sov. Phys.—Solid State *12*, 1157 (1970)
4.92 Adolfson, W.F., Schlecht, R.G., Morton, J.B.: Appl. Phys. *14*, 49 (1977)
4.93 Hills, G.W., Jones, W.J.: J. Chem. Soc., Trans. Faraday II, *71*, 812 (1975)
4.94 Bridges, N.J., Buckingham, A.D.: Proc. R. Soc. A*295*, 334 (1966)
4.95 Hills, G.W., Foster, R.B., Jones, W.J.: Mol. Phys. *33*, 1571 (1977)
4.96 Foster, R.B., Hills, G.W., Jones, Mol. Phys. *33*, 1589 (1977)
4.97 Murphy, W.F.: Mol. Phys. *33*, 1701 (1977)
4.98 Escribano, R., Mills, I.M., Brodersen, S.: J. Mol. Spectrosc. *61*, 249 (1976)
4.99 Warsop, P.A., Montero, S.: J. Raman Spectrosc. *7*, 115 (1978)
4.100 Clark, R.J.H., D'Urso, N.R.: J. Chem. Soc. (Dalton) , 170 (1978)
4.101 Abe, N., Ito, M.: J. Raman Spectrosc. *7*, 161 (1978)
4.102 Eliseev, A.A., Popova, T.N., Churakov, C.P.: Experimental Determination of Differential Raman Scattering Cross Sections for Molecules of some Gases (in Russian), in *Spektroskopiya Kombinatsionnogo Rasseyaniya Sveta*, Materialy II Vsesoyuznoj Konferentsii (Akademiya Nauk SSSR, Moskva 1978) p.111

5. Intermolecular Forces Revealed by Raman Scattering

R. P. Srivastava and H. R. Zaidi

With 4 Figures

In this chapter, the recent experimental and theoretical developments in the general area of Raman spectral shapes are surveyed. In particular, the dependence of the spectral shapes on intermolecular forces is emphasized. Information about intermolecular forces that has been obtained from the study of Raman spectra in gases and liquids is summarized. It is pointed out that quantitative information has been obtained from the Raman line shapes in many low density gases.

5.1 Introductory Remarks

In general, the study of Raman scattering from molecular systems (gases, liquids and solids) yields information about the internal structure of the molecule, its dynamical behavior, and the interaction with its surrounding [5.1]. For example, the intensity and polarization measurements of the scattered light determines the polarizability of the molecular system [5.2], and the frequency measurements at low pressures furnish information about the molecular structure ([5.1,3]; see also Chaps.2, 3 and 4 of this volume). The line width, shift, and band shape measurements in the Raman spectra provide information about the intermolecular forces. Our interest, here, is in the latter aspect of the Raman scattering. This subject was earlier reviewed by GRAY and WELSH [5.4] in 1971. Significant amount of experimental and theoretical work has been done since then. We shall emphasize the developments since the last review, although the work prior to that is also cited wherever it is necessary.

There are several other sources of information about the intermolecular forces [5.5], For example, the following spectroscopic methods also yield related information: 1) microwave and infrared line and band shapes [5.6], 2) depolarized Rayleigh line shapes [5.7] and, 3) NMR relaxation [5.8]. The formal theories of all the above processes are interrelated, as pointed out by GORDON [5.8]. Therefore, the consistency and reliability of the results obtained from different methods can be checked against each other. The traditional methods (virial coefficients, transport properties, crystal structure, etc.) usually provide information about the isotropic forces only [5.5]. However, in some cases, the study of transport phenomena in polyatomic gases [5.9a] and structure of molecular crystals [5.9b] can also provide information about anisotropic forces. Molecular beam scattering has been employed to learn about anisotropic forces between molecules [5.8b]. Indirect information about intermolecular forces is contained in collision induced absorption [5.10] and scattering [5.11,12]; but these subjects will not be discussed here. We will also omit the discussion of the resonance Raman effect (Chap.6) and the nonlinear phenomena such as stimulated light scattering [5.13].

The properties of low density gases can be directly related to the interaction between two molecules. In dense gases and liquids the many-body interactions complicate the situation considerably. Therefore, the main thrust of this article is toward the Raman line and band shapes in gases, but relevant work in liquids will also be summarized. However, a discussion of Raman scattering in solids is beyond the scope of this article. This is a vast subject, and information about intermolecular forces can be obtained in molecular crystals (see, e.g., [5.14]). We will start with a brief review of some of the well known features of Raman line and band shapes.

In general, the polarizability of a molecule consists of an isotropic as well as an anisotropic part. Light scattering originates in both parts, but the two types of scattering behave differently with respect to intermolecular forces. Also, the characteristics of the Raman spectrum depends on the density of the molecular system. At very low gas densities, where all the collision effects are negligible, an isolated Raman line has its natural width superimposed with the Doppler width. However, as the density increases, the collisions become more frequent. In some cases, the Dicke narrowing of the Doppler profile [5.15] is observed, At still higher densities the collision broadening takes over, and the line width increases with gas density. In the case where many broadened lines overlap, the phenomenon of frequency degeneracy occurs, and the broadening ceases. When the density of the gas is further increased the band spectrum shows collisional narrowing (e.g., isotropic Q-branch of totally symmetric vibrational bands). The frequencies of the Raman lines are also shifted by the collisions. Most of these effects can be related to the intermolecular forces through collision parameters. The measurement of line widths, shifts, and band shapes can determine some of these parameters.

In the case of dense gases or liquids, the Raman spectrum is mostly in the form of bands of overlapping lines (except for very light molecules such as H_2 and D_2). Because of the complexity of the liquid state, the spectrum is generally not analyzed in the same manner as for gases. In many cases, the band moments and correlation functions are determined, which provide information about molecular dynamics in the fluids. Enormous amounts of work have been done in recent years on scattering of light from liquids. There seem to be two main reasons for this: 1) The liquid state is not very well understood, and therefore, attracts more attention; new qualitative effects are being discovered, 2) the scattering is usually stronger than from gases, and thus experimental observation is easier. However, the information obtained so far about intermolecular forces is mostly qualitative. The relevant work will be summarized in Sect.5.6.

5.2 Spectral Function for Raman Scattering

In this section we will briefly outline a derivation of the spectral function $F(\omega)$ for non-resonant Raman scattering. To keep the discussion simple, we will consider the case of single active molecule ("system") surrounded by a thermal bath of perturbers. This excludes some well known effects, e.g., correlations in scattering, collision induced scattering, and local field corrections. More general formulations of the scattering theory exist which take into account these effects [5.11,12,16]. We will briefly discuss some of these effects at appropriate places. In the single scatterer approximation the total Hamiltonian of the system, the bath, and the radiation field is given by

$$\mathcal{H} = H_0 + H' \tag{5.1a}$$

$$H_0 = H + H_r \quad , \tag{5.1b}$$

where H is the Hamiltonian of the system and the bath including all the intermolecular interactions, H_r is the Hamiltonian of the radiation field, and H' represents the interaction of radiation with the system. We assume that the system and the bath is in a thermal equilibrium, and the radiation field is not too strong (linear response regime).

The eigenstates of H satisfy the time independent equation

$$H|n\rangle = E_n|n\rangle \tag{5.2}$$

where E_n is the energy eigenvalue in the state $|n\rangle$. We further assume that the incident radiation field consists of a monochromatic beam of frequency ω_0, wave vector $\underline{k}_0$, and polarization $\hat{e}_0$. Under the influence of the radiation field, a molecule of the system scatters a photon in state $|\omega_0\underline{k}_0\hat{e}_0\rangle$ to a photon in state $|\omega_s\underline{k}_s\hat{e}_s\rangle$. Using the second-order transition probability formula, one obtains an expression for the differential scattering cross section [5.17]. For the non-resonant Raman scattering, the differential scattering cross section per molecule can be written as [5.18,19]

$$d^2\sigma/d\Omega d\omega_s = (\omega_0\omega_s^3/c^4)\sum_n \rho_n \sum_{n'} |\langle n|\hat{e}_0\cdot\underline{\underline{\alpha}}\cdot\hat{e}_s|n'\rangle|^2 \times \delta(\omega_s-\omega_0+E_{n'}/\hbar-E_n/\hbar) \tag{5.3}$$

$\hbar = h/2\pi$ (normalized Planck's constant).

where $d\omega_s$ and $d\Omega$ represent the frequency range and the solid angle element, respectively, for the scattered photon. The symbol $\underline{\underline{\alpha}}$ represents the polarizability tensor [5.19], $|n\rangle$ and $|n'\rangle$ represent the initial and final states, and ρ_n represents the thermal equilibrium density in the state $|n\rangle$

$$\rho_n = \exp(-E_n/kT)/\sum_{n'} \exp(-E_{n'}/kT) \quad . \tag{5.4}$$

Here an assumption has been made that the system returns to a thermal equilibrium after each photon has been scattered. The delta function in (5.3) represents the energy conservation in the scattering process: $\hbar\omega_0 + E_n = \hbar\omega_s + E_{n'}$.

Using the Fourier integral representation of delta function, (5.3) can be written in the following form

$$\begin{aligned} d^2\sigma/d\Omega d\omega_s &= (\omega_0\omega_s^3/c^4) \sum_n \rho_n \sum_{n'} \langle n|\underline{\underline{\alpha}}|n'\rangle\langle n'|\underline{\underline{\alpha}}|n\rangle \\ &\quad \times (2\pi)^{-1} \int_{-\infty}^{\infty} dt \exp[i(\omega_s-\omega_0+E_{n'}/\hbar-E_n/\hbar)t] \\ &= (\omega_0\omega_s^3/2\pi c^4) \int_{-\infty}^{\infty} \exp[i(\omega_s-\omega_0)t]dt \\ &\quad \times \sum_n \rho_n \sum_{n'} \langle n|\underline{\underline{\alpha}}|n'\rangle\langle n'|\exp(iE_{n'}t/\hbar)\underline{\underline{\alpha}} \exp(-iE_nt/\hbar)|n\rangle \end{aligned} \tag{5.5}$$

where $\hat{e}_0 \cdot \underline{\underline{\alpha}} \cdot \hat{e}_s$ is replaced by $\underline{\underline{\alpha}}$ for simpler notation.

Since $|n\rangle$ and $|n'\rangle$ are the eigenstates of the Hamiltonian H, one can write (5.5) as

$$\begin{aligned} d^2\sigma/d\Omega d\omega_s &= (\omega_0\omega_s^3/2\pi c^4) \int_{-\infty}^{\infty} \exp[i(\omega_s-\omega_0)t]dt \\ &\quad \times \sum_n \rho_n \sum_{n'} \langle n|\underline{\underline{\alpha}}|n'\rangle\langle n'|\exp(iHt/\hbar)\underline{\underline{\alpha}} \exp[t(-iHt/\hbar)]|n\rangle \\ &= (\omega_0\omega_s^3/2\pi c^4) \int_{-\infty}^{\infty} dt \exp[i(\omega_s-\omega_0)t]\, C(t) \end{aligned} \tag{5.6}$$

$$C(t) = \mathrm{Tr}\{\rho\underline{\underline{\alpha}}(0)\underline{\underline{\alpha}}(t)\} = \sum_n \rho_n \langle n|\underline{\underline{\alpha}}(0)\underline{\underline{\alpha}}(t)|n\rangle \tag{5.7}$$

where $\underline{\underline{\alpha}}(t)$ is the Heisenberg operator satisfying the equation of motion

$$d\underline{\underline{\alpha}}(t)/dt = i\hbar^{-1}[H,\underline{\underline{\alpha}}(t)] \quad . \tag{5.8}$$

Equation (5.6) gives the Raman spectrum of the system. The slowly varying factor $(\omega_0\omega_s^3/c^4)$ in (5.6) is not important for the spectral function. Therefore, we define the Raman spectral function $F(\omega)$ by dropping this factor

$$F(\omega) = (2\pi)^{-1} \int_{-\infty}^{\infty} dt \exp(i\omega t)\, C(t) \tag{5.9a}$$

where

$$\omega = \omega_s - \omega_0 \quad . \tag{5.9b}$$

Thus the spectrum can be obtained by Fourier transforming the correlation function C(t) of the polarizability with itself. The frequency ω, here, represents the difference between the frequencies of the scattered and incident photons.

Since $F(\omega)$ is real, one can express it in the form

$$F(\omega) = \pi^{-1} \mathrm{Re}\{\int_{0}^{\infty} dt \exp(i\omega t)\, C(t)\} \quad . \tag{5.10}$$

In the case of an absorption spectrum, the correlation function similar to (5.7) is obtained by replacing $\underline{\underline{\alpha}}$ by the dipole moment operator $\underline{\mu}$ (see, e.g., [5.8]). The main difference between the two cases results from the different symmetry properties and the selection rules for $\underline{\underline{\alpha}}$ and $\underline{\mu}$. However, the general problem of calculation of C(t) is essentially the same for the absorption and the non-resonant Raman scattering.

Various methods—classical, semiclassical [5.20-25,8], and quantum mechanical [5.26-29]—have been proposed for the calculation of C(t) or $F(\omega)$ for absorption or Raman scattering. Alternative formulations of the problem using kinetic theory [5.9, 30-34], Greens functions, and other many-body methods [5.35-39] have also been employed for a variety of atomic and molecular systems. In some of the limiting cases, most of the approaches lead to essentially the same result. For example, one of the limiting results is the well known impact theory of collision broadening [5.21,26-28]. To keep the discussion simple, we will neglect the Doppler broadening. The Doppler broadening and Dicke narrowing [5.15] of spectral line shapes has been investigated by several workers [5.30,34,36,40-42]. This narrowing has also been observed in Raman scattering [5.43,44]. It gives information about the diffusion constant for low density gases. A related, but different effect is the diffusional narrowing in collision induced spectra, which provides information about diffusion constant in high density gases [5.45]. We will not discuss such effects further, since only indirect information about intermolecular forces is contained in the observed results.

In the absence of Doppler broadening and related phenomena, the Liouville approach of FANO [5.29] is quite convenient. We will follow it to reduce (5.10) to more usable forms. The impact limit and the application to gases will be considered in the next section. Some of these results are also applicable to liquids, to be discussed in Sect.5.6.

According to (5.7), the correlation function for the Raman scattering is

$$C(t) = \mathrm{Tr}\{\rho\underline{\underline{\alpha}}(0)\underline{\underline{\alpha}}(t)\} = \mathrm{Tr}\{\underline{\underline{\alpha}}(t)\rho\underline{\underline{\alpha}}(0)\} \quad . \tag{5.11}$$

At this stage we define the Liouville super operator L, which enables one to obtain a formal solution of the correlation function and the spectral function without specifying further details of the system and the bath. The action of L on any operator X is defined by

$$LX = \hbar^{-1}(HX-XH) = \hbar^{-1}[H,X] \tag{5.12}$$

and the matrix elements are defined by

$$(LX)_{mn} = \sum_{m'n'} L_{mn.m'n'} X_{m'n'} \tag{5.13}$$

$$L_{mn,m'n'} = \hbar^{-1}(H_{mm'}\delta_{nn'}-\delta_{mm'}H_{n'n}) = \hbar^{-1}(H_{mm'}\delta_{nn'}-\delta_{mm'}H^{*}_{nn'}) \tag{5.14}$$

where the asterisk denotes the complex conjugate. The Liouville operator is Hermitian and its eigenvalues represent the transition frequency between the two states

$$L_{mn,mn} = \hbar^{-1}[H_{mm}-H_{nn}] = \hbar^{-1}(E_m-E_n) = \omega_{mn} \quad . \tag{5.15}$$

By using the definition (5.12) one can write the Heisenberg equation of motion, (5.8), for $\underline{\underline{\alpha}}(t)$ as

$$d\underline{\underline{\alpha}}(t)/dt = iL\underline{\underline{\alpha}}(t) \tag{5.16}$$

which has the solution

$$\underline{\underline{\alpha}}(t) = \exp(iLt)\underline{\underline{\alpha}}(0) = \underline{\underline{\alpha}}(0)\,\exp(-iLt) \tag{5.17}$$

Here, the Hermitian property of $\underline{\underline{\alpha}}$ has been used.

From (5.10), (5.7) and (5.17) one obtains

$$F(\omega) = -\pi^{-1}\,\mathrm{Im}\,\mathrm{Tr}\{\underline{\underline{\alpha}}(0)(\omega-L)^{-1}\rho\underline{\underline{\alpha}}(0)\} \quad . \tag{5.18}$$

This formal expression is still quite general and it is applicable to solids, liquids, and gases; but further progress is not possible in general. However, by making proper approximations, depending on the system, one can study the spectral function (5.18) in more detail.

We now condider the specific case in which the thermal bath consists of the foreign atoms or molecules. The total Hamiltonian can be written as

$$H = H_0^{(s)} + H_0^{(b)} + V \tag{5.19}$$

where $H_0^{(s)}$ and $H_0^{(b)}$ are the unperturbed Hamiltonians of the system (scattering molecule) and the bath (perturbers), respectively, and V contains all the intermolecular interactions (bath-system and bath-bath). The system-system correlations are neglected. The resulting Liouville operators are defined as

$$L = L_0^{(s)} + L_0^{(b)} + L_1$$
$$L_0^{(s)}X = \hbar^{-1}\left[H_0^{(s)},X\right]$$
$$L_0^{(b)}X = \hbar^{-1}\left[H_0^{(b)},X\right]$$

and

$$L_1X = \hbar^{-1}[V,X] \quad . \tag{5.20}$$

The operators $L_0^{(s)}$ and $L_0^{(b)}$ are diagonal in the unperturbed system and bath states, respectively. Thus their matrix elements can be written as

$$\left(L_0^{(s)}\right)_{if,i'f'} = \hbar^{-1}(E_i-E_f)\delta_{ii'}\delta_{ff'} = \omega_{if}\delta_{ii'}\delta_{ff'} \tag{5.21}$$

and

$$\left(L_0^{(b)}\right)_{ab,a'b'} = \hbar^{-1}(E_a-E_b)\delta_{aa'}\delta_{bb'} - \omega_{ab}\delta_{aa'}\delta_{bb'} \tag{5.22}$$

where i, f, i', and f' represent the eigenstates of the $H_0^{(s)}$ and a, b, a', and b' represent the eigenstates for $H_0^{(b)}$.

With the assumption that no initial correlation exists between the system and bath molecules one can write the density matrix of the sample gas as a product of the density matrices of the system and bath molecules

$$\rho = \rho^{(s)}\rho^{(b)} \quad . \tag{5.23}$$

This assumption allows one to separate, at least formally, the bath variables in (5.18), and write the trace only over the system variables, provided one also assumes that the distribution of bath molecules among the eigenstates of $H_0^{(b)}$ is constant in time, i.e.

$$\rho_{aa'}^{(b)} = \rho_a^{(b)}\,\delta_{aa'} \tag{5.24}$$

(Note that the operator $\underline{\underline{\alpha}}$ acts only on the system variables). Thus one can write the spectral function (5.18) as [5.29]

$$F(\omega) = -\pi^{-1} \text{ Im } Tr_S\left\{\underline{\underline{\alpha}}(0)\left[\omega - L_0^{(s)} - \bar{R}(\omega)\right]^{-1} \rho^{(s)} \underline{\underline{\alpha}}(0)\right\} \qquad (5.25)$$

where Tr_S stands for the trace over the system variables and $\bar{R}(\omega)$ for the relaxation operator (matrix) which is averaged over the bath variables

$$\bar{R}(\omega) = Tr_b\{\rho^{(b)} R(\omega)\} \quad . \qquad (5.26)$$

Neglect of initial correlations is a standard approximation, and its effect on spectral shapes has been discussed, e.g., in [5.31,38]. Moreover, effects such as collision induced and nonlinear phenomena are excluded in making this approximation.

The relaxation operator, $\bar{R}(\omega)$ in (5.26) is still an operator with respect to the system coordinates. It is determined by the interaction, V or L_1, and contains information about the dynamics of the system molecule [5.29]. We will not write down the full expression here. However, it should be pointed out that $\bar{R}(\omega)$ still includes most of the many-body collision and memory effects. In the binary collision regime, the ω-dependence represents the duration of a collision effect . Therefore, (5.25) still contains the impact as well as the two body static limits [5.38]. In recent years, the duration of collision effects have been investigated in atomic spectra [5.31,46,47]. In general, memory effects should be important in dense gases and liquids.

5.3 Impact Theory

Impact limit is the most commonly used approximation in the theory of spectral shapes. It is usually valid for low density gases; but surprisingly, it is also applicable, in some cases, to dense gases and liquids (sometimes, impact theory is used simply because of want of anything better!).

The two main assumptions of impact limits are:

1) Binary collision regime (i.e., $\tau_c \gg \tau_d$, where τ_c is the average time between the collisions, and τ_d is the average duration of a collision).

2) Instantaneous collisions i.e., τ_d is small compared to the characteristic time of the relevant radiative process (absorption, emission or scattering of light) . In this limit, $\bar{R}$ is linear in the density of perturbers, N, and its ω-dependence is negligible [5.26-29,38]. In addition, the classical path approximation is usually valid for the translational motion of molecules. In that case, $\bar{R}$ can be expressed in terms of S, the semi-classical two-body scattering matrix as follows [5.6a,21-23,38]

$$\overline{R}_{i'f',if} = -i2\pi N \int_0^\infty dv\ vf(v) \sum_a \rho_a \int_0^\infty db\ b <a|[\delta_{i'i}\delta_{f'f}$$
$$-<i'|S(b)|i><f'|S(b)|f>^*]|a> \tag{5.27}$$

where v is the relative velocity of the colliding molecules, f(v) the Boltzmann velocity distribution function, N the density of the perturber, and b the impact parameter. The internal states $|i>$ and $|f>$ (or $|i'>$ and $|f'>$) are the initial and final radiative states of the system connected by the operator $\underline{\underline{\alpha}}$, whereas $|a>$ denotes an internal state of a perturber. More general, quantum mechanical, forms for $\overline{R}$ are given in [5.27-29,38,48]; these expressions involve the quantum two body collision operators.

Thus the impact theory result for the spectral function is given by

$$F(\omega) = -\pi^{-1}\ \mathrm{Im}\left\{\sum_{\substack{if\\i'f'}} <i'|\underline{\underline{\alpha}}(0)|f'>^*\left[\omega-L_0^{(s)}-\overline{R}\right]^{-1}_{i'f',if}\ \rho_i^{(s)}<i|\underline{\underline{\alpha}}(0)|f>\right\} \quad . \tag{5.28}$$

This expression is valid for both isolated and overlapping lines. In the case of isolated lines, the collisions do not mix different transitions. Therefore, the total spectrum consists of a superposition of Lorentzian lines without any interference. That is

$$\overline{R}_{i'f',if} = \overline{R}_{if}\delta_{ii'}\delta_{ff'} \tag{5.29a}$$

and (5.28) reduces to

$$F(\omega) = -\pi^{-1}\ \mathrm{Im}\left\{\sum_{if} \rho_i^{(s)}|<i|\underline{\underline{\alpha}}(0)|f>|^2(\omega-\omega_{if}-\overline{R}_{if})^{-1}\right\} \tag{5.29b}$$

where the real and imaginary parts of $\overline{R}_{if}$ determine, respectively, the shift and the width of the line $|i>\rightarrow|f>$. This case will be further discussed in Sect.5.3.

The general case of overlapping lines is quite complicated because of the presence of the inverse matrix $\left[\omega-L_0^{(s)}-\overline{R}\right]^{-1}$. This will be discussed further in Sect.5.4. However, overlapping of lines due to the degeneracy of the magnetic sublevels, can be treated easily by transforming (5.28) into a rotationally invariant form which is discussed below.

In general, the internal states of the system (scattering molecule) can be written in terms of the electronic and vibrational-rotational wave functions. Since the electronic states usually do not change in the non-resonant Raman scattering, we express the internal states just by $|vjm>$ where v, j, and m are the vibrational, rotational, and magnetic quantum numbers, respectively. In the absence of external fields, the states with different m values for a given value of v and j are degenerate, since the total Hamiltonian (5.19) is invariant under any rotation provided the sample is isotropic and no external field is present. For such a system, one can write the rotationally invariant form of (5.28) as [5.23,32,48]

$$F(\omega) = -\pi^{-1}\,\mathrm{Im}\left\{\sum_{\substack{kj_i j_f\\ j_i' j_f'}} \langle v_i j_i' || \underline{\underline{\alpha}}^{(k)}(0) || v_f j_f' \rangle^* \times\left[\omega - L_0^{(s)} - \overline{R}^{(k)}\right]^{-1}_{v_i j_i' v_f j_f', v_i j_i v_f j_f} \rho_{j_i}^{(s)} \langle v_i j_i || \underline{\underline{\alpha}}^{(k)}(0) || v_f j_f \rangle \right\} \qquad (5.30)$$

where the polarizability tensor $\underline{\underline{\alpha}}(0)$ has been resolved into its irreducible components of rank k, and $\langle v_i j_i || \underline{\underline{\alpha}}^{(k)}(0) || v_f j_f \rangle$ represents the reduced matrix element. Since no overlapping of vibrational bands is expected, only the matrix elements of $\overline{R}$ with $v_i' = v_i$ and $v_f' = v_f$ appear in (5.30). It is also clear from (5.30) that there is no interference between the spectrum obtained from different irreducible components of $\underline{\underline{\alpha}}$. For Raman scattering, the allowed values of k are: 0, 1 and 2. For most systems of interest, however, the component of $\underline{\underline{\alpha}}$ with $k = 1$ is negligible. On the other hand, if $\underline{\underline{\alpha}}$ is replaced by the electric dipole moment, $\underline{\mu}$, there is only one value, $k = 1$, allowed. In that case (5.30) gives the spectral function for the electric dipole absorption and emission processes. Therefore, normally, the infrared and Raman isolated transitions are mutually exclusive. However, for bands in dense systems, the normal selection rules are sometimes broken because of the overlap of many lines.

The invariant form of the spectral function, (5.30), is quite useful in the sense that it has separate non-interfering contributions from $\underline{\underline{\alpha}}^{(0)}$ and $\underline{\underline{\alpha}}^{(2)}$. The first component (k=0) gives the isotropic scattering and the second one (k=2), the anisotropic scattering. The reduced matrix element of $\underline{\underline{\alpha}}^{(0)}$ is non-zero only when $j_i = j_f$, i.e., when $\Delta j = 0$. In general, there are two parts of $\underline{\underline{\alpha}}^{(0)}$: one is diagonal in v, i.e., $v_i = v_f$, and the other has non-vanishing value only for $v_i \neq v_f$. The first part gives rise to the isotropic Rayleigh scattering (Brillouin scattering also emerges if system-bath correlations are retained [5.7]). The second part yields the isotropic Raman Q-branch. Similarly, $\underline{\underline{\alpha}}^{(2)}$ has non-vanishing matrix elements only when $j_i = j_f$ or $j_i = j_f \pm 2$. The component of $\underline{\underline{\alpha}}^{(2)}$, which is non-zero for $\Delta v = 0$, yields the depolarized Rayleigh scattering [5.7,8] for $\Delta j = 0$, and pure rotational Raman lines (S- and O-branches) for $\Delta j = \pm 2$. The part which is non-vanishing for $\Delta v \neq 0$ gives rise to the anisotropic Raman Q-branch ($\Delta j=0$) and S- and O-branches ($\Delta j=\pm 2$).

So far we have discussed only the case of foreign broadening. However, the results can be generalized for self-broadening by taking into account the excitation transfer processes [5.38]. In that case there are two kinds of contributions to the relaxation matrix; one results from collisions in which the excitation is transferred from one molecule to the other, and the other contribution results from the direct collisions. The related problem of resonance broadening and shift in atomic absorption has received considerable attention during recent years [5.49-52]. For example, it has been pointed out that the excitation transfer contribution to $\overline{R}$ is related to the Lorentz-Lorenz local field correction [5.49,52]. Very recently, the theory has been extended to dense gases [5.53].

5.4 Isolated Lines

The spectral function for isolated lines is given by (5.29b). For the line $|i\rangle \to |f\rangle$, the two validity conditions for the impact theory reduce to: 1) $\tau_c \gg \tau_d$, and 2) $|\omega-\omega_{if}|^{-1} \gg \tau_d$. Because of the condition 2), $\overline{R}^{(k)}$ in (5.30) is independent of ω. Thus the isolated lines would have Lorentzian shapes and their widths (half width at half maximum) and shifts (in wave number units) are given by

$$\gamma_{if}^{(k)} = -(2\pi c)^{-1}\,\mathrm{Im}\left\{\overline{R}_{if}^{(k)}\right\} \tag{5.31}$$

and

$$\Delta_{if}^{(k)} = (2\pi c)^{-1}\,\mathrm{Re}\left\{\overline{R}_{if}^{(k)}\right\} \tag{5.32}$$

respectively, where c is the speed of light in vacuum.

5.4.1 Calculation of Line Widths and Shifts

Within the impact theory and classical path approximation, the relaxation matrix, in rotationally invariant form, can be written as [5.21,22,54]

$$\overline{R}_{if}^{(k)} = -i2\pi\, N \int_0^\infty dv\, vf(v) \sum_a \rho_a \int_0^\infty db\, b\underset{\sim}{S}_a(b) \quad , \tag{5.33a}$$

$$\begin{aligned}\underset{\sim}{S}_a(b) = &\langle a|[1-(2j_f+1)^{-1} \textstyle\sum C(j_i k j_f; m_i m m_f)\; C(j_i k j_f; m_i' m m_f')\\ &\times \langle v_i j_i m_i'|S(b)|v_i j_i m_i\rangle\langle v_f j_f m_f'|S(b)|v_f j_f m_f\rangle^*]|a\rangle\end{aligned} \tag{5.33b}$$

where C is a Clebsch-Gordan coefficient, and the other quantities are the same as in (5.27). The summation here is over all the magnetic quantum numbers. The semi-classical S-matrix is given by

$$S(b) = T\exp[-i\hbar^{-1}\int_{-\infty}^{\infty} dt\; V_I(b,t)] \tag{5.34}$$

where T is the Dyson time-ordering operator and $V_I(b,t)$ is the interaction representation of the semi-classical potential evaluated along a classical trajectory. In the quantum expression, the semi-classical S-matrix is replaced by the quantum mechanical S-matrix [5.38,48].

In general, an exact evaluation of the S-matrix is quite difficult. Therefore, several approximate methods have been developed. These methods vary from fully quantum to purely classical treatments, depending on the system under study. For example, in the case of lighter molecules and atoms, such as H_2 perturbed by He, the quantum mechanical effects are expected to be relatively more important, and a fully quantum treatment would be desirable. Moreover, for long range potentials, a perturbation approach may be valid, whereas a non-perturbative calculation may be required for short range interactions. We will discuss the two cases separately.

a) *Non-Perturbative Approaches*

A quantum calculation of the S-matrix, for H_2 perturbed by He, was performed by SHAFER and GORDON [5.48], using the close coupling scheme. This requires solutions of coupled equations of motion. Comparing their results with the measured Raman line widths [5.55-57] and other data, they deduced a H_2-He interaction potential; it will be discussed in Sect.5.4.2.

A detailed quantum calculation becomes quite time consuming if many rotational levels are involved. The computational problem is especially difficult for molecule-molecule collisions. To reduce the dimensionality of the coupled equations of motion, other approximate schemes have been proposed. For example, RABITZ and co-workers [5.58] have developed the "effective Hamiltonian method", for quantum or semi-classical calculations. In this approach, essentially an averaging is done over the rotational projection quantum numbers before calculating the S-matrix. Recently, GOLDFLAM and KOURI [5.59] have developed a theory for relaxation phenomena in gaseous mixtures, using the "centrifugal decoupling (averaged j_z-conserving coupled states) approximation", and used it to analyze the Raman and other data for H_2 perturbed by He [5.60]. NEILSEN and GORDON [5.61a] have solved the Schrödinger equation numerically, using the classical trajectories for the translational motion, and applied it to the Raman and other data for HCl perturbed by Ar. An alternative semi-classical approach has been used by FITZ and MARCUS for CO, HCl, and OCS [5.61b]. Another classical path approach has been developed by SMITH et al. [5.62a]. Using this approach, SMITH and GIRAUD [5.62b] analyzed the rotational line widths in HCl perturbed by Ar, and concluded that the linewidths are not sufficiently sensitive to the details of the anisotropies of the interaction, to be able to deduce a unique potential function from a set of line broadening data.

Another non-perturbative approach is the purely classical one, proposed by GORDON [5.24]; the rotational motion is also treated classically. This approximation is expected to be valid for high j-values; its application will be discussed in Sect. 5.5. The non-perturbative methods have been used mainly for atom-molecule collisions. A general survey is given in [5.6a]. For a comparison of different theories, see e.g.,

[5.63]. For molecule-molecule collisions, most of the calculations so far, have used perturbation theory for determining the S-matrix. A discussion of such perturbative methods will be taken up now.

b) *Perturbative Approaches*

In the traditional approach, the S-matrix is expanded in powers of the interaction potential ($S(b)=1+S^{(1)}(b)+S^{(2)}(b)+\ldots$), and the classical trajectories are used for the translational motion. Usually, the expansion is truncated at the second order term, and the classical trajectory is approximated by a straight line path (the latter approximation is expected to be valid for heavier molecules although it has been used even for lighter molecules such as H_2-H_2, because of its simplicity). These approximations cause S(b) to diverge at small impact parameter, b. To avoid this problem, several procedures have been proposed where either $\underset{\sim}{S}_a(b)$ is approximated by a simpler form, or some kind of cut off is used. The most frequently used procedure is by ANDERSON [5.21, approximation #2] where $\mathrm{Re}\{\underset{\sim}{S}_a(b)\}$ assumes the following form

$$\mathrm{Re}\{\underset{\sim}{S}_a(b)\} = \begin{cases} 1 & b \leqq b_0 \quad ; \quad \underset{\sim}{S}_a^{(2)}(b_0) = 1 \quad \text{and} \quad b_0 > d_0 \\ \underset{\sim}{S}_a^{(2)}(b) & b \geqq b_0 \end{cases} \tag{5.35}$$

The superscript on $\underset{\sim}{S}_a^{(2)}(b)$ in the above equation implies the contributions from the second order term only and d_0 represents the kinetic diameter. In general $\underset{\sim}{S}_a^{(2)}(b)$ would be complex; however, within the approximation that the interactions commute at different times [5.21], it is real. This latter approximation is fairly good for the width calculations, but it eliminates the shift [5.64]. The shift is determined by $\mathrm{Im}\{\underset{\sim}{S}_a\}$, according to (5.32). For the case where $b_0 < d_0$, VAN KRANENDONK's [5.65] approach has been commonly used; here the condition (5.35) is replaced by

$$\mathrm{Re}\{\underset{\sim}{\bar{S}}(b)\} = \begin{cases} \underset{\sim}{\bar{S}}^{(2)}(d_0) & ; \quad b \leqq d_0 \\ \underset{\sim}{\bar{S}}^{(2)}(b) & ; \quad b \geqq d_0 \end{cases} \tag{5.36}$$

where

$$\underset{\sim}{\bar{S}}^{(2)}(b) = \sum_a \rho_a \, \underset{\sim}{S}_a^{(2)}(b) \tag{5.37}$$

Here again $\underset{\sim}{S}_a^{(2)}(b)$ is real, as in (5.35). Therefore, only the line broadening can be calculated in the Anderson or Van Kranendonk theories. We will consider the calculation of the line shift later. Following FIUTAK and VAN KRANENDONK [5.54], one can write $\underset{\sim}{\bar{S}}^{(2)}(b)$ as

$$\underset{\sim}{\bar{S}}^{(2)}(b) = \sum_{j_a} \rho_a \Bigg[\frac{1}{2} \sum_{m_i m_a} \frac{\langle j_i m_i j_a m_a | P^2 | j_i m_i j_a m_a \rangle}{(2j_i+1)(2j_a+1)} + \frac{1}{2} \sum_{m_f m_a} \frac{\langle j_f m_f j_a m_a | P^2 | j_f m_f j_a m_a \rangle}{(2j_f+1)(2j_a+1)}$$

$$- \sum_{\text{all } m\text{'s}} C(j_i k j_f; m_i m m_f)\, C(j_i k j_f; m_i' m m_f')$$

$$\times \sum_{j_a'} \frac{\langle j_i m_i j_a m_a | P | j_i m_i' j_a' m_a' \rangle \langle j_f m_f j_a m_a | P | j_f m_f' j_a' m_a' \rangle}{(2j_f+1)(2j_a+1)} \Bigg] \tag{5.38}$$

$$P = \hbar^{-1} \int_{-\infty}^{\infty} dt\, \exp[iH_0 t/\hbar]\, V(t)\, \exp[-iH_0 t/\hbar] \tag{5.39}$$

where H_0 is the *internal* Hamiltonian of the colliding pair, and $|j_a m_a\rangle$ represents the internal state of the perturber. Further progress requires the specification of $V(t) = V[\underline{r}(t)]$, the intermolecular potential along the classical trajectory, $\underline{r} = \underline{r}(t)$, for the relative motion of the colliding pair.

In general, V contains induction, dispersion, and overlap interactions along with the multipolar terms such as dipole-dipole, dipole-quadrupole, etc. To illustrate the calculation of line width, let us consider a simple case: collision of axially symmetric molecules, where one multipole term is dominant (generalization will be considered later). The multipole interaction $V_{\ell_1 \ell_2 \ell}$ of order ℓ_1, ℓ_2 and ℓ ($\ell = \ell_1 + \ell_2$ for multipoles, but for induction, dispersion, and overlap forces, this does not hold necessarily [5.66]) can be written as

$$V_{\ell_1 \ell_2 \ell}(\underline{r}) = 4\pi\varepsilon(\ell_1 \ell_2 \ell)\, r^{-(\ell+1)} \times \sum_{m_1 m_2 m} C(\ell_1 \ell_2 \ell; m_1 m_2 m)\, Y_{\ell_1 m_1}(\Omega_1)\, Y_{\ell_2 m_2}(\Omega_2)\, Y^*_{\ell m}(\Omega) \tag{5.40}$$

where

$$\varepsilon(\ell_1 \ell_2 \ell) = (-1)^{\ell_2} \{4\pi(2\ell+1)!/[(2\ell_1+1)!(2\ell_2+1)!]\}^{1/2} Q_{\ell_1} Q_{\ell_2}/(2\ell+1) \quad . \tag{5.41}$$

The orientations of the internuclear axes of molecule 1 and 2 are represented by Ω_1 and Ω_2, respectively, and the intermolecular separation, $\underline{r} \equiv (r,\theta,\phi) = (r,\Omega)$, has an orientation Ω, with respect to a space fixed coordinate system; Q_ℓ is the scalar multipole moment of order ℓ. The matrix elements of operator P have been calculated by GRAY and VAN KRANENDONK [5.67] for a general multipole potential; substitution in (5.38) yields

$$\underset{\sim}{\bar{S}}^{(2)}(b) = \sum_{j_a} \rho_a\, \lambda(a,v,\tau)/b^m \quad ; \quad m = 2\ell \tag{5.42}$$

$$\lambda(a,v,\tau) = (\eta^2 v^{-2}) \sum_{j_a'} C(j_a \ell_2 j_a';000)^2 \Big\{ \sum_{j_i'} C(j_i \ell_1 j_i';000)^2 g_\ell(|\omega'\tau|)$$

$$+ \sum_{j_f'} C(j_f \ell_1 j_f';000)^2 g_\ell(|\omega'\tau|)$$

$$-2[(2j_i+1)(2j_f+1)]^{1/2}(-1)^{j_i+j_f-k-\ell_1} W(j_i j_f j_i j_f;k\ell_1)$$

$$\times C(j_i \ell_1 j_i;000) C(j_f \ell_1 j_f;000) g_\ell(|\omega'\tau|) \Big\} \qquad (5.43)$$

$$g_\ell(|\omega'\tau|) = 2(|\omega'\tau|/2)^{2\ell}(2\ell)![(\ell-1)!]^{-2} \sum_{\mu=-\ell}^{\ell} \{(\ell+\mu)!(\ell-\mu)!\}^{-1} K_\mu^2(|\omega'\tau|) \qquad (5.44)$$

$$\omega' = 2\pi c\{B[j'(j'+1)-j(j+1)]+B_a[j_a'(j_a'+1)-j_a(j_a+1)]\} \qquad (5.45)$$

$$\eta = 2^\ell(\ell-1)!Q_{\ell_1}Q_{\ell_2}/\{\hbar[(2\ell_1+1)!(2\ell_2+1)!]^{1/2}\} \qquad (5.46)$$

where W is a Racah coefficient, $\tau = b/c$, and K_μ is a modified Bessel function. The constants B and B_a are the rotational constants of the scatterer and perturber, respectively, in the respective rotational-vibrational states, and τ represents the duration of a collision. It should be pointed out that the argument of $g_\ell(|\omega'\tau|)$ in (5.43) is j-dependent. If there are more than one term in the potential, the line width may not necessarily be the sum of the contributions from the different terms. Interference terms may also arise depending on the orders of the potentials [5.68]. The function $g_\ell(|\omega'\tau|)$ is commonly known as resonance function and has been tabulated by TSAO and CURNUTTE [5.22] for $\ell = 2, 3, 4$.

Now, within the framework of this theory, we can make some general comments about the line width contribution of various collision processes arising from different potentials. For example, Clebsch-Gordan coefficients in (5.43) yield the following selection rules for the transitions due to collisions: 1) $j_i + j_i' + \ell_1$, $j_f + j_f' + \ell_1$, and $j_a + j_a' + \ell_2$ should be even; 2) $(j_i j_i' \ell_1)$, $(j_f j_f' \ell_1)$, and $(j_a j_a' \ell_2)$ should satisfy the triangular relation, e.g., $j_i' = j_i + \ell_1, \dots |j_i - \ell_1|$. This implies that the collisions which are elastic ($j_i = j_i'$ or $j_f = j_f'$) with respect to the scatterer, do not contribute to the broadening, in general, if ℓ_1 is odd. In fact, the elastic term contains both the phase shift and reorientation effects. For isotropic (k=0) lines in a Q-branch ($j_i = j_f$), these contributions are absent and only the inelastic collisions contribute to the width. But for anisotropic lines (k=2) of the Q-, S- and O-branches ($\Delta j = 0, \pm 2$) both elastic and inelastic contributions are non-zero. This effectively makes the anisotropic lines broader than the isotropic ones. Note that (5.43) reduces to the Anderson result when $k = 1$; normally, this applies to absorption and emission spectra only.

The AFVK (Anderson-Fiutak and Van Kranendonk) approach, outlined above, is reasonable if effects of the short range interaction are small or negligible. SRIVASTAVA and ZAIDI [5.69] have recently investigated the effects of short range interaction on collision broadening using impact theory and classical straight path approximations. In this approach, the total interaction is assumed to be: $V(\underline{r}) = V_0(r) + V_1(\underline{r})$, where $V_0(r) = A/r^n$ is a short range term (core potential), and $V_1(\underline{r})$ is the long range (multipolar) part. The core potential is treated to all orders whereas contributions up to second order in V_1 are retained. This yields two kinds of contributions to $\mathrm{Re}\{\underset{\sim}{S}_a(b)\}$ in (5.33b): one arises from the 'core' potential only, and the other is the result of interference between the core and the long range term. Both of these terms give a finite result after b-integration in (5.33a). Thus inclusion of a core potential with the multipolar part, and treating it to all orders, eliminates the problem of divergence at the lower limit of b. A similar approach was adopted by JARECKI and HERMAN [5.70] in the study of infrared line width and shift, but they employed a cut off procedure. In the absence of V_0, the first term vanishes, and the second one reduces to (5.42) if $V_1(\underline{r})$ is given by (5.40).

The function λ, given by (5.43), is usually a slowly varying function of τ; therefore it can be replaced by $\bar{\tau}$, the average duration of a collision. Then (5.42) gives

$$\bar{\underset{\sim}{S}}^{(2)}(b) = \sum_{j_a} \rho_a\, \lambda(a,v,\bar{\tau})/b^m \quad . \tag{5.47}$$

Now the total width arising from V_0 and V_1 can be written as [5.69]

$$\gamma = \sum_{j_a} \rho_a \int_0^\infty dv\; f(v)\Delta\tilde{\nu}(a,v) \tag{5.48}$$

where

$$\begin{aligned}\Delta\tilde{\nu}(a,v) = (Nv/2c)\Big\{&\alpha^{2/(n-1)}\Gamma[(n-3)/(n-1)]\,\sin[(n-3)\pi/2(n-1)] \\ &+[2\lambda/(m-2)]\alpha^{-(m-2)/(n-1)}\Gamma[1+(m-2)/(n-1)]\,\cos[(m-2)\pi/2(n-1)]\Big\}\end{aligned} \tag{5.49}$$

and m must satisfy the condition

$$0 < m - 2 < n - 1 \quad . \tag{5.50}$$

In (5.49) α represents the phase shift produced by the core potential [5.69] and Γ is the gamma function. All the information about the internal states is contained in λ. The indices i, j, and k have been suppressed in (5.48).

The first term in (5.49) represents the direct 'core' contribution, and the second one is the interference term. In practice, both the terms are not always important. Therefore, one can distinguish between two cases.

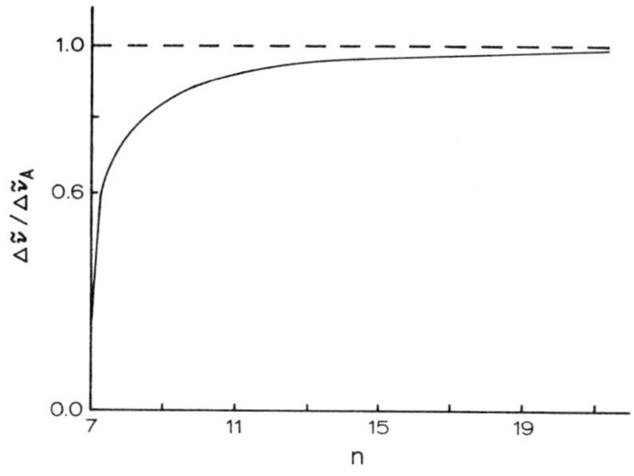

Fig. 5.1. Variation of the ratio $\Delta\tilde{\nu}/\Delta\tilde{\nu}_A$ with the steepness parameter, n, of the core potential for m = 8 (quadrupole-quadrupole interaction) where $\Delta\tilde{\nu}_A$ is the Anderson-Fiutak and Van Kranendonk result. (From [5.69])

Case a: If only the interference effect of the core is important, the phase shift parameter, α, can be eliminated through a self-consistent minimization procedure. This minimization represents a generalization of the Anderson procedure [5.69] and it reduces (5.49) to

$$\Delta\tilde{\nu}(a,v) = (Nv/2c)[m/(m-2)]\{\Gamma[(n-3)/(n-1)]\ \sin[(n-3)\pi/2(n-1)]\}^{(m-2)/m} \times \{\Gamma[1+(m-2)/(n-1)]\ \cos[(m-2)\pi/2(n-1)]\lambda\}^{2/m} \quad . \tag{5.51}$$

In the limit $n\to\infty$, (5.51) gives the same result as the AFVK theory [5.69]. Thus the above minimization method leads to a gradual, n-dependent, cut off. A comparison of $\Delta\tilde{\nu}$ for m = 8 (quadrupole-quadrupole interaction) with the AFVK result, $\Delta\tilde{\nu}_A$, is presented in Fig.5.1. It shows that the AFVK result is the upper limit of (5.51), as $n\to\infty$.

Case b: If the direct contribution from the core is also important, a more suitable form of (5.49) can be obtained by replacing the first term by $Nvd^2/2c$, where d is the effective collision diameter of the core, for the broadening collisions. (In general, $d \neq d_0$, the kinetic collision diameter.) Thus one obtains [5.69]

$$\Delta\tilde{\nu} = (Nvd^2/2c)\Big[1+[2\lambda/(m-2)]d^{-m}\Gamma[1+(m-2)/(n-1)] \times \cos[(m-2)\pi/2(n-1)]\{\Gamma[(n-3)/(n-1)]\ \sin[(n-3)\pi/2(n-1)]\}^{(m-2)/2}\Big] \quad . \tag{5.52}$$

Again, the AFVK result is the upper limit of (5.52), as $n\to\infty$.

To summarize, (5.51) and (5.52) represent two possible generalizations of the AFVK theory, if the effects of the core potential are appreciable. The core potential, V_0, represents an average of the actual short range interaction. Such an approach is useful, because very little is known about core potentials. Therefore, (5.51) and (5.52) can be used to determine the average core potential parameters, through comparison with experiments [5.71]; the details will be discussed in Sect. 5.4.2.

MORAAL [5.72] has also derived expressions for the widths of the rotational Raman lines of homonuclear diatomic molecules, using the distorted wave Born approximation. He considered an anisotropic short range potential together with the quadrupole-quadrupole interaction. He has also analyzed the recent data on H_2 [5.73,74]. Another approach was used by BONAMY et al. [5.39c], who have calculated the line widths and shifts in HD using curved collision trajectories. This work will be discussed in Sect.5.5. RABITZ and GORDON [5.75] have used the straight path approximation to derive general expressions for the first and second order contributions to the S-matrix for multipolar potential; the results were applied to broadening in HCN. Some investigators [5.6,76] have assumed special forms of $\mathrm{Re}\{\underset{\sim}{S}_a^{(2)}(b)\}$ such that the divergence at low b is eliminated; but such approaches neglect the effect of the *actual* short range interaction which seems to be important in many cases.

c) *Line Shift*

From (5.32) it follows that the shift is determined by the imaginary part of the S-matrix. An example of a non-perturbative quantum calculation is the work by SHAFER and GORDON [5.48] on Raman line shifts for H_2 perturbed by He. The interaction potential deduced from a comparison of the theory and experiment will be discussed in Sect.5.4.2.

For molecule-molecule collisions, only perturbative approaches have been used. In the second order perturbation expansion of the S-matrix, the shift could arise from the first order as well as the second order terms. However, the first order contribution vanishes for anisotropic potentials because of the averaging over orientations [5.23b]. On the other hand, the first order contribution of an isotropic potential does not vanish if the potential is different in the two vibrational states [5.55,77-80]. The second order term is quite complicated because of the time ordered integrals. Several investigators have evaluated it [see, e.g., 5.23b,39c,70,75,81]. In most cases, the divergence problem is encountered, and the usual cut off procedure is employed. However, this problem can be avoided if more realistic collision trajectories are used [5.39c].

Most of the reported works on shifts are for infrared lines. Raman line shifts have been reported for H_2 [5.44,55,77,85], HD [5.84], and D_2 [5.44]. Theoretical work and the analysis of the data on H_2 line shifts have been discussed in [5.4]. The only recent theoretical work which has been compared with the experiment is on HD [5.39c]. This work will be discussed in Sect.5.5.

In general, the line shifts are more sensitive to the short range isotropic part of the intermolecular potential compared with the line widths. Therefore, a combined investigation of the line widths and shifts can provide more complete information about the potentials.

5.4.2 Comparison of Theory with Experiment and Information Obtained About Intermolecular Potentials

Here we will consider selected cases in which information about the interaction potential has been obtained from a comparison of measured Raman line widths and shifts with the theoretical calculations. The emphasis will be on the recent works, reported since the last review [5.4]. Even then, a discussion of all the works is not possible. However, we have tabulated the relevant works, experimental as well as theoretical, in Table 5.1.

For H_2 perturbed by He, SHAFER and GORDON [5.48] calculated the S-matrix quantum mechanically, and compared the results with the relevant data on Raman line widths and shifts, sound absorption, and NMR. They concluded that the following potential agrees with experimental data as well as with the theoretical calculations of the interaction:

$$V(\underline{r}) = V_0(r) + V_2(r)P_2(\cos\theta) = \sum_{i=0,2} V_i(\underline{r})P_i(\cos\theta)$$

where

$$V_i(\underline{r}) = \begin{cases} \varepsilon A_i\{\exp[2\beta(1-r/r_m)]-2\gamma_i \exp[\beta(1-r/r_m)]\} \quad , & r \leqq r_1 \\ \sum_{n=0}^{3} a_n r^n \quad , & r_1 \leqq r \leqq r_2 \\ -B_i\left(c_8/r^8+c_6/r^6\right) \quad , & r_1 \geqq r_2 \end{cases} \qquad (5.53)$$

The values of all the potential parameters are given in [5.48]; the parameters a_n were obtained by matching V_i and dV_i/dr at r_1 and r_2.

The self broadening and shift of H_2 lines has also been investigated in some detail [5.55-57,13,14,73,74,86-92]. The shift was discussed in [5.4], and there is no new calculation since then. KEIJSER et al. [5.73,74] have reported the widths for the largest number of lines, the first five members of the S-branch. The results exhibit sharp variations of the widths with the rotational quantum number, j ($j=j_i$, the initial state quantum number). The lines with odd j-values have relatively higher widths than the ones with even j-values. This is due to the dominant contribution from the resonant collisions. Though these collisions contribute to all the lines, they are more significant for $j=1$ and $j=3$ states because more than two-thirds of the molecules at room temperature are in the $j=1$ state. KEIJSER et al. [5.74] also studied the variation of width with ortho-para concentration and found the results to agree with the above explanation. No alternation of width was present in the theoretical calculation by VAN KRANENDONK [5.65] for H_2 where only the quadrupole-quadrupole interaction was considered, and the relative velocity spread was neglected. KEIJSER et al. [5.74] applied the above approach to ortho and para H_2 and found a

Table 5.1. List of experimental and theoretical works on Raman spectral shapes in gases

System	Perturber	Experimental Reference	Theoretical Reference
H_2	H_2	[43,44,55-57,73,74,77, 85a,85b,86,88-92,142]	[44,55,65,71a,72,74,77,87,143]
	He	[55-57,85b]	[48,55,60]
	Ar	[55,56,85b]	[55]
	Ne	[57]	-
	N_2, CH_4	[85b]	-
HD	HD	[57,74,84,108]	[39c,84,108]
D_2	D_2	[44,57,74]	[44]
O_2	O_2	[86,87,88,94,137a,137b]	[67]
	He, Ar	[86,87,94]	-
N_2	N_2	[86,87-89,93,94,107, 110,137b,139,140]	[65,67,71a]
	He	[94]	-
	Ar	[86,87,94]	-
CO	CO	[87,94,107,110,138]	[67,138]
	H_2	[138]	[138]
	Ar, CH_4	[87]	-
CO_2	CO_2	[86,87,94,95,140]	[67,71b]
	He, Ne	[95]	[95]
	Ar	[86,87,95]	[95]
HCl	HCl	[99-101,141]	[23b,97,98]
	DCl	-	[98]
	Ar	[141]	[23b,61a]
	C_2F_6, SF_6	[101]	-
DCl	DCl	[101]	[98]
	HCl	-	[98]
	C_2F_6	[101]	-
HBr	HBr	[101]	-
DBr	DBr, C_2F_6	[101]	-
HF	HF, SF_6	[112]	-
C_2H_2	C_2H_2, He, Ar	[87]	
N_2O	N_2O, Ar	[113]	
CH_4	CH_4	[86,88,142]	

poor agreement with the results. However, by increasing the average duration of collision, used in the resonance function $g_4(\omega'\tau)$, by about 20% the agreement was much better. This effectively increased the resonant contribution and decreased the nonresonant and elastic contributions, yielding a better result. MORAAL [5.72] calculated the line widths for an anisotropic short range potential combined with the quadrupole-quadrupole term, and found a good agreement for high frequency lines: $S_0(2)$, $S_0(3)$, and $S_0(4)$. However, $S_0(0)$ and $S_0(1)$ had large discrepancies which were attributed to the approximate treatment of the anisotropic short range potential.

SRIVASTAVA and ZAIDI [5.71a] also analyzed the H_2 self broadening data, and concluded that the Case b result, represented by (5.52), could also explain it; a core potential with parameters, $d = 0.18$ Å and $n \simeq 7$, together with the long range quadrupole-quadrupole (q-q) term, gave the best fit. Therefore, it appears that the available data is not sensitive to the details of the short range interaction, although the core contribution is essential in this case. Hopefully, data for higher j-values will be able to discriminate between the various forms of the core potentials.

KEIJSER et al. [5.74] have also determined the self broadened widths of the S-branch Raman lines in D_2 and HD. A monotonic decrease in width with increasing j was observed in HD, whereas the D_2 lines showed a less prominent alternation in widths.

The self broadening of rotational Raman lines of the S-branch in N_2 has been investigated by several workers [5.93,65,94,67,71a]. Figure 5.2 shows a comparison of different theoretical approaches with the experimental data. VAN KRANENDONK [5.65]

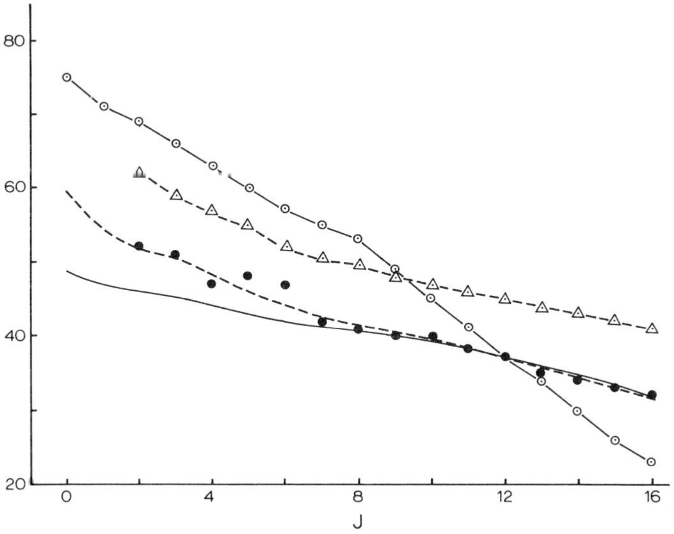

Fig. 5.2. Variation of the broadening constant, <a> (where $\langle a\rangle = \gamma/p$; p is the pressure) for the pure rotational Raman lines in N_2, with the rotational quantum number J, of the initial state. ● Experimental data [5.94], Δ theory [5.65], o theory [5.67], —— theory (case a) and ---- theory (case b) [5.71a]. (From [5.71a])

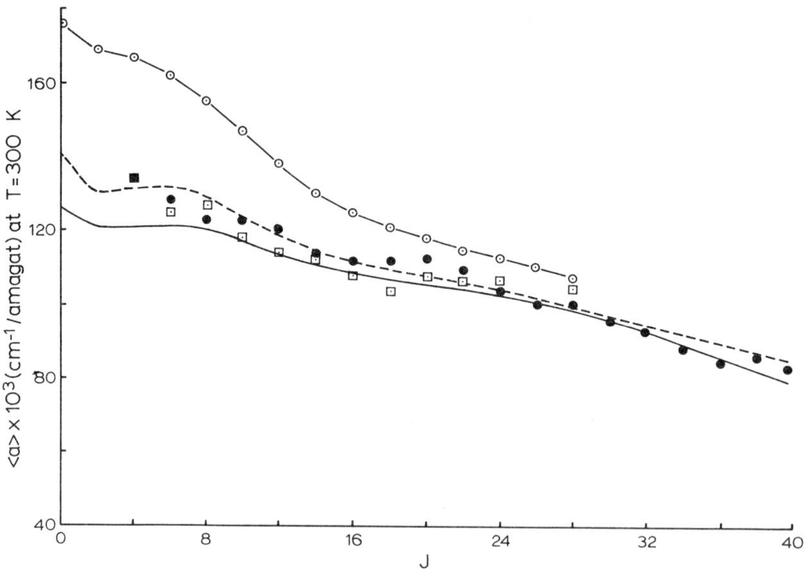

Fig. 5.3. Variation of collision broadened widths of pure rotational Raman lines (S-branch) in CO_2 with the rotational quantum number, J, of the initial state. ● Experimental data [5.95], □ experimental data [5.94], o theory [5.67], —— theory (case a), and --- theory (case b) [5.71b]. (From [5.71b])

calculated the widths using the q-q potential, and neglecting the relative velocity spread. His results were 18-30% higher than the ones obtained experimentally [5.94]. GRAY and VAN KRANENDONK [5.67] calculated the widths using anisotropic dispersive interaction together with the q-q term. Their calculation yielded a steeper j-dependence and the agreement with the data became worse. SRIVASTAVA and ZAIDI [5.71a] using the q-q interaction together with the core potential, showed that their results (both Case a and b) can explain the experimental results much better (see Fig.5.2). In case b, the core potential parameters, $n = 7.6$ and $d = 2.9$ Å, were obtained from the comparison. Thus the core contribution seems to be important for N_2N_2 collisions.

In the case of CO_2 Raman lines (S-branch), the width data [5.94,95] is available for a much larger range of j-values; the calculated results of GRAY and VAN KRANENDONK [5.67], based on the q-q potential and anisotropic dispersive interaction show a poor agreement. SRIVASTAVA and ZAIDI [5.71b], by including a short range potential of the form r^{-n} with the q-q interaction, obtained a much better agreement as shown in Fig.5.3. The resulting core potential parameters for case b are: $n = 8$ and $d = 5.6$ Å. Using the same approach, good agreement was also obtained with the infrared data for CO_2.

The importance of short range interaction in line broadening has also been demonstrated by HERPIN and LALLEMAND [5.95], and, BONAMY and ROBERT [5.96]. These works, however, employ the second order perturbation theory in the total interaction. While trying to interpret their data on broadening of rotational Raman lines (S-branch) perturbed by He, Ne, and Ar, on the basis of the FIUTAK and VAN KRANENDONK [5.23a,54] theory, HERPIN and LALLEMAND [5.95] realized the need of a shorter range interaction compared with the anisotropic London dispersive potential of the form r^{-6}. In the case of CO_2 perturbed by Ne, they showed that a short range potential of the form r^{-14} will explain the data better. BONAMY and ROBERT [5.96] used a detailed short range atom-atom potential, in addition to the multipolar part, for self and foreign gas broadening of CO.

GRAY [5.97,98] has calculated the self-broadened widths of infrared and pure rotational Raman lines in HCl and DCl; the foreign broadening by DCl and HCl, respectively, was also calculated. Fiutak and Van Kranendonk theory and the following four multipole potentials were used in the calculations: dipole-dipole, dipole-quadrupole, quadrupole-dipole and quadrupole-quadrupole. No adjustable parameters were used and the dipole and quadrupole moments were obtained from other sources. The results were compared with available data for HCl [5.99-101]. The agreement is generally good. However, for the interesting region of high j-values, where the width oscillates with j, no experimental data are yet available. The oscillations in widths with j result from the resonant collisions. This effect is also present in the Raman line broadening of DCl perturbed by HCl, but not in the reverse case [5.98]. A similar effect is observed in the H_2 lines, but for low j-values, as mentioned earlier.

To summarize the situation, the linewidths are generally not very sensitive to the details of the short range part of the interaction. However, the j-dependence of the widths furnishes additional detail, which can provide information about an effective core potential, if the long range potential parameters are known. This is specially true if data are available for low as well as high rotational quantum numbers. Of course, the combined analysis of the line shifts and widths can provide more complete information. If a single multipole term is dominant, the line width measurements can determine the multipole moment. However, this procedure is not reliable for higher multipoles, or if short range interactions are also present [5.102].

5.5 Overlapping Lines and Band Shapes

At low pressures, the lines are usually well separated and their widths are proportional to the density. However, when the pressure increases, they start to overlap and finally at high pressure, they collapse into a single band whose width decreases with the density. The narrowing (usually known as "collisional narrowing") is a result of the interference between different coherent contributions from all the individual transitions. The spectral function, in general, is given by (5.25) which reduces to (5.30) in the impact limit. The various approaches for the calculation of the relaxation matrix, $\overline{R}$, have been discussed in detail in Sect.5.4.2. For the overlapping lines, the diagonal as well as the off diagonal $\overline{R}$ matrix elements are required. The band shape is obtained by inverting the non-Hermitian matrix, $\omega-L_0-\overline{R}$. For a large number of lines, this can be a difficult problem. However, GORDON and McGINNIS [5.103] have suggested a method of inversion through diagonalization, which is claimed to be much faster.

Several qualitative discussions [5.104,105,83] have been presented for the overlapping lines, but not many detailed calculations have been reported in the literature. GORDON and McGINNIS [5.106], using the classical treatment of the S-matrix, have calculated the band shape of infrared spectra in CO perturbed by He. Comparing the results with their measurements, they estimated an intermolecular potential for the system. Their calculation yields agreement over a wide range of pressure, except at low pressure where the long range part of the potential used is thought to be too strong. The agreement is good even in high pressure regions, where impact theory is not rigorously applicable. No such calculation has been reported as yet for Raman band shapes.

For the case of molecule-molecule collisions, the non-perturbative calculation of the S-matrix becomes computationally very difficult. Therefore, the perturbative approaches, discussed in Sect.5.4.1, are normally used. Recently, CHERKASOV [5.83] has formally extended the ANDERSON-TSAO-CURNUTTE [5.21,22] formalism to overlapping lines for a general multipole interaction potential; but no specific case was treated in detail.

The experimental data on the collisional narrowing of the vibrational Raman bands in N_2 and CO by MAY et al. [5.107] were analyzed by them in a semiquantitative manner, and the results were summarized in the previous review [5.4]. DION and MAY [5.108] have studied the broadening of Raman Q-branch lines in HD at low and high densities and at room temperature. They qualitatively analyzed their results using the ALEKSEEV and SOBELMAN [5.105] approach. Later WITKOWICZ and MAY [5.84] studied the collisional effects in compressed HD at room temperature and at 85 K. They have also calculated the width of Raman Q-branch lines at low pressures, using the GRAY and VAN KRANENDONK [5.67] approach and a potential which consists of the quadrupole-quadrupole and dispersive interactions. The agreement with experiment is not satisfactory; the calculated widths vary much faster, with increasing j, compared to the experimental results. Their shift data for Q(0) and Q(2) at room temperature show a weak quadratic dependence with density whereas the other shift measurements exhibit a linear dependence. They also analyzed the band shapes at higher pressures, using the ALEKSEEV and SOBELMAN [5.105] theory, but an agreement was obtained only after modifying the theory. This discrepancy was partly blamed on the neglect of the correlated collisions.

Recently, non-impact theories have been used for calculations in gases and liquids [5.39b,39c,109]. BONAMY et al. [5.39c] have employed such a theory for investigation of overlapping lines. (This theory is based on the "cumulant" approach of KUBO [5.39a], which will be discussed in the next section.) The treatment includes the vibrational degrees of freedom and the interaction potential up to second order. The classical path is not assumed to be a straight line; rather, it is approximated by a parabolic trajectory. This improvement eliminates the divergence in the perturbation expansion of the S-matrix.

As an application of the theory, BONAMY et al. [5.39c] analyzed the broadening data [5.84] of isotropic Raman Q-branch lines in HD perturbed by HD. In general, the widths and shifts, calculated at low density, have very good agreement with the experimental results at room temperature as well as at 85 K, except for the shift in the Q(0) line at 300 K. This discrepancy was attributed to the higher order terms in V which have been neglected. The quadratic terms in the gas density were not included in the calculation, although the experimental shifts for Q(0) and Q(2) at 300 K show a small nonlinear dependence. For a comparison with the high density results, the band shapes were computed and plotted with the experimental contours. At 300 K, the two profiles are in good agreement up to 100 amagat. However, at higher densities, some discrepancies begin to appear until there is no agreement at very high densities. At these densities, many body collisions should be quite important.

The low temperature (85 K) band shapes have excellent agreement even up to 300 amagat (see Figs.5.4a,b). Some deviation in the contour is found at very high density, although the two peaks are still in good agreement (see Figs.5.4c,d). The net result of this detailed comparison is to give some confidence in the intermolecular potential used in the calculations for HD. The general form of the potential is

$$
\begin{aligned}
V &= V_{iso} + V_{aniso} \\
V_{aniso} &= V_{qq} - 6\left(C_6/r^7 - 2C_{12}/r^{13}\right)d\cos\theta \\
V_{iso} &= \xi_1(\partial V/\partial\xi_1)_0 + \xi_2(\partial V/\partial\xi_2)_0 + \frac{1}{2}\,\xi_1^2\left(\partial^2 V/\partial\xi_1^2\right)_0 \\
&\quad + \frac{1}{2}\,\xi_2^2\left(\partial^2 V/\partial\xi_2^2\right)_0 + \xi_1\xi_2\left(\partial^2 V/\partial\xi_1\partial\xi_2\right)_0
\end{aligned}
\qquad (5.54)
$$

where θ is the angle between the axes of the two molecules; ξ_1 and ξ_2 are the vibrational coordinates of the two molecules. All the parameters in the above potential were obtained from independent sources [5.39c].

PERCHARD et al. [5.101] have investigated the self-broadened Raman lines in HCl, DCl, HBr, and DBr gases and their foreign broadening with SF_6 and C_2F_6 gases. The Q-branch band shapes have been analyzed qualitatively, and no information about intermolecular potential was obtained. WANG and WRIGHT [5.110] have investigated the collisional narrowing in the isotropic and anisotropic Raman Q-branches in N_2, and they also reported some preliminary results on CO. Using the GORDON [5.111] approach (to be discussed in Sect.5.6) for band moments, they observed some discrepancy between the expected theoretical results and the experiment. No quantitative calcula-

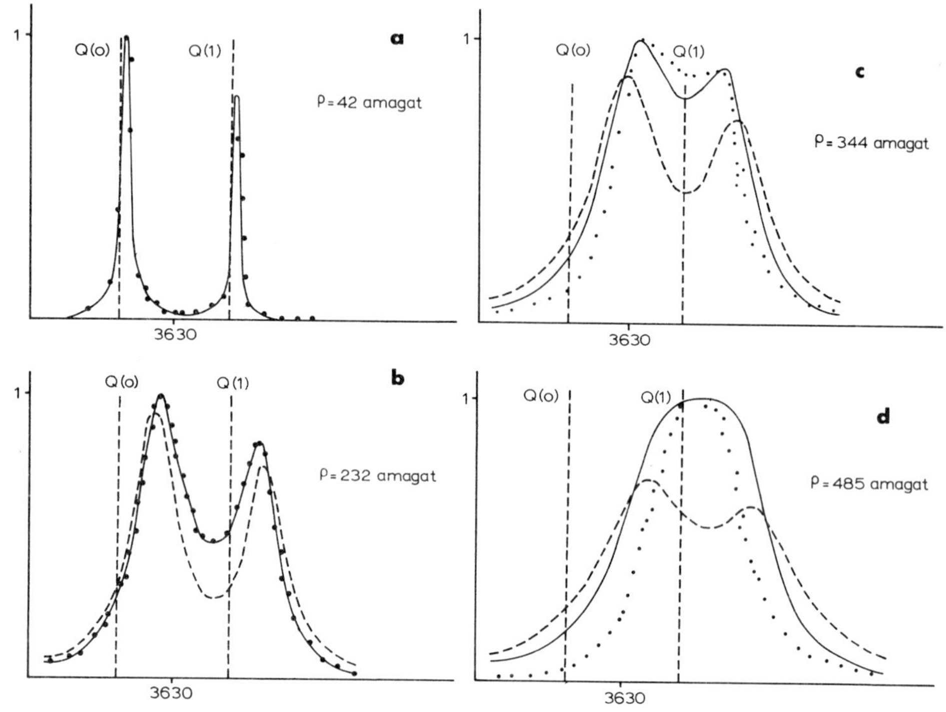

Fig. 5.4a-d. Contour of the Q-branch of HD at 85 K for various gas densities, ρ. The solid curves refer to the spectrum calculated and include the cross-correlation terms. The dashed curves refer to the spectrum calculated under the same conditions without the cross-correlation terms contribution. The experimental results of [5.84] are represented by the dots. The vertical scale is in arbitrary units and the maximum calculated intensity (solid curves) has been adjusted on the corresponding experimental one (From [5.39c]).

tion was performed using intermolecular potentials. Some other experimental works [5.112,113] on Raman band shapes and collisional narrowing have been reported, without making a connection with intermolecular forces.

5.6 Statistical Theories

In this section a brief survey of the non-impact theories of spectral shapes will be given. As mentioned earlier, (5.25) contains the many body and the memory effects. Similar results and generalizations can be obtained from other many body formulations [5.31,33,35-39,42,47,53,81b]. However, the actual calculation of $F(\omega)$ or $C(t)$ is a very difficult problem of statistical physics; detailed calculations have been performed only in some special cases (see, e.g., [5.53]).

An approximate statistical procedure, in which the initial and perturber-perturber correlations are neglected, is sometimes used to construct a many body correlation function [5.42,47]; it leads to the following result

$$C(t) = \mathrm{Tr}_s\left\{\underline{\underline{\alpha}}(0)\ \exp\left(-iL_0^{(s)}t\right)T\ \exp[N\langle U(t)-1\rangle_p]\underline{\underline{\alpha}}(0)\right\} \tag{5.55}$$

$$U(t) = T\ \exp\left[-i\int_0^t dt'\ \exp(iL_0t')L_{sp}\right] \tag{5.56}$$

where T is the time ordering operator, $L_0 = L_0^{(s)} + L_0^{(p)}$, L_{sp} is the Liouville operator for the system-perturber interaction and $\langle\ldots\rangle_p$ denotes an average over a single perturber variable. The well known statistical theory of MARGENAU [5.114], and its generalizations [5.115], are special cases of (5.55). The two body static limit of such statistical theories is the opposite extreme, compared with the impact theory. For an isolated line $|i\rangle \to |f\rangle$, the static limit is valid if $|\omega-\omega_{if}| \gg \tau_d^{-1}$, a region usually lying in the far wings of the line. The resulting line wing shape is generally sensitive to the short range part of the interaction potential [5.46,47,116]. Attempts have been made, using the static limit, to explain the non-Lorentzian line wings in the CO_2 infrared spectra [5.17].

Another related approximation can be obtained through the "cumulant" approach of KUBO [5.39a], on replacing the $\langle\ldots\rangle_p$ by the cumulant average. A second order cumulant expansion represents a non-impact generalization of the ANDERSON theory [5.21-22]. Such an approach has been used recently for gases, liquids, and solids [5.39b,c]; the application to HD was considered in Sect.5.4. For band shapes, moment analysis and the expansion of C(t) in powers of t, can also be useful in dense systems; this will be discussed in the next section.

5.7 Liquids

The present knowledge of the liquid state is rather limited. Therefore, while Raman spectroscopy of liquids is more exciting simply because there is more to learn, the information obtained about intermolecular forces is generally qualitative. The more immediate task facing the investigators of liquids is to understand the structure, molecular dynamics, and other basic properties of liquids. Most of these properties, in turn, are related to the intermolecular forces. However, in many cases, quantitative theories of such relationships are in early formative stages [5.118-122]. Therefore, we will focus our attention here on a brief discussion of the main features, rather than on a complete literature survey. The literature on light scattering in liquids has experienced an explosive growth in recent years, and the reader is referred to several recent review articles for details [5.118-122,11].

The Raman spectrum of molecules in liquids usually consists of bands of overlapping lines; only for light molecules, like H_2, HD, etc., are the individual lines resolved [5.120]. The spectrum of pure molecular liquids involves further complications because of the correlation effects. Therefore, the spectrum of a dilute molecular solution, in a solvent of different species, generally gives more direct information about molecular dynamics. The most striking feature in liquids is the appearance of forbidden transitions, e.g., the Q-branch in the infrared spectra [5.120]. Raman band narrowing studies provide interesting information about solute-solvent interactions [5.123].

The general theory of Sect.5.1 applies to liquids as well. In particular, the band narrowing and the forbidden transitions can be understood as a coupling effect between the overlapping lines. However, as suggested by GORDON [5.8,111], the time domain study of the correlation function C(t) also gives useful information about the short time behavior of the system. This supplements the information on the long time behavior contained in the width of the spectral function $F(\omega)$. Gordon's work, later extended by McCLUNG and others [5.118], neglects the vibrational motion and focuses attention on the rotational (or orientational) motion of the molecules. Only the anisotropic part of the polarizability contributes to the orientational part of the correlation function $C_0(t)$. For example, for a linear or symmetric top molecule, one obtains the following result[1] in the classical limit [5.4,8]

$$\hat{C}_0(t) = \langle P_2[\cos\theta(t)]\rangle \tag{5.57}$$

where θ is the angle through which the symmetry axis of the molecule has rotated in time t, and P_2 is the second order Legendre polynomial. (For generalizations of (5.57), see. e.g., [5.118]). Using a moments analysis, GORDON [5.8] obtained the following expansion for a classical linear molecule, on neglecting the frequency fluctuations

$$\hat{C}_0(t) = \sum_n M_n(it)^n/n! = 1 - (3kT/I)t^2 + [4(kT)/I)^2+(8I^2)^{-1}\langle(\underline{O}V)^2\rangle]t^4+ \dots \tag{5.58}$$

$$M_n = \int_{-\infty}^{+\infty} \omega^n \hat{F}(\omega)d\omega \tag{5.59}$$

where M_n is the nth order band moment, I is the moment of inertia, and $\underline{O}V$ is the angular derivative of the intermolecular potential V, so that $\langle(\underline{O}V)^2\rangle$ is the mean squared torque. In this approximation the second moment, 6kT/I, is independent of V since frequency fluctuations are neglected. Therefore, the mean squared torque can be obtained from the second and fourth band moments (M_2 and M_4); this provides

---

[1] In this section we will use the normalized correlation function $\hat{C}(t)$ defined such that $\hat{C}(0) = 1$; the Fourier transform will be denoted by $\hat{F}(\omega)$.

information about the strength of the intermolecular forces. (For an explicit expression for $<(\underline{O}V)^2>$ for multipolar forces; see [Ref. 5.4, (7.17)].) Analysis of Raman and infrared experimental data along these lines has been discussed, e.g., in [5.4,111,120,124].

Expansion (5.58) is useful only for short times t. The complete correlation function $\hat{C}(t)$ can be obtained by inverse Fourier transform of $\hat{F}(\omega)$. The study of molecular dynamics through the correlation functions obtained from the inverse Fourier transform of the experimental band shapes has been vigorously pursued in recent years [5.118-121]. At the same time, the observed band width usually provides direct information on the long time behavior of the system.

The theoretical study of orientational motion in liquids is based on various models. The earliest is the rotational diffusion model of DEBYE [5.125], which involves weak collisions perturbing the rotation in small steps. This model is valid if $\tau_0 \gg \tau_j$, where τ_0 is the average orientational relaxation time and τ_j is the average time between collisions perturbing the rotation. GORDON [5.126] introduced the strong collision model in which each collision randomizes either the complete rotational state (j-diffusion model) or only the orientation of the rotational angular momentum (m-diffusion model). This model is valid even when $\tau_0 \sim \tau_j$; it was later extended and has been studied in great detail by McCLUNG and others [5.118, 127]. The band shape function can be described by (5.28), where the matrix elements of $\overline{R}$ are proportional to τ_j^{-1}; the actual form depends on the molecule and the model chosen for rotational motion [5.118,128]. Measured band shapes and derived correlation functions in Raman as well as infrared spectra have been compared with the theory in numerous cases. The value of τ_j is the basic information obtained from such comparisons. In most cases, the extended j-diffusion model appears to be satisfactory for spherical top and symmetric top molecules, but not so satisfactory for linear molecules [5.118]. This model is based on impact theory and, therefore, should break down in dense systems where many body and memory effects are also important. However, the above discussion of liquids, as well as the behavior of dense gases discussed in Sect.5.4, suggests that an impact theory is applicable to band shapes in some cases. This could be partly due to the fact that a lot of detailed information is averaged out in a band of a large number of overlapping lines. Some non-impact theories were discussed in Sect.5.5. The memory function approach has been used to investigate such theories in liquids [5.129].

BRATOS and MARECHAL [5.130] further developed the theory for diatomic molecules by including the effects of vibrational motion in the correlation function $\hat{C}(t)$. They expressed the polarized and the depolarized Raman spectra in terms of the isotropic ($\hat{C}_i$) and the anisotropic ($\hat{C}_a$) parts of the correlation function. For example, for 90° scattering geometry, it was shown that the polarized and the depolarized spectra are given by the Fourier transforms of

$$\hat{C}_p(t) = \hat{C}_i(t) + (4/3)\hat{C}_a(t) \quad , \tag{5.60}$$

and

$$\hat{C}_d(t) = \hat{C}_a(t) \quad , \tag{5.61}$$

respectively. Furthermore, if rotational and vibrational motions are uncorrelated, it follows that

$$\hat{C}_a(t) = \hat{C}_v(t)\, \hat{C}_0(t) \tag{5.62}$$

where $\hat{C}_v(\hat{C}_0)$ is the vibrational (orientational) part of the correlation function. (The theory was later extended to polyatomic molecules [5.131].) Thus the Raman spectrum is particularly suitable for a study of vibrational and rotational relaxation in liquids, since the polarized and the depolarized parts determine $\hat{C}_v$ and $\hat{C}_0$ completely. However, the analysis of experimental results is usually complicated because of the presence of rotation-vibration interaction, the correlations between neighboring molecules, and the uncertainties in the spectral measurements [5.121,132]. Still, vibrational and orientational relaxation has been studied in many liquids, and relaxation times have been measured. Combined studies of infrared, Raman, and depolarized Rayleigh scattering provide more complete information. (For a sampling of experimental works in this area, see [5.121,133].)

An interesting question in the study of a liquid is whether its behavior is closer to a solid (structure-dominated) or to a dense gas (collision-dominated). Most of the discussion in this article has revolved around the collision-dominated regime. However, structure-dominated models (e.g., cell model) are also appropriate for many liquids. For example, hindered rotations have been analyzed through an energy level approach, and information about the hindering potential was obtained from the band shapes [5.120]. Unified treatments of collision-dominated and structure-dominated effects (e.g., [5.39b,109,134]) should be helpful in the analysis of data which throws light on the structure of liquids.

To summarize, Raman spectrum studies in liquids in recent years have provided considerable information about molecular dynamics. However, reliable quantitative theories for the measured quantities (e.g., relaxation times and diffusion constants) in terms of the intermolecular forces are not fully developed. Some of these quantities may not be very sensitive to the details of the forces. However, attempts have been made in this direction. (See, e.g., [5.39b,132,135].) A promising recent development is the use of computer simulation of many body dynamics and study of correlation functions [5.136]; comparison with light scattering and other data should lead to more quantitative information in future.

5.8 Conclusion

The recent theoretical and experimental developments in the general area of Raman spectral shapes have been surveyed. It has been shows that significant quantitative information about intermolecular forces can be obtained from a comparison of theory and experiment, specially in low density gases. Semi-quantitative and qualitative information about interactions in dense systems can also be obtained from analysis of data in high density gases and liquids. In fact, the information provided by the gas phase studies sometimes helps in the understanding of the complicated many body interactions and dynamics in liquids [5.118,120].

Table 5.1 summarizes the experimental and theoretical papers on Raman spectrum studies in gases. Selected works in liquids were also briefly discussed in Sect.5.6.

Acknowledgments. The authors would like to acknowledge, with thanks, the assistance provided by several workers in the field, in the literature survey. Special thanks are due to Dr. C.G. Gray in this connection.

References

5.1 Cherlow, J.M., Porto, S.P.S.: In *Topics in Applied Physics*, Vol.2, ed. by H. Walther (Springer, Berlin, Heidelberg, New York 1976) Chap.4, pp.253-282

5.2 Barron, L.D., Buckingham, A.D.: Ann. Rev. Phys. Chem. *26*, 381 (1975)

5.3 Weber, A.: In *The Raman Effect*, ed. by A. Anderson (Marcel Dekker, New York 1973) Vol.2, Chap.9, pp.543-757

5.4 Gray, C.G., Welsh, H.L.: In *Essays in Structural Chemistry*, ed. by Long,D.A., Downs,A.J., Stravely,L.A.K. (MacMillan, London 1971) Chap.7, pp.163-188

5.5 Buckingham, A.D., Utting, B.D.: Ann. Rev. Phys. Chem. *21*, 287 (1970)

5.6a Rabitz, H.: Ann. Rev. Phys. Chem. *25*, 155 (1974)

5.6b Krishnaji: J. Sci. Ind. Res. *32*, 168 (1973)

5.7 Knaap, H.F.P., Lallemand, P.: Ann. Rev. Phys. Chem. *26*, 59 (1975)

5.8a Gordon, R.G.: Adv. Mag. Resonance *3*, 1 (1968)

5.8b Gordon, R.G., Klemperer, W., Steinfeld, J.I.: Ann. Rev. Phys. Chem. *19*, 215 (1968). For a general reference on magnetic resonance, see Slichter, C.P. : *Principles of Magnetic Resonance*, Springer Series in Solid-State Sciences, Vol.1, 2nd ed. (Springer, Berlin, Heidelberg, New York 1978)

5.9a Moraal, H.: Phys. Reports (Phys. Lett. C) *17*, 225 (1975)

5.9b Kihara, T., Koide, A.: Adv. Chem. Phys. *33*, 51 (1975)

5.10 Rich, N.H., McKellar, A.R.W.: Can. J. Phys. *54*, 486 (1976)

5.11 Gelbart, W.M.: Adv. Chem. Phys. *26*, 1 (1974)

5.12 Samson, R., Pasmanter, R.A., Ben-Reuven, A.: Phys. Rev. A*14*, 1224, 1238 (1976)

5.13 Flytzanis, C., Bloembergen, N.: In *Progress in Quantum Electronics*, Vol.4, Pt.3, ed. by Sanders,J.H., Stenhom,S. (Pergamon Press, New York 1976)

5.14 Van Kranendonk, J., Karl, G.: Rev. Mod. Phys. *40*, 531 (1968)

5.15 Dicke, R.H.: Phys. Rev. *89*, 472 (1953)

5.16 Fulton, R.L.: J. Chem. Phys. *50*, 3355 (1969)

5.17 Baym, G.: *Lectures in Quantum Mechanics* (Benjamin, New York 1969) Chap.13

5.18 Gordon, R.G.: J. Chem. Phys. *42*, 3658 (1965)

5.19 Placzek, G.: Z. Phys. *70*, 84 (1931); Placzek, G.: "Rayleigh-Streuung und Raman Effekt", in *Handbuch der Radiologie*,

Vol.VI, Pt.2, ed. by E. Marx (Akademische Verlagsgesellschaft, Leipzig 1934) pp.205-374. English translation by A. Werbin: "Rayleigh and Raman Scattering", UCRL Transl. No. 526L, available from National Technical Information Service, U.S. Department of Commerce, Springfield, Va.;
Born, M., Huang, K.: *Dynamical Theory of Crystal Lattices* (Clarendon Press, Oxford 1954)
5.20 Foley, H.M.: Phys. Rev. *69*, 616 (1946)
5.21 Anderson, P.W.: Phys. Rev. *76*, 647 (1949)
5.22 Tsao, C.J., Curnutte, B.: J. Quant. Spectrosc. Radiat. Transf. *2*, 41 (1962)
5.23a Fiutak, J., Van Kranendonk, J.: Can. J. Phys. *40*, 1085 (1962)
5.23b Gray, C.G.: Ph.D. Thesis, University of Toronto (1967)
5.24 Gordon, R.G.: J. Chem. Phys. *44*, 3083 (1966)
5.25 Gordon, R.G.: J. Chem. Phys. *45*, 1649 (1966)
5.26 Kolb, A.C., Griem, H.: Phys. Rev. *111*, 514 (1958)
5.27 Baranger, M.: Phys. Rev. *111*, 481, 494 (1958)
5.28 Baranger, M.: Phys. Rev. *112*, 855 (1958)
5.29 Fano, U.: Phys. Rev. *131*, 259 (1963)
5.30 Hess, S.: Physica *61*, 80 (1972)
5.31 Chappell, W.R., Cooper, J., Smith, E.W., Dillon, T.: J. Stat. Phys. *3*, 401 (1971)
5.32 Moraal, H.: Physica *73*, 379 (1974)
5.33 Hussey, T., Dufty, J.W., Hooper, C.F.: Phys. Rev. *12*, 1084 (1975)
5.34 Berman, P.R.: Phys. Rev. A*5*, 927 (1972)
5.35 Ross, D.W.: Ann. Phys. (N.Y.) *36*, 458 (1966)
5.36 Zaidi, H.R.: Can. J. Phys. *50*, 2792 (1972)
5.37 Bezzerides, B.: Phys. Rev. *181*, 379 (1969)
5.38 Ben-Reuven, A.: Adv. Chem. Phys. *33*, 235 (1975)
5.39a Kubo, R.: In *Fluctuations, Relaxation and Resonance in Magnetic Systems*, ed. by D. Ter Haar (Oliver and Boyd, Edinburgh 1962) p.23
5.39b Robert, D., Galatry, L.: J. Chem. Phys. *64*, 2721 (1976)
5.39c Bonamy, J., Bonamy, L., Robert, D.: J. Chem. Phys. *67*, 4441 (1977)
5.40 Galatry, L.: Phys. Rev. *122*, 1218 (1961)
5.41a Rautian, S.G., Sobelman, I.I.: Sov. Phys.-Usp. *9*, 701 (1967)
5.41b Fiutak, J., Paul, E.: Acta Phys. Polonica A*49*, 773 (1976)
Fiutak, J., Horodecki, R.: Acta Phys. Polonica A*51*, 301 (1977)
5.42 Smith, E.W., Cooper, J., Chappel, W.R., Dillon, T.: J. Quant. Spectrosc. Radiat. Transf. *11*, 1547, 1567 (1971)
5.43 Cooper, V.G., May, A.D., Hara, E.H., Knaap, H.F.P.: Can. J. Phys. *46*, 2019 (1968);
Gupta, B.K., Hess, S., May, A.D.: Can. J. Phys. *50*, 778 (1972)
5.44 Murray, J.R., Javan, A.: J. Mol. Spectrosc. *42*, 1 (1972)
5.45 Zaidi, H.R., Van Kranendonk, J.: Can. J. Phys. *49*, 385 (1971)
5.46 Srivastava, R.P., Zaidi, H.R.: Can. J. Phys. *53*, 84 (1975)
5.47 Futrelle, R.P.: Phys. Rev. A*5*, 2162 (1972)
5.48 Shafer, R., Gordon, R.G.: J. Chem. Phys. *58*, 5422 (1973)
5.49 Bezzerides, B.: Phys. Rev. *159*, 3 (1967)
5.50 Vdovin, Y.A., Galiskii, V.M.: Sov. Phys. - JETP *25*, 894 (1967)
5.51 Zaidi, H.R.: Can. J. Phys. *53*, 76 (1975)
5.52 Srivastava, R.P., Zaidi, H.R.: J. Quant. Spectrosc. Radiat. Transf. *16*, 301 (1976)
5.53 Zaidi, H.R.: Can. J. Phys. *55*, 1229, 1243, 1252 (1977)
5.54 Fiutak, J., Van Kranendonk, J.: Can. J. Phys. *41*, 21 (1963)
5.55 May, A.D., Degen, V., Stryland, J.C., Welsh, H.L.: Can. J. Phys. *39*, 1769 (1961
5.56 Cooper, V.G., May, A.D., Gupta, B.K.: Can. J. Phys. *48*, 725 (1970)
5.57 Gupta, B.K., May, A.D.: Can. J. Phys. *50*, 1747 (1972)
5.58a Rabitz, H.: In *Modern Theoretical Chemistry*, ed. by W.H. Miller (Plenum Press, New York 1976) Vol.III and references therein
5.58b Fisanick-Englot, G., Rabitz, H.: J. Chem. Phys. *63*, 1547 (1975)
5.59 Goldflam, R., Kouri, D.J.: J. Chem. Phys. *66*, 542 (1977)
5.60 Goldflam, R., Kouri, D.J.: J. Chem. Phys. *66*, 2452 (1977)
5.61a Neilsen, W.B., Gordon, R.G.: J. Chem. Phys. *58*, 4131, 4149 (1973)
5.61b Fitz, D.E., Marcus, R.A.: J. Chem. Phys. *62*, 3788 (1975)

5.62a Smith, E.W., Giraud, M., Cooper, J.: J. Chem. Phys. *65*, 1256 (1976)
Smith, E.W., Giraud, M., Cooper, J.: J. Chem. Phys. *66*, 376 (1977)
5.62b Smith, E.W., Giraud, M.: J. Chem. Phys. *66*, 1762 (1977)
5.63a Green, S., Monchick, L., Goldflam, R., Kouri, D.J.: J. Chem. Phys. *66*, 1409 (1977)
5.63b Tarr, S.M., Rabitz, H., Fitz, D.E., Marcus, R.A.: J. Chem. Phys. *66*, 2854 (1977)
5.64 Herman, R.M.: Phys. Rev. *132*, 262 (1963)
Herman, R.M.: J. Quant. Spectrosc. Radiat. Transf. *3*, 449 (1963)
5.65 Van Kranendonk, J.: Can. J. Phys. *41*, 433 (1963)
5.66 Gray, C.G.: Can. J. Phys. *46*, 135 (1968)
5.67 Gray, C.G., Van Kranendonk, J.: Can. J. Phys. *44*, 2411 (1966)
5.68 Gray, C.G.: J. Chem. Phys. *53*, 457 (1970)
5.69 Srivastava, R.P., Zaidi, H.R.: Can. J. Phys. *55*, 533 (1977)
5.70 Jarecki, J., Herman, R.M.: J. Quant. Spectrosc. Radiat. Transf. *15*, 707 (1975)
5.71a Srivastava, R.P., Zaidi, H.R.: Can. J. Phys. *55*, 542 (1977)
5.71b Srivastava, R.P., Zaidi, H.R.: Can. J. Phys. *55*, 549 (1977)
5.72 Moraal, H.: Physica *73*, 379 (1974)
5.73 Keijser, R.A.J., Lombardi, J.R., Van den Hout, K.D., DeGroot, M., Sanctuary, B.C., Knaap, H.F.P.: Phys. Lett. *45*, 3 (1973)
5.74 Keijser, R.A.J., Lombardi, J.R., Van den Hout, K.D., Sanctuary, B.C., Knaap, H.F.P.: Physica *76*, 585 (1974)
5.75 Rabitz, H.A., Gordon, R.G., J. Chem. Phys. *53*, 1815, 1831 (1970)
5.76a Murphy, J.S., Boggs, J.E.: J. Chem. Phys. *47*, 691, 4152 (1967)
Murphy, J.S., Boggs, J.E.: J. Chem. Phys. *49*, 3333 (1968)
Murphy, J.S., Boggs, J.E.: J. Chem. Phys. *50*, 3320 (1969)
Murphy, J.S., Boggs, J.E.: J. Chem. Phys. *51*, 3891 (1969)
Murphy, J.S., Boggs, J.E.: J. Chem. Phys. *54*, 2443 (1971)
5.76b Johri, G.K., Srivastava, S.L.: Chem. Phys. Lett. *45*, 364 (1977)
Johri, G.K., Srivastava, S.L.: Chem. Phys. Lett. *39*, 579 (1976)
5.77 May, A.D., Varghese, G., Stryland, J.C., Welsh, H.L.: Can. J. Phys. *42*, 1058 (1964)
5.78 Boulet, C., Robert, D., Galatry, L.: J. Chem. Phys. *65*, 5302 (1976)
5.79 Sharma, R.D.: Chem. Phys. Lett. *8*, 428 (1971)
Sharma, R.D.: Chem. Phys. Lett. *13*, 261 (1972)
5.80 Gray, C.G., Tipping, R.H.: Chem. Phys. Lett. *13*, 260 (1972)
5.81a Frost, B.S.: J. Phys. B*9*, 1001 (1976)
5.81b Davies, R.W.: Phys. Rev. A*12*, 927 (1975)
5.82 Herman, R.M., Tipping, R.H.: J. Quant. Spectrosc. Radiat. Transf. *10*, 897 (1970)
5.83 Cherkasov, M.R.: Opt. Spectrosc. *40*, 3 (1976)
5.84a Witkowicz, T., May, A.D.: Can. J. Phys. *54*, 575 (1976)
5.84b Witkowicz, T.: Master Thesis, University of Toronto (1975)
5.85a Lallemand, P., Simova, P., Bret, G.: Phys. Rev. Lett. *17*, 1239 (1966)
5.85b Lallemand, P., Simova, P.: J. Mol. Spectrosc. *26*, 262 (1968)
5.86 Bazhulin, P.A., Lazarev, Yu.A.: Opt. Spectrosc. *8*, 106 (1960)
5.87 Lazarev, Yu.A.: Opt. Spectrosc. *13*, 373 (1962)
5.88 Bazhulin, P.A.: Sov. Phys.-Usp. *5*, 661 (1963)
5.89 Allin, E.J., May, A.D., Stoicheff, B.P., Stryland, J.C., Welsh, H.L.: Appl. Opt. *6*, 1597 (1967)
5.90 Murray, J., Javan, A.: Bull. Am. Phys. Soc. *12*, 113 (1967)
5.91 Murray, J.R., Javan, A.: J. Mol. Spectrosc. *29*, 502 (1969)
5.92 Clements, W.R.L., Stoicheff, B.P.: J. Mol. Spectrosc. *33*, 183 (1970)
5.93 John, G.E. St.: Ph.D. Thesis, University of Toronto (1952)
5.94 Jammu, K.S., John, G.E.St., Welsh, H.L.: Can. J. Phys. *44*, 797 (1966)
5.95 Herpin, M.C., Lallemand, P.: J. Quant. Spectrosc. Radiat. Transf. *15*, 779 (1975)
5.96 Bonamy, J., Robert, D.: J. Quant. Spectrosc. Radiat. Transf. *16*, 185 (1976)
5.97 Gray, C.G.: Chem. Phys. Lett. *8*, 527 (1971)
5.98 Gray, C.G.: J. Chem. Phys. *61*, 418 (1974)
5.99 Rich, N.H., Welsh, H.L.: Chem. Phys. Lett. *11*, 292 (1971)
5.100 Fabre, D., Widenlocher, G., Vu, H.: Opt. Commun. *4*, 421 (1972)

5.101 Perchard, J.P., Murphy, W.F., Bernstein, H.J.: Mol. Phys. *23*, 535 (1972)
5.102a Varanasi, P., Sarangi, S.: J. Quant. Spectrosc. Radiat. Transf. *14*, 845 (1974)
5.102b Levy, A., Boulet, C.: J. Quant. Spectrosc. Radiat. Transf. *15*, 1067 (1975)
5.103 Gordon, R.G., McGinnis, R.P.: J. Chem. Phys. *49*, 2455 (1968)
5.104 Burshtein, A.I., Naberukhin, Yu.I.: Sov. Phys. - JETP *25*, 799 (1967)
5.105 Alekseev, B.A., Sobelman, I.I.: Acta Phys. Polonica *34*, 579 (1968)
Alekseev, B.A., Sobelman, I.I.: IEEE J. QE-*4*, 654 (1968)
5.106 Gordon, R.G., McGinnis, R.P.: J. Chem. Phys. *55*, 4898 (1971)
5.107 May, A.D., Stryland, J.C., Varghese, G.: Can. J. Phys. *48*, 2331 (1970)
5.108 Dion, P., May, A.D.: Can. J. Phys. *51*, 36 (1973)
5.109 Talin, B., Galatry, L., Klein, L., J. Chem. Phys. *66*, 2789 (1977)
5.110 Wang, C.H., Wright, R.B.: J. Chem. Phys. *59*, 1706 (1973)
5.111 Gordon, R.G.: J. Chem. Phys. *43*, 1307 (1965)
5.112 Le Duff, Y., Holzer, W.: J. Chem. Phys. *60*, 2175 (1974)
5.113 Ouillon, R.: Chem. Phys. Lett. *35*, 63 (1975)
5.114 Margenau, H.: Phys. Rev. *48*, 755 (1935)
5.115 Atakan, A.K., Jacobson, H.C.: J. Quant. Spectrosc. Radiat. Transf. *12*, 289 (1972)
5.116 Szudy, J., Baylis, W.: J. Quant. Spectrosc. Radiat. Transf. *15*, 641 (1975)
5.117 Ben Aryeh, Y., Sorgen, A.: Phys. Rev. A*4*, 2170, 2178 (1971)
Fowler, B.W., Sung, C.C.: Phys. Rev. A*13*, 2318 (1976)
5.118 McClung, R.E.D.: Adv. Mol. Ralaxation and Interaction Processes *10*, 83 (1977)
5.119 Steele, W.A.: Adv. Chem. Phys. *34*, 1 (1976)
Berne, B.J., Pecora, R.: *Dynamic Light Scattering* (Wiley and Son, New York 1976)
Fleury, P.A., Boon, J.P.: Adv. Chem. Phys. *24*, 1 (1973)
Gray, C.G.: In *Specialist Periodical Reports, Statistical Mechanics*, Vol.2, ed. by K. Singer (Chemical Society, London 1975) Chap.5, pp.300-323
Egelstaff, P.A., Gray, C.G., Gubbins, K.E.: In *International Review of Science*, Physical Chemistry Series 2, Vol.2, ed. by A.D. Buckingham (Butterworths, London 1975) Chap.8, pp.299-347
5.120 Marabella, L.J.: Appl. Spectrosc. Rev. *7*, 313 (1974)
5.121 Griffiths, J.E.: In *Vibrational Spectra and Structure*, ed. by J.R. Durig (Elsevier Publishing, New York 1977) Vol.6, pp.273-375
5.122 Peticolas, W.L.: Ann. Rev. Phys. Chem. *23*, 93 (1972)
5.123 Altmann, K., Holzer, W., Le Duff, Y.: Chem. Phys. Lett. *36*, 259 (1975)
Le Duff, Y., Holzer, W.: Chem. Phys. Lett. *24*, 212 (1974)
5.124 Briquet, J.C., Soussen-Jacob, J.: Mol. Phys. *28*, 921 (1974)
Twu, C.H., Gray, C.G., Gubbins, K.E.: Mol. Phys. *27*, 1601 (1974)
Twu, C.H., Gubbins, K.E., Gray, C.G.: Mol. Phys. *30*, 1607 (1975)
5.125 Debye, P.: *Polar Molecules* (Dover, New York 1945)
5.126 Gordon, R.G.: J. Chem. Phys. *44*, 1830 (1966)
5.127 Lynden-Bell, R.M.: Mol. Phys. *31*, 1653 (1976)
5.128 Eagles, T.E., McClung, R.E.D.: J. Chem. Phys. *59*, 435 (1973)
5.129 Biot, F., Constant, E.: Chem. Phys. Lett. *29*, 618 (1974)
Marsault, J.P., Marsault-Harail, F., Levi. G.: J. Chem. Phys. *62*, 893 (1975)
5.130 Bratos, S., Marechal, E.: Phys. Rev. A*4*, 1078 (1971)
5.131 Bartoli, F., Litovitz, T.A.: J. Chem. Phys. *56*, 404, 413 (1972)
Nafie, L.A., Peticolas, W.L.: J. Chem. Phys. *57*, 3145 (1972)
5.132 Wang, C.H.: Mol. Phys. *33*, 207 (1977)
5.133 Perchard, J.P., Murphy, W.F., Bernstein, H.J.: Mol. Phys. *23*, 499, 519 (1972)
Wang, C.H., Wright, R.B.: J. Chem. Phys. *59*, 1706 (1973)
Wang, C.H.: J. Chem. Phys. *62*, 747 (1975)
Wang, C.H., Jones, D.R., Christensen, D.H.: J. Chem. Phys. *64*, 2820 (1976)
Amorim da Costa, A.M., Norman, M.A., Clarke, J.H.R.: Mol. Phys. *29*, 191 (1975)
Griffiths, J.E., Clerc, M., Rentzepis, R.M.: J. Chem. Phys. *60*, 3824 (1974)
Griffiths, J.E., Clerc, M., Rentzepis, R.M.: J. Chem. Phys. *63*, 2262 (1975)
Neuman, M.N., Tabisz, G.C.: Chem. Phys. *15*, 195 (1976)
Soussen-Jacob, J., Dervil, E., Vincent-Geisse, J.: Mol. Phys. *28*, 935 (1974)
5.134 Keys, T., Kivelson, D.: J. Chem. Phys. *56*, 1057 (1972)
Kivelson, D., Keys, T.: J. Chem. Phys. *57*, 4599 (1972)
Kivelson, D.: Mol. Phys. *28*, 321 (1974)

5.135 Madden, P.A., Lynden-Bell, R.M.: Chem. Phys. Lett. *38*, 163 (1976)
Diestler, D.J.: Chem. Phys. Lett. *39*, 39 (1976)

3.136 Berne, B., Bishop, M., Rahman, A.: J. Chem. Phys. *58*, 2696 (1973)
Adler, B.J., Young, D.A., Mark, M.A.: J. Chem. Phys. *56*, 3013 (1972)

5.137a Fletcher, W.H., Rayside, J.S.: J. Raman Spectrosc. *2*, 3 (1974)

5.137b Mikhailov, G.V.: Sov. Phys. - JETP *9*, 974 (1959)
Mikhailov, G.V.: Sov. Phys. - JETP *10*, 1114 (1960)

5.138 Compaan, A., Langer, W.D., Eden, D., Swinney, H.L.: Astrophys. J. *185*, L105 (1973)

5.139 Le Duff, Y.: J. Chem. Phys. *59*, 1984 (1973)

5.140 Pinter, F.: Opt. Spectrosc. *17*, 428 (1964)

5.141 Fabre, D., Widenlocher, G., Thiery, M.M., Vu, H., Vodar, B.: In *Advances in Raman Spectroscopy*, ed. by J.P. Mathieu, Vol.1 (Heyden and Son, London 1973) pp.497-503

5.142 May, A.D., Stryland, J.C., Welsh, H.L.: J. Chem. Phys. *30*, 1099 (1959)

5.143 May, A.D., Poll, J.D.: Can. J. Phys. *43*, 1836 (1965)

6. The Resonance Raman Effect

D. L. Rousseau, J. M. Friedman, and P. F. Williams

With 26 Figures

The application of Raman scattering to molecular systems, as discussed in the previous chapters, yield a great deal of information about *ground state* structural, spectroscopic, and physical properties. In contrast when the incident frequency is in resonance with a real transition of a system, (bringing about the resonance Raman effect) *excited electronic state* properties may be studied. In addition due to the phenomena of resonance enhancement, specific parts of complicated molecules may be probed without interference from other parts of the molecule or from solvents. These features make the application of resonance Raman scattering to molecules complementary to non-resonance investigations and they make it a very important tool in the study of properties of molecular systems.

In this chapter we discuss the basic theory of resonance Raman scattering and its application to a few specific systems. Our goal is, at the expense of oversimplifying certain aspects of the theory, to give a clear physical understanding of the basic aspects of resonance Raman scattering, to demonstrate the origin of many of the phenomena currently being studied, and to point out a few areas in which our qualitative understanding of the phenomena is still lacking. We shall emphasize the fundamental unity of light scattering theory as it is applied to a wide range of situations to enable the student in the field to grasp the general principles with a minimal amount of familiarity with specialized theory. It is hoped that this chapter will serve to complement the treatment of Raman scattering in solids given in [Ref.6.1, Chap.3]. It is also hoped that differences in our approach will make this chapter complementary to the many excellent review articles that have recently been written about theories of resonance Raman scattering [6.2-6], the techniques of resonance Raman scattering [6.7], and its applications to a variety of systems [6.8-16]. Finally we apologize to those who are very familiar with our earlier work since in some sections of this chapter we have made extensive excerpts from them, especially from our paper [6.17] which dealt with resonance Raman scattering from homonuclear diatomic molecules.

Over the past several years a vast amount of work has been done on resonance Raman scattering from molecules from which significant contributions about molecular properties have been made and the basic scattering phenomena have been demonstrated. In this chapter we make no attempt to include a discussion of all the important contributions. Instead we describe in detail a few systems which we believe best illustrate the basic phenomena. Therefore in the examples most attention will be given to diatomic molecules which exhibit scattering via the Franck-Condon terms in the nuclear coordinate expansion of the Raman amplitude [6.17], and to porphyrins which clearly exhibit scattering via the Herzberg-Teller terms in the expansion [6.18]. When appropriate, other less generally important terms in the expansion will also be presented. Some examples of the types of molecular property information available from resonance Raman scattering will also be presented.

6.1 Resonance Raman Scattering

In a Raman process from a molecular system, an incident photon is destroyed, a scattered photon is created, and the molecular system undergoes a transition from an initial state to a final state. This process is coherent and therefore these three events are not independent and resolvable, i.e., we may not view this process as a sequential absorption then emission of a photon but instead as a process in which the three events occur indivisibly. If the excitation frequency is close to or in resonance with an available excited state, the process is called *Resonance Raman Scattering* (RRS). In varying the excitation frequency from off resonance to on resonance striking changes in the intensity, temporal behavior, and frequency distributions generally occur. In spite of these frequency dependent differences, the same theoretical treatment is applicable independent of excitation frequency. It is not necessary to arbitrarily separate out any special regions of incident frequency for separate consideration.

As the excitation frequency is scanned through resonance with a separated discrete transition smooth continuous changes are observed in the intensity of the re-emission. The temporal behavior, however, is somewhat more confusing. For excitation frequencies well off resonance, the scattering time is very fast in that the re-emission intensity envelope accurately reproduces variations in the excitation intensity. If the excitation frequency is exactly on resonance with the transition, the scattering time can be much slower, reflecting the natural lifetime of the intermediate state involved. For intermediate frequency shifts away from resonance, the behavior is much more complex, reflecting fundamental uncertainty principle limitations. These variations are well explained by the Raman process, and a much more detailed discussion, in addition to a comparison with experimental results, is presented in Sect.6.2.

In real systems other radiative processes are often observed simultaneously with the Raman re-emission. Redistribution scattering [6.19], hot luminescence [6.20,21], and relaxed fluorescence [6.22] are such processes that may occur when the system can exchange energy with its environment. In redistribution scattering the phase of the process is interrupted by an interaction with the environment so that the process becomes incoherent and typically displays very different spectral and temporal behavior. For relaxed fluorescence or hot luminescence the state of the system is changed through either radiative or non-radiative interaction with the environment. These processes are also incoherent. They will only be considered in more detail in this chapter when they either interfere with or influence the resonant Raman reemission.

The basic theories of Raman scattering have been described in numerous reports [6.2-6,12,23,24]. Here we introduce the general expressions from which the Raman intensity may be derived and then present various approximations to the electronic state wavefunctions which appropriately may be made in order to be able to interpret Raman spectra from various molecular systems. The total scattering intensity I_S in photons per molecule per second for a transition from the initial ground state, $|G>$, to the final state, $|F>$, may be given by the standard expression [6.2,23]

$$I_S = \frac{8\pi\omega_S^4 I_L}{9c^4} \sum_{\rho\sigma} |(\underline{\underline{\alpha}}_{\rho\sigma})_{GF}|^2 \quad . \tag{6.1}$$

Here c is the velocity of light, I_L is the incident intensity at frequency ω_L, and ω_S is the scattered frequency. $(\underline{\underline{\alpha}}_{\rho\sigma})_{GF}$ is the polarizability tensor component for the transition from $|G>$ to $|F>$ with incident and scattered polarizations indicated by σ and ρ, respectively. The second-order perturbation expression for $(\underline{\underline{\alpha}}_{\rho\sigma})_{GF}$ is

$$(\underline{\underline{\alpha}}_{\rho\sigma})_{GF} = \frac{1}{\hbar} \sum_I \left(\frac{<F|r_\rho|I><I|r_\sigma|G>}{\omega_{GI}-\omega_L-i\Gamma_I} + \frac{<I|r_\rho|G><F|r_\sigma|I>}{\omega_{IF}+\omega_L-i\Gamma_I} \right) \quad . \tag{6.2}$$

between an excited rotational-vibrational (rovibronic) state $|I>$ and the ground state $|G>$; Γ_I is a damping term reflecting the homogeneous width of state $|I>$; and r is the electron position operator. The notation used in these and in subsequent equations is listed in Table 6.1. For simplicity we have not included the effects of damping of the initial and final states, $|G>$ and $|F>$.

It is very difficult to make direct interpretations from (6.2) because the exact molecular eigenstates in principle depend on *both* nuclear and electronic coordinates in a complicated way and are very difficult to utilize. To simplify the matrix elements in (6.2) we introduce the Born-Oppenheimer approximation where the wavefunction ψ representing the state I may be expressed as

$$\psi_{i,k}(\underline{r},\underline{R}) = \theta_i(\underline{r},\underline{R})\phi_{i,k}(\underline{R}) \tag{6.3}$$

where $\underline{r}$ represents a set of electronic coordinates and $\underline{R}$ a set of nuclear coordinates. In this expression $\theta_i(\underline{r},\underline{R})$ is the electronic wavefunction for the electronic state i

---

$\hbar = h/2\pi$ (normalized Planck's constant).

Table 6.1. Notation used in Chap.6

G,F	Complete rovibronic ground electronic states	$\underline{r}$	Electron coordinate
I	Complete rovibronic excited electronic states	$\underline{R}$	Nuclear coordinate
g	Born-Oppenheimer ground electronic state	ξ	Normal vibrational mode
i,j	Born-Oppenheimer excited electronic state	R_ξ	Normal coordinate of mode ξ
Q,B	Born-Oppenheimer excited electronic states in porphyrins	M	Electronic matrix element
u,v	Vibrational levels of a normal mode of state g	h	Herzberg-Teller perturbation term
k,t	Vibrational levels of a normal mode of state i (or Q)	Γ_k	linewidth of state k
m,n	Vibrational levels of a normal mode of state j (or B)	$\hbar\omega_L$	Incident photon energy
$\underline{\underline{\alpha}}$	Polarizability tensor	$\hbar\omega_S$	Scattered photon energy
σ,ρ	Polarization directions	$\hbar\omega_{vk}$	Energy difference between states v and k
$\underline{r}$	Electron coordinate	Q_0	v=0 level of excited electronic state Q in porphyrin macrocycle
		Q_1	v=1 level of excited electronic state Q in porphyrin macrocycle
		B	Soret state of porphyrin macrocycle
		$Q(\underline{r},R_0)$	Crude Born-Oppenheimer state of Q

and it depends parametrically on the nuclear coordinate $\underline{R}$. $\phi_{i,k}(\underline{R})$ is the vibrational wavefunction with quantum number k. (For molecules with more than two atoms k will represent a set of quantum numbers.) For most cases the Born-Oppenheimer approximation accurately describes the molecular states. However, in cases of near degenerate levels of different vibronic manifolds, breakdowns in the Born-Oppenheimer approximation may occur. These may be introduced in the scattering formalism by expansion of the exact molecular eigenstates using the Born-Oppenheimer states as a basis. This point will be discussed in greater detail in Sect.6.4.7. For the present we will assume that the adiabatic (Born-Oppenheimer) approximation is valid.

In writing (6.3) in the above form we implicitly ignore the dependence of the wavefunctions on molecular orientation (the angular coordinates of a molecular-fixed axis). Inclusion of this dependence in the matrix elements of (6.2) only results in a multiplicative factor which depends on the rotational quantum numbers and on the polarization of the incident and scattered light. This factor is responsible for the dipole selection rule on rotational transitions, but in most of the following discussion it plays no other role, and for clarity of presentation it will be ignored. Detailed evaluation of these rotational matrix elements for homonuclear, diatomic molecules has been carried out most recently by SILVERSTEIN and St. PETERS [6.25], by JACON and VAN LEBEKE [6.26], and by WILLIAMS et al. [6.27].

By substituting (6.3) in (6.2) and taking the initial and final states to belong to the ground electronic state manifold the matrix elements may be rewritten. For example, the first matrix element is

$$\langle F|r_\rho|I\rangle = \langle\phi_{g,v}(\underline{R})\theta_g(\underline{r},\underline{R})|r_\rho|\theta_i(\underline{r},\underline{R})\phi_{i,k}(\underline{R})\rangle \tag{6.4}$$

Here $\phi_{g,v}(\underline{R})$ is the Born-Oppenheimer nuclear wavefunction of state $|F\rangle$ and $\theta_g(\underline{r},\underline{R})$ is the electronic wavefunction. The v and k designate the vibrational modes of the electronic states g and i, respectively. The integration over electronic coordinates may be carried out to yield the $\underline{R}$ dependent electronic dipole matrix element

$$M_{ig}(\underline{R}) = \int \theta_g^*(\underline{r},\underline{R})r_\rho\theta_i(\underline{r},\underline{R})d\underline{r} \quad . \tag{6.5}$$

These matrix elements are generally weakly varying functions of the internuclear coordinates $\underline{R}$, so that they may be expanded in a multidimensional Taylor series about the equilibrium configuration $\underline{R}_0$. In order to maintain clarity of presentation, here we will consider cases in which only one normal mode coordinate R_ξ of the ξth normal mode is important in this expansion. The extension of these ideas to more general cases is trivial, but leads to much bulkier notation. Also to keep the notation simple we have not included the polarization subscript on the matrix element $M_{ig}(\underline{R})$. For the Taylor series expansion we obtain the expression

$$M_{ig}(\underline{R}) = M_{ig}(\underline{R}_0) + \left(\frac{\partial M_{ig}}{\partial R_\xi}\right)_{R_0} R_\xi + \dots \quad . \tag{6.6}$$

Using the Taylor series expansion and considering only one excited electronic state, with vibrational levels k for a Raman transition from initial state v to final state u, we obtain for the zero and first order terms the polarizability

$$(\underline{\underline{\alpha}})_{u,v} = |M_{ig}(\underline{R}_0)|^2 \frac{1}{\hbar}\sum_k \frac{\langle u|k\rangle\langle k|v\rangle}{\omega_{vk}-\omega_L-i\Gamma_k}$$

$$+ M_{ig}(\underline{R}_0)M'_{ig}(\underline{R}_0)\frac{1}{\hbar}\sum_k \frac{\langle u|R_\xi|k\rangle\langle k|v\rangle+\langle u|k\rangle\langle k|R_\xi|v\rangle}{\omega_{vk}-\omega_L-i\Gamma_k} + \dots \quad . \tag{6.7}$$

In this expression $M'_{ig}(\underline{R}_0)$ represents $(\partial M_{ig}/\partial R_\xi)_{R_0}$; the vibrational states u, v belong to the ground electronic manifold g, and the vibrational state k belongs to the intermediate electronic manifold i (all of the notation is summarized in Table 6.1). In writing (6.7) we have dropped the polarization subscripts and ignored the second term in (6.2)—since very near resonance it is much smaller than the first term owing to the large energy denominator. Also here and in subsequent discussions we have neglected the antiresonance term in (6.2). In many cases more than one electronic state may contribute to the sum over states in (6.7). It is straight forward to include other states and examples will be given in Sect.6.2.2.

In the first term of (6.7) the matrix elements are simply vibrational overlap factors (the squares of such factors are called Franck-Condon factors). The term corresponds to ALBRECHT's A term [6.23] but Raman intensity derived from it is also referred to as "Franck-Condon scattering" or just "Condon scattering". The second term in (6.7) corresponds to ALBRECHT's B term [6.23]. However, in the formulation written here the dependence of the electronic moment on internuclear separation is described in general terms. The ALBRECHT formulation [6.23] specifically treats the coordinate dependence of the electronic moment as a Herzberg-Teller perturbation, i.e., indicates a partial derivative

$$M'_{ig}(\underline{R}_0) = \sum_j M_{jg}(\underline{R}_0)h_{ij} \quad , \quad h_{ij} = \frac{1}{\hbar}\frac{\langle i|\partial H/\partial R_\xi|j\rangle}{\omega_i-\omega_j} \quad . \tag{6.8}$$

In addition to being referred to as the ALBRECHT B term this term is also called the Herzberg-Teller term. The ALBRECHT C term results from inclusion of coupling to the ground state which we have dropped from (6.8) for simplicity. In detailed treatments of some complex systems, such as porphyrins for example, it becomes necessary to include non-adiabatic interactions (i.e., breakdown of the Born-Oppenheimer approximation). Such interactions may be included by expanding the molecular eigenstates defined in (6.3) and lead to additional terms in the polarizability expansion (6.7). These effects will be discussed in Sect.6.4.6.

The Herzberg-Teller interaction (6.8) results from the mixing of two electronic states by the coordinate operator. This electronic state mixing distorts both the upper and lower states that are mixed, giving "extra" intensity to the weaker one thereby creating a strong resonance Raman enhancement. The modes that are group theoretically capable of coupling the electronic states may be directly determined by examining the matrix elements in (6.8).

For most systems the basic features of the Raman re-emission may be explained by either Franck-Condon scattering or by Herzberg-Teller scattering. We will therefore devote most of our attention to these two scattering terms. First, however, it is instructive to consider the effect on the re-emission spectra of having a nearly constant denominator in (6.7). For example, this may occur in situations in which the laser frequency is far from resonance with any state with a large transition matrix element or where the damping constant Γ_k is large and approximately constant for all states $|k\rangle$ with significant transition matrix elements. Under these conditions the summation in the first term in (6.7) reduces to $\langle u|v\rangle$ as a result of completeness of the set of states $|k\rangle$. Because of the orthogonality of the states $|u\rangle$ and $|v\rangle$ this term then allows only elastic (Rayleigh) scattering. Inelastic (vibrational Raman) scattering must then arise from the higher-order terms in (6.7) or from corrections to the first term arising from small variations in the constancy of the energy denominator. Similar considerations of the second term in (6.7) result in the replacement of the summation in it by $\langle u|R_\xi|v\rangle$. If we assume harmonic oscillator wavefunctions this term yields first order Raman scattering only, i.e., $\langle u|R_\xi|v\rangle = \delta_{u,v\pm1}$. Higher-order terms which are expected to be weaker yet result in the vibrational overtones. Thus, far from resonance or in situations in which the excited state damping constant Γ_k is large and relatively constant, we expect strong Rayleigh scattering, with weaker fundamental Raman scattering, and with very weak overtone scattering.

This behavior is in sharp contrast to that expected and observed when the laser frequency is close to resonance with some well defined excited electronic state. In this case, the energy denominator is not constant in any approximation, and completeness may not be used to collapse the sum. If the transition matrix element is large and the excited state potential function has a different shape and internuclear equilibrium position than the ground state, then the Franck-Condon term (A term) is expected to dominate the scattering. If this is the case there is no restriction on the vibrational quantum number change in the Raman process and overtone progressions are observed. The relative intensities of the bands in these progressions is governed only by the magnitude of the relevant overlap factors, i.e., $\langle u|k\rangle$. For the matrix elements in the first term in (6.7) to be non-zero they must be invariant to all symmetry operations of the molecular symmetry group. Since the inital vibrational level is generally totally symmetric only totally symmetric vibrations may serve as intermediate states and only totally symmetric vibrations become enhanced.

In cases in which the incident frequency is in resonance with a very weak transition in a system in which there is a nearby very strong transition, the electronic matrix elements in the second term in (6.7) through a Herzberg-Teller interaction may be substantially larger than those in the first: $M_{ig}(\underset{\sim}{R}_0)M_{jg}(\underset{\sim}{R}_0) \gg |M_{ig}(\underset{\sim}{R}_0)|^2$. If this is the case the resonant behavior may be dominated by the second term in the equation and harmonics of asymmetric modes may then be observed since due to the matrix element $\langle k|R_\xi|v\rangle$, the intermediate vibrational state no longer needs be totally symmetric. In addition to the influence of a dominant nearby electronic state, in many systems the excited state potential surface is nearly the same as that of the ground state. Hence the vibrational wavefunctions are orthogonal so the Franck-Condon term results only in Rayleigh scattering and the Herzberg-Teller term dominates. When the two potentials are similar, but not identical, this consideration still has the effect of favoring the Herzberg-Teller term.

In the remaining sections of this chapter we consider in some detail resonance Raman scattering from specific systems. To illustrate Franck-Condon scattering a great deal of attention will be given to homonuclear diatomic molecules because they are simple enough to clearly demonstrate many of the principles involved without undue interference from unwanted complications, and because we are quite familiar with these molecules. To illustrate Herzberg-Teller scattering most attention will be given to porphyrins, again because we are quite familiar with them and also because they have been very well studied and clearly demonstrate the general principles.

6.2 Diatomic Molecules

For ease of discussion of diatomic molecules it is convenient to classify the range of incident frequencies into three distinct regions [6.17]. This classification is illustrated in Fig.6.1. On the left, when the incident frequency is far from resonance with a real transition we label the re-emission as normal Raman scattering. In the center, when the incident frequency is in the region of discrete vibronic transitions we term the re-emission discrete resonance Raman scattering independent of whether or not the incident frequency is at exact resonance. To the right, when the incident frequency is above the excited state dissociation limit in a continuum region, we label the re-emission as continuum resonance Raman scattering. The differences in the scattering behavior in these three excitation frequency regions is very great. The halogen molecules have served as examples in which experimental observation could be made in each of these three regions.

In the case of ordinary nonresonance Raman scattering, the reradiation is characterized by the following: 1) The Rayleigh line is strong and the fundamental vibrational transition ($\Delta v=1$) is weak. The higher overtones are weaker yet and are generally not observed at all in the spectrum; 2) The scattered band envelope is relatively broad, may have some structure, and exhibits no dependence on the excitation frequency; 3) The scattering intensity varies slowly and smoothly (an ω^4 dependence) with excitation frequency; 4) The Stokes-anti-Stokes ratio may be calculated from the Boltzmann factor; 5) The scattering time is very fast; 6) Addition of foreign

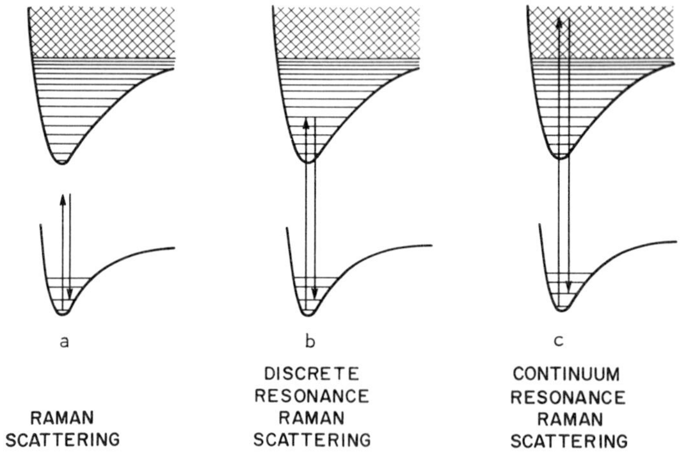

Fig. 6.1a-c. Classification of Raman scattering from diatomic molecules according to laser frequency. a) The incident laser frequency is far from resonance with any real electronic transition so *Normal Raman Scattering* is observed. b) The incident laser frequency is in the region of discrete levels of a single electronic intermediate state. We term this process *Discrete Resonance Raman Scattering*. c) The incident frequency is in the range of dissociative continuum levels. We label this process *Continuum Resonance Raman Scattering*. (From [6.17])

gas does not quench the scattering intensity; 7) The Q-branch depolarization ratio is small, with the exact value determined by the anisotropy in the polarizability.

At higher incident laser frequencies in the region of discrete resonance Raman scattering the properties of the re-emission are significantly different: 1) The overtones may be of comparable intensity and, in fact, may be as strong as the Rayleigh line. The ratio of the intensity of one overtone to the next varies erratically; 2) With sufficiently narrow excitation each overtone consists of only a few very sharp lines. Their widths are limited by the combination of the natural, collisional, and Doppler broadening in addition to possible nuclear hyperfine splitting. The structure within each overtone changes erratically with small incident frequency shifts; 3) The scattered intensity varies greatly with excitation frequency; 4) The Stokes-anti-Stokes ratio is not simply related to the ground state population factors and in fact the anti-Stokes intensity may be greater than the Stokes intensity in some cases; 5) At exact resonance the scattering time can be long and is typically in the range of 10^{-5}-10^{-8} s, although as the incident frequency is moved away from exact resonance the scattering time becomes very short; 6) Addition of foreign gas strongly quenches the intensity; 7) Generally scattering from the Q-branch is depolarized, with a depolarization ratio of 3/4.

In the region of continuum resonance Raman scattering the observed spectra are substantially different from those in the discrete resonance case: 1) Again there are numerous overtones with comparable intensity to the fundamental and the Rayleigh line. The intensity variation from one overtone to the next is now very regular; 2) The band envelope within each overtone region appears broad with considerable structure which smoothly changes as the excitation frequency is varied; 3) The scattering intensity varies with the excitation frequency in a smooth systematic way; 4) The Stokes-anti-Stokes ratio is not solely related to the ground state population factors but varies systematically with incident frequency; 5) The scattering time is fast; 6) Addition of foreign gas does not quench the intensity; 7) The depolarization ratio is small and may be calculated from the electronic structure of the molecule. Despite these differences, the one Raman scattering process properly accounts for the behavior in all three regions [6.17].

In off resonance Raman scattering, as discussed earlier, only elastic, Rayleigh scattering is allowed in lowest order, with fundamental Raman lines arising from first-order correction terms. For diatomic molecules where there is only one degree of freedom this results in a single vibrational line at the resonance frequency of the oscillator in addition to the Rayleigh line.

For resonance Raman scattering where the incident frequency coincides with some real transition ($\omega_L \approx \omega_{vk}$) a situation completely different from the non-resonance case arises. Since the damping factor for a low pressure gaseous system is quite small, for the resonance condition in the sum over states in (6.7) the denominators are not constant, so the completeness argument fails and all terms may contribute to both the Rayleigh and the Raman scattering. However, because of the rapidly converging

Taylor series most of the Rayleigh and the Raman scattering intensity will now result from the first term alone, i.e.,

$$(\underline{\underline{\alpha}})_{uv} = |M_{ig}(\underline{R}_0)|^2 \frac{1}{\hbar} \sum_k \frac{\langle u|k\rangle\langle k|v\rangle}{\omega_{vk}-\omega_L-i\Gamma_k} \quad . \tag{6.9}$$

No longer is there a selection rule discriminating against overtones, but instead the overtones are expected to be strong and dependent only on the magnitude of the appropriate Franck-Condon overlap factors. By using (6.9) the difference between discrete resonance Raman scattering and continuum resonance Raman scattering now may be understood.

6.2.1 Discrete Resonance Raman Scattering

When the incident frequency is in the discrete resonance region and resonant with a specific initial to intermediate state transition then that transition can dominate the summation in which case all other transitions are weak in comparison. Equation (6.9) may then be written

$$(\underline{\underline{\alpha}})_{uv} = |M_{ig}(\underline{R}_0)|^2 \frac{1}{\hbar} \frac{\langle u|k\rangle\langle k|v\rangle}{\omega_{vk}-\omega_L-i\Gamma_k} \quad . \tag{6.10}$$

From this equation we see that the structure in the spectrum of the re-emitted light then will depend on the specific vibrational and rotational quantum numbers of the selected excited state. The intensities of the successive overtones will vary erratically depending on the magnitudes of the Franck-Condon overlap factors connecting the intermediate and final states. A slight change in incident frequency will bring about resonance with a different discrete state of the complicated excited state vibration-rotational manifold, with different vibrational and rotational quantum numbers so that a completely different appearing re-emission spectrum may be observed.

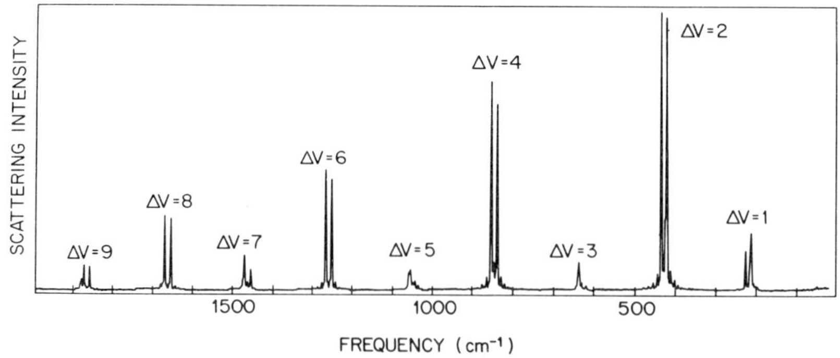

Fig. 6.2. Survey spectrum of discrete resonance Raman scattering from I_2. The laser wavelength was 5145 Å and an etalon was used for single mode operation. The doublets result from the $\Delta J = \pm 1$ dipole selection rule for changes in the rotational quantum number. The frequency scale is the wavenumber shift from the excitation frequency (From [6.17])

Extensive data of discrete resonance Raman scattering has been obtained on molecular iodine by using narrow line laser excitation [6.28]. As predicted a series of overtones of erratically varying intensity are obtained. Such a spectrum [6.17] is shown in Fig.6.2. This spectrum was obtained on a gaseous sample at an I_2 vapor pressure of 0.25 torr and a single mode argon ion laser oscillating in the 5145 Å region was used as the excitation source. The vibrational fundamental appears at about 213 cm^{-1} and each overtone is separated by about this frequency. Each of the overtones results from a transition to a specific intermediate excited state, and then a re-emission to the ground state with the overall change in the vibrational quantum number indicated in the figure. The doublets result from the $\Delta J = \pm 1$ dipole selection rule for rotational transitions.

Often for discrete resonance Raman scattering only one intermediate state dominates the scattering, resulting in the re-emission doublets seen here. The intensity variation between overtones results from this variation in the Franck-Condon overlap factors between this intermediate state and the final state, i.e., $\langle u|k\rangle$. The additional structure seen near some of the overtones results from collisional transfer lines and re-emission lines from additional intermediate states that are only slightly out of resonance.

By changing laser modes it is possible to change from resonance with one intermediate state transition to resonance with another. In Fig.6.3 a higher resolution spectrum of the vibration fundamental region of Fig.6.2 is shown (labeled Mode #20)

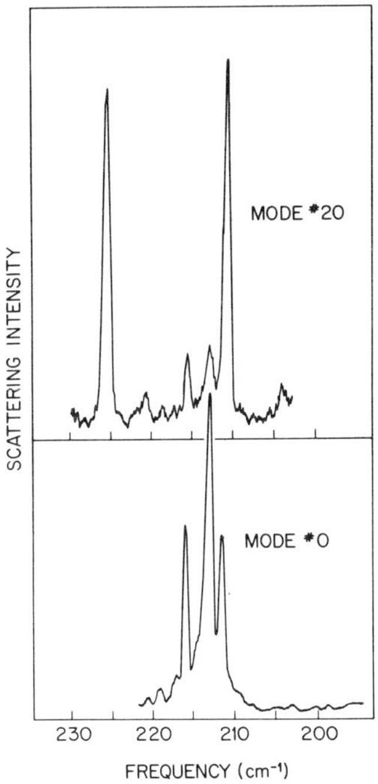

Fig. 6.3. Re-emission spectra of the vibrational fundamental region of I_2 excited in the region of discrete resonance Raman scattering with 5145 Å. The laser mode spacing is about 130 MHz. The top spectrum, labeled Mode 20, was taken at the same laser frequency as the spectrum in Fig.6.2 (From [6.17])

and compared to a spectrum obtained with a laser mode shifted by ~2.5 GHz (labeled Mode #0). In shifting the frequency by only 2.5 GHz, the emission doublet has disappeared entirely and a triplet is now observed. This triplet results from the dipole selection rule when the laser is accidentally in resonance with two transitions.

An interesting feature of the discrete resonance Raman scattering is a consideration of the re-emission lifetime. In normal off-resonance Raman scattering the re-emission time is, of course, very fast. An upper limit for the scattering time in this case may be given by $(\Delta\omega^2+\Gamma^2)^{-1/2}$ where Γ is the natural width. Since in off-resonance experiments $\Delta\omega$ is very large the scattering time is very small and hence may be considered to be instantaneous. For continuum resonance Raman scattering when the incident frequency is above a dissociation limit the re-emission time must be limited by the time needed for the atoms to fly apart ($<10^{-12}$ s). However, in discrete Raman scattering, when the excitation is on resonance the re-emission time is expected to be characteristic of the excited state involved, and both theoretical and experimental considerations of its variation in moving away from resonance is intriguing [6.17,29].

Consider a simple three level system such as that shown in Fig.6.4. The scattering process consists of the absorption of an excitation photon (angular frequency ω_L) and the emission of a Raman shifted photon (angular frequency ω_S) with a consequent change in the system state from $|v\rangle$ to $|u\rangle$. Although throughout this section we use Raman scattering as our model in the lifetime calculations, the arguments are equally applicable to Rayleigh scattering. We now apply a pulse of light, such as that shown in Fig. 6.5, of frequency ω_L to the system and ask what the time response of the Raman shifted re-emission is. The answer to this question may be readily determined by use of time dependent perturbation theory. Let S(t) be the envelope function for the incident light pulse, i.e., let the incident field be given by $E_L(t) = S(t)\exp(i\omega_L t)$. We will assume $S(t) = 0$ for all $t \leq 0$, but this restriction is not necessary. Then the time dependent intensity of the Raman shifted re-emission is given by [6.17]

$$I_S(t) = \frac{8\pi\omega_S^4 I_L}{9c^4\hbar^2}\,|\langle F|r|I\rangle\langle I|r|G\rangle|^2\,|R(t)|^2 \qquad (6.11)$$

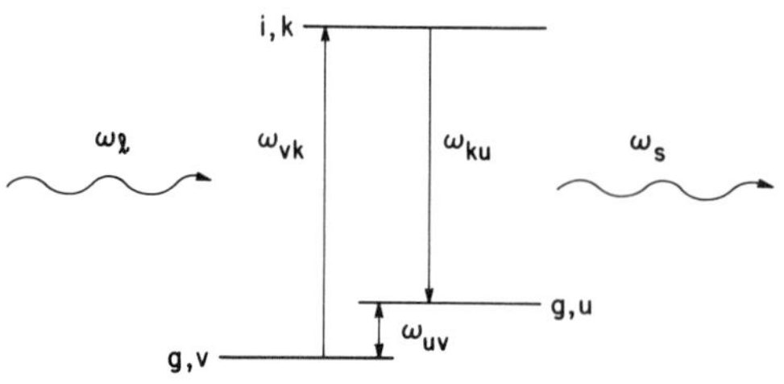

Fig. 6.4. Model three level system for considering discrete resonance Raman scattering. ω_L and ω_S are the incident and scattered frequencies, respectively. g is the electronic ground state in which a Raman transition from vibrational level v to u is occurring. i is an excited electronic state with a vibrational level k such that the energy $\hbar\omega_L$ is at resonance with $\hbar\omega_{vk}$. ω_{uv} is the Raman shift

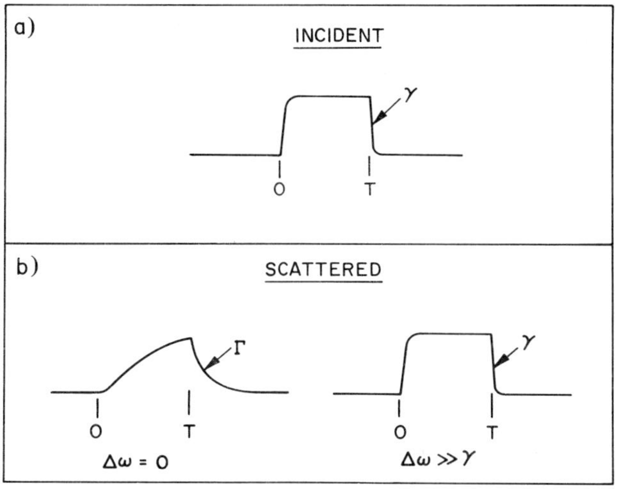

Fig. 6.5a and b. Incident and scattered theoretical pulse shapes. a) Rectangular laser pulse with finite rinse and fall times used in the calculations. b) Calculated scattered pulse shapes for resonant excitation ($\Delta\omega=0$) and off-resonant excitation ($\Delta\omega>>\gamma$). (From [6.17])

where the time dependent response function R(t) is given in terms of the envelope function S(t) by

$$R(t) = \exp(-\Gamma_k t)\ \exp(-\omega_{GI} t) \int_0^t \exp[i(\omega_{GI}-i\Gamma_k)\tau]\ \exp(-\omega_L \tau) S(\tau) d\tau \quad . \tag{6.12}$$

In deriving (6.11) it was assumed that the spectrometer bandpass was sufficient to pass essentially all of the Stokes-shifted emission.

To examine this expression in more detail, we take the incident pulse envelope S(t) to have the shape shown in Fig.6.5a. Although many pulse shapes have been proposed [6.30], we find this pulse shape to be the most convenient one for interpreting time response behavior. The exponential rise and fall times minimize transient effects from rapid turn-on and turn-off of the pulse and the excitation has not been turned on for an infinite time in the past. Finally, by choosing T sufficiently large, the deconvolution of the input pulse from the output is relatively easy.

Specifically, the function S(t) is defined by

$$\begin{aligned} S(t) &= 0 && t \leq 0 \\ &= (1-\exp[-\gamma t]) && 0 \leq t \leq T \\ &= (1-\exp[-\gamma T])\exp[-\gamma(t-T)] && T \leq t \quad , \end{aligned} \tag{6.13}$$

where T is the pulsewidth and γ^{-1} gives the rise and fall time of the electric field of the pulse. If we assume the pulse rise and fall times are much faster than the natural lifetime of the excited state $\gamma >> \Gamma$, then for exact resonance $\Delta\omega = \omega_{GI} - \omega_L = 0$, we get

$$|R(t)|^2 \simeq \frac{1}{\Gamma^2}(1-\exp[-\Gamma t])^2 \qquad 0 \leq t \leq T$$

$$\simeq \frac{1}{\Gamma^2}(1-\exp[-\Gamma T])^2 \exp[-2\Gamma(t-T)] \quad T \leq t \quad . \tag{6.14}$$

$|R(t)|^2$ for this case, as shown in Fig.6.5, has a rise and fall time characteristic of the natural response time of the system $(2\Gamma)^{-1}$. However, far off-resonance, the response is "fast" in that it follows the excitation pulse. For this case in which the shift from resonance is much greater than the Fourier broadening of the pulse, i.e., $\Delta\omega >> \gamma$, we get

$$|R(t)|^2 \simeq \frac{1}{\Delta\omega^2}(1-\exp[-\gamma t])^2 \qquad 0 \leq t \leq T$$

$$\simeq \frac{1}{\Delta\omega^2}(1-\exp[-\gamma T])^2 \exp[-2\gamma(t-T)] \quad T \leq t \quad . \tag{6.15}$$

The behavior of $|R(t)|^2$ under this condition is shown in Fig.6.5 also.

Actual measurements of the re-emission lifetime were made by WILLIAMS et al. [6.29] on molecular I_2 at 0.03 torr in the 215 cm^{-1} region of the fundamental Raman mode. These results are shown in Fig.6.6. The laser pulse width in these measure-

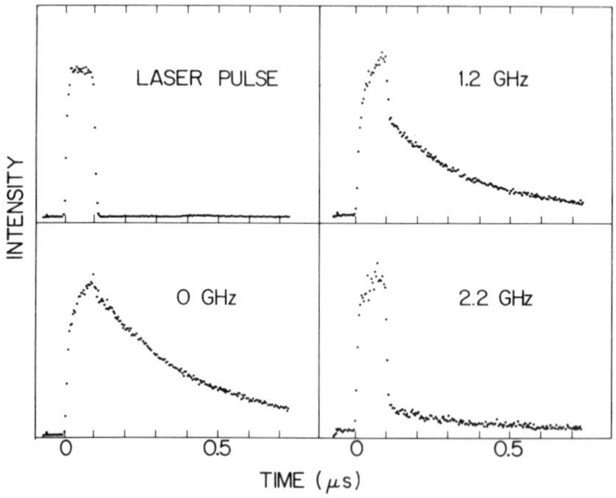

Fig. 6.6. Time decay data for pulsed excitation of molecular iodine. The spectrometer was set to observe the re-emission from a single rotational component in the region of the Raman fundamental (~216 cm^{-1} shift). The laser pulse is shown in the upper left. In the lower left is the time decay data obtained for resonant excitation and on the right are data for off-resonant excitation with frequency shifts of 1.2 GHz and 2.2 GHz. The accumulation time was substantially different for each plot ranging from a few minutes for the 0 GHz decay plot to several hours for the 2.2 GHz decay plot. (From [6.29])

ments was 100 ns. The 0.0 GHz spectrum was obtained by adjusting the laser frequency to obtain the maximum intensity of the re-emission, i.e., to be at the center of the Doppler broadened and hyperfine split lines. The other time decay measurements were made by shifting the laser modes away from this position.

As predicted for on-resonance excitation, a long lifetime is observed although it is somewhat shorter than the I_2 natural lifetime due to a spatial diffusion effect caused by the narrow slits used in the experiment. At 1.2 GHz off-resonance there are two different lifetime contributions one short-lived re-emission, decaying in a time limited by the 10 ns temporal resolution, and one long-lived component with the same decay constant as the 0.0 GHz measurement. This behavior results from the complicated line shape of the I_2 absorption bands consisting of natural as well as Doppler broadening. Due to the Doppler broadening the excitation frequency is exactly on-resonance with some molecules giving the long-lived contribution and is well off-resonance for other molecules giving the short-lived contribution. Further away from the line center at 2.2 GHz a negligible number of molecules are resonant with the laser frequency and the re-emission is dominated by the fast response.

The conclusions from both theory and experiment for the lifetime measurements are clear. On-resonance the lifetime is long and off-resonance it is limited by the rise time (or fall time) of the excitation pulse. The theory shows intermediate frequency shifts to have a very complex response with sinusoidally oscillating terms. In a real system due to inhomogeneous broadening mechanisms these oscillations are not normally observed. Although one might naively expect to be able to observe a continuous variation in lifetime by smoothly shifting the incident frequency away from resonance, in practice fundamental uncertainty principle limitations prevent the direct observation of intermediate lifetimes.

Many workers use the term resonance fluorescence to describe the re-emission following on-resonance excitation. Although it is only a semantic argument we prefer to use the term discrete resonance Raman scattering to describe re-emission following excitation with frequencies throughout the region of well defined discrete transitions in order to emphasize the underlying unity of the basic physical process. The distinction between resonance fluorescence and Raman scattering becomes in many cases somewhat arbitrary and considerable confusion has resulted. As in any Raman process, there is no loss in phase information in discrete RRS. Only when there is a phase interruption in the excited state does it become possible to consider the absorbed and emitted photons independently. These processes are more appropriately separated from the undisturbed processes by terming them fluorescence, and indeed when collisions interrupt the phase of the excited state, the re-emission properties, i.e., the frequency width and position, and the lifetime, are dramatically altered.

6.2.2 Continuum Resonance Raman Scattering

For excitation into the continuum [6.31], just as in the case of discrete RRS, we must consider (6.9)

$$(\underline{\underline{\alpha}})_{uv} = |M_{ig}(\underline{R}_0)|^2 \frac{1}{\hbar} \sum_k \frac{\langle u|k\rangle\langle k|v\rangle}{\omega_{vk}-\omega_L-i\Gamma_k} \tag{6.16}$$

and again we would expect overtones to be strong and depend only on the magnitude of the appropriate Franck-Condon factors. Here, because we are dealing with a continuum, we may assume the limit of infinitely narrow states and it is not necessary to include the damping term. However, as will be discussed later, it is necessary to accommodate the pole in the denominator which occurs for those continuum states at exact resonance.

For excitation into a continuum unlike the discrete case the resonant denominator no longer picks out a single vibrational-rotational state transition. Instead, all populated vibrational rotational levels of the ground electronic state are at resonance with some continuum state. The re-emission observed in a given overtone now, rather than resulting from a single vibrational-rotational transition, is composed of a summation of lines resulting from Raman transitions from all the populated levels. This large number of superimposed bands has an averaging effect on the overtone intensities and frequency dependence. An additional difference exists between the discrete and continuum resonance cases. In the discrete case a given line in the re-emission spectrum results from a single excited state energy level. Because of the $\Delta J = \pm 1$ dipole transition selection rule, the next closest intermediate state which could contribute to the intensity of that line is one separated by the excited state vibrational spacing. Because of the resonance denominator this "off-resonance" scattering does not contribute significantly to the intensity of the line in question. In contrast, though, in the continuum region the spacing of vibrational levels is effectively infinitesimal, and a given re-emission line gets intensity not only from the continuum state it is in exact resonance with but also from other nearby continuum states that are only slightly off-resonance. This serves as an additional averaging effect on the intensity. In the continuum resonance case the various overtones, rather than varying in a haphazard manner as in the discrete case, vary in a systematic manner because of those two averaging effects. Similarly, when the excitation frequency is changed, the form of the spectra changes in a very continuous and systematic manner as certain vibrational-rotational states become weighted more or less heavily by the combination of the Franck-Condon factors and the resonance denominator.

Continuum resonance Raman scattering from molecular iodine was first reported by HOLZER et al. [6.32]. Shortly after that KIEFER and BERNSTEIN [6.33] reported high-resolution spectra of I_2 with 4880 Å excitation and were able to interpret the complex fine structure. A survey spectrum of I_2 with 4880 Å excitation is shown in Fig.6.7. As in the discrete case many overtones are observed. However, in contrast to the dis-

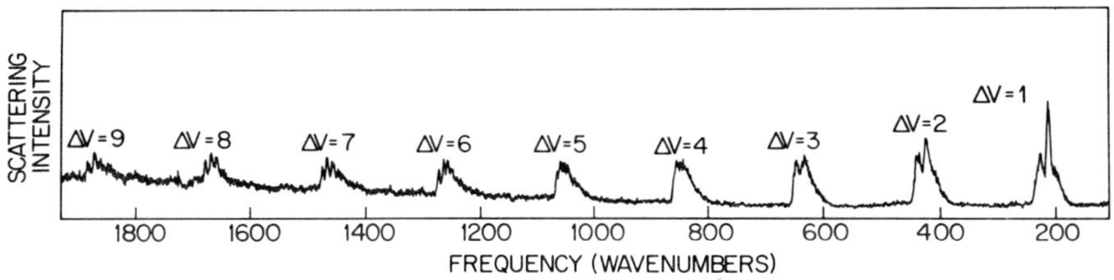

Fig. 6.7. Survey spectrum of continuum resonance Raman scattering from I_2. The laser excitation wavelength was 4880 Å. (From [6.17])

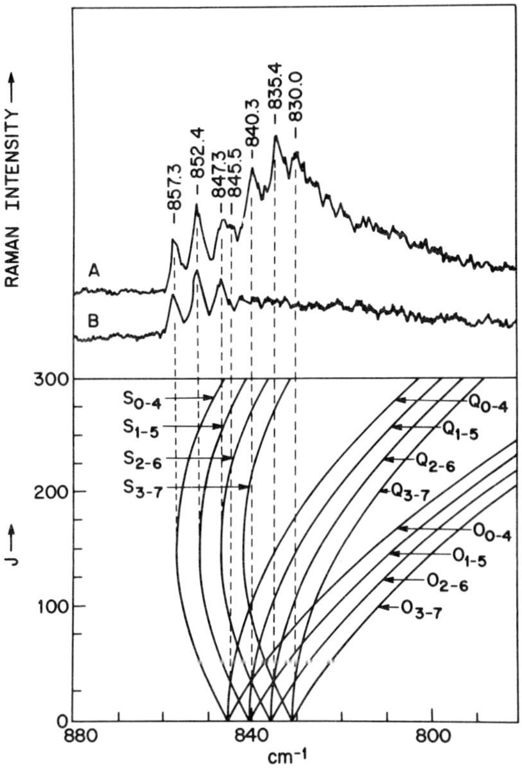

Fig. 6.8. Third overtone of resonance Raman scattering from molecular iodine obtained with 4880 Å excitation. In the top are the experimental data obtained in polarized (A) and depolarized (B) orientations. On the bottom are the Fortrat diagrams for each of the rotational branches for the four most populated levels. The dashed lines show the relation of the Raman peaks to the band heads. (From [6.33])

crete case, the relative intensity of one overtone to the next varies in a systematic way. The structure within each overtone is real and results from the O, Q, and S branches of the multitude of vibrational and rotational transitions that are occurring. This structure may be better seen under high resolution conditions as shown in Fig.6.8 taken from [6.33] for the $\Delta v = 4$ Raman line. As was pointed out by them, consideration of the Fortrat diagram for each of the rotational branches readily explains the origin of the structure in the spectra. As may be seen in Fig.6.8, where the spectra are displayed above and the Fortrat diagrams below, the strong sharp features result from

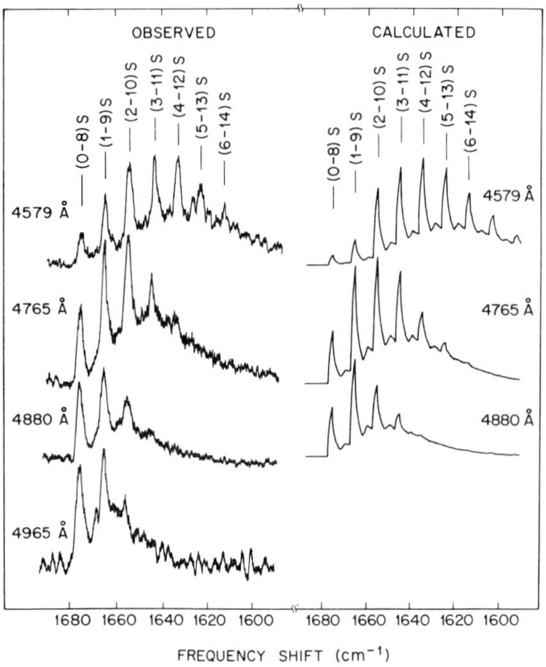

Fig. 6.9. Experimentally observed (left) and theoretically calculated (right) iodine resonant Raman spectra for the $\Delta v=8$ transition. The sharp lines in the spectra are S-branch transitions and the initial and final vibrational quantum numbers for each of these lines are assigned in the upper spectra. (From [6.31])

the S-branches. Each of the strong lines consists of several rotational transitions within the S-branch manifold in the region of its band head. Owing to the energy dispersion of the Q and O transitions as seen in the Fortrat diagrams, a well defined band head is absent for these branches, and therefore they do not lead to well-defined discrete features in the spectrum, but instead to the broad underlying band.

In the region of discrete RRS a change in the laser frequency of 1 GHz resulted in a dramatic change in the scattering spectrum. In Fig.6.9 on the left, in changing the laser frequency from 4965 to 4579 Å over 1000 cm^{-1} the appearance of the spectrum has changed in a very continuous and systematic way. In particular, the relative intensity of the various hot bands has changed from a condition at long wavelengths where low v" hot bands dominate to a condition at shorter wavelengths were higher v" hot bands dominate. This systematic variation in which higher hot bands dominate at higher laser energies is true of all the overtones. Such a smooth variation in structure with incident frequency is expected on the basis of (6.9) and occurs because of the continuous variation in magnitude of the overlap factors as a function of excited state energy.

To quantitatively interpret continuum resonance Raman spectra numerical calculations of the Raman intensities have been made. In initial calculations [6.31] only transitions via the continuum $B(^3\Pi_{0^+u})$ intermediate states were included (see Fig.6.10). The discrete states of the B potential well and the states of the dissociative $^1\Pi_{1u}$ potential were not included in the calculation. Since damping is not

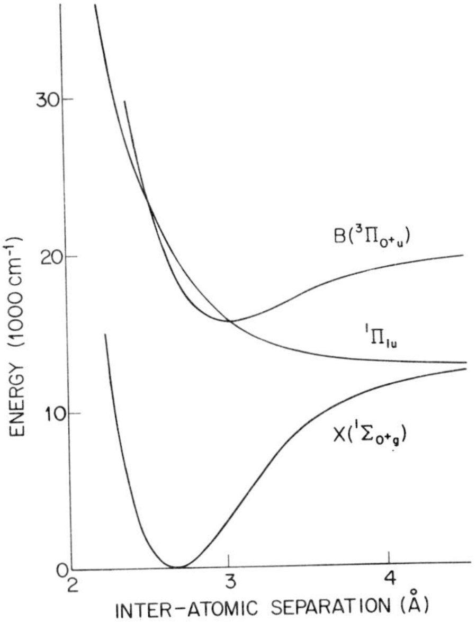

Fig. 6.10. Potential energy curves of I_2. (From [6.17])

relevant to continuum states the Γ_k in (6.9) is taken to be infinitesimal, and the pole in the denominator at exact resonance is accommodated by rewriting the equation as

$$(\underline{\underline{\alpha}})_{uv} = |M_{ig}(\underline{R}_0)|^2 \frac{1}{\hbar}\left\{ P\int_0^\infty \frac{<u|k><k|v>\rho(\omega_k)}{\omega_{vk}-\omega_L} d\omega_k - i\pi <u|k><k|v>\rho(\omega_k)\Big|_{\omega_k=\omega_L}\right\} \quad (6.17)$$

where $\rho(\omega_k)$ is the density of continuum states appropriate to the chosen normalization for the excited state wavefunctions. By using this expression for $(\underline{\underline{\alpha}})_{uv}$, in addition to the rotational and vibrational population factors as well as the rotational branch intensity factors, (6.1) may be then used to obtain the Raman intensity.

A comparison [6.31] of the observed and calculated spectra for the $\Delta v = 8$ overtone is shown in Fig.6.9. The only parameter in these calculations was that of scaling the absolute intensity of the calculated spectrum for each laser line to the experimental intensity. In the calculated spectra, as in the observed spectra, the trend toward the dominance of higher hot bands at higher laser frequencies may be seen. This results from changes in the magnitude of the Franck-Condon factors with incident laser frequency.

In all the calculated spectra there is a general weakness of transitions originating from the lowest vibrational states (i.e., v"=0,1) as compared with the observed spectra. It was argued that these systematic quantitative discrepancies resulted from some of the approximations used in the calculations. Specifically the sum over electronic states was restricted to the B state only; the sum over rotational-vibrational levels was restricted to the continuum states only, and the dependence of the Franck-Condon matrix elements on rotational quantum numbers was handled in an approximate way.

The importance of including all the available intermediate states in Raman intensity calculations was made clear in recent experiments by BAIERL et al. [6.34] on Br_2. Like the electronic structure of I_2, Br_2 has a bound potential function overlapping a dissociative potential. However, unlike I_2 in which the bound potential dominates the visible absorption, in Br_2 both potentials make nearly equal contributions. To properly treat this case for two potentials we must rewrite (6.17) to include contributions from both potentials. Since the form of the rotational branch factors b depend on the electronic state symmetry it is necessary to include them explicitly in the polarizability expression. Doing this we get

$$(\underline{\alpha})^2_{uv} = 1/\hbar^2\Big[|M_{XB}(\underline{R}_0)|^4\big(b_J^B\big)^2\big(R_B^2+I_B^2\big)+|M_{1uX}(\underline{R}_0)|^4\big(b_J^{1u}\big)^2\big(R_{1u}^2+I_{1u}^2\big)$$
$$+2|M_{BX}(\underline{R}_0)|^2|M_{1uX}(\underline{R}_0)|^2 b_J^B b_J^{1u}\,(R_B R_{1u}+I_B I_{1u})\Big] \quad . \tag{6.18}$$

The R's correspond to the real parts of (6.17) and the I's to the imaginary parts. The terms X, B and 1u represent the $X(^1\Sigma_{0+g})$, $B(^3\Pi_{0+u})$ and $^1\Pi_{1u}$ states. Note that

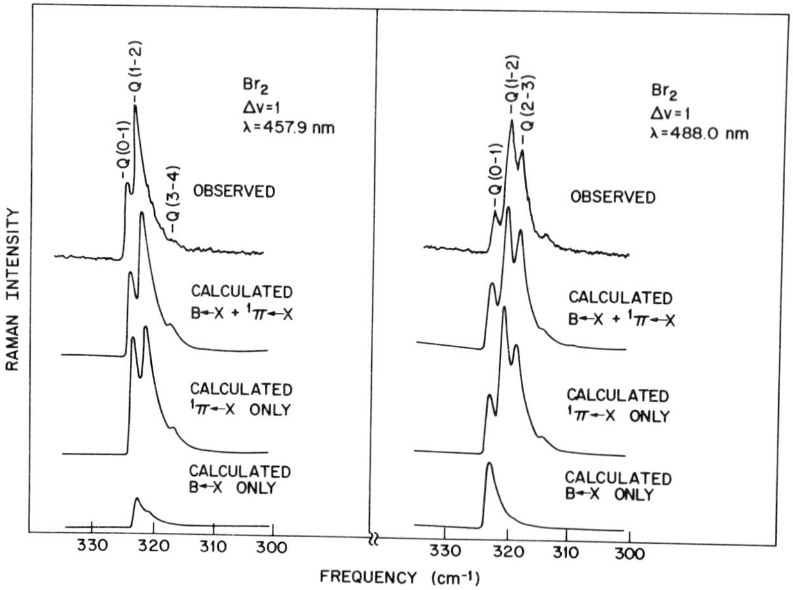

Fig. 6.11. Experimentally observed and theoretically calculated resonance Raman spectrum of $^{79}Br_2$ vapor for $\Delta v = 1$ with 488.0 and 457.9 nm excitation frequencies. The lines are Q branch transitions and the initial-final vibrational quantum numbers are assigned to the observed peaks. The calculated spectra demonstrate the importance of interference effects. (From [6.34])

if scattering events occurring via each intermediate state B and 1u were physically distinguishable, only the first two terms (corresponding to the sum of the probabilities) would be present. The third, interference, term is required because these events are fundamentally indistinguishable in the experiment.

Calculations were carried out on Br_2 in a fashion similar to those done on I_2 [6.34]. In Fig.6.11 the observed spectra and calculated spectra for three different cases are shown for both 488.0 and 457.9 nm excitation. The two lower traces for each excitation frequency represent calculated spectra where only the labeled transition (either $B \leftarrow X$ or $1u \leftarrow X$) contributes to the resonance Raman intensity. The second trace from above was obtained by including both electronic states in the summation. The absolute intensity of these spectra was scaled to the intensity of the experimentally observed spectra. The spectra arising from contributions of the pure $B \leftarrow X$ or $1u \leftarrow X$ transitions alone are scaled relative to the combined spectra.

The importance of inclusion of both states in interpreting the resonance Raman data points to the possibility that when an excited state potential is not accurately known, resonance Raman scattering may be used to accurately position the potential function. Currently in KIEFER's laboratory such calculations are being carried out for the Br_2 to accurately determine its potential functions. Also WILLIAMS et al. [6.35] recently reported on such a positioning of the $^1\Pi_{1u}$ state in I_2.

In order to include all the discrete state transitions of the B state as well as to accurately treat the rotational dependence of the matrix elements a more efficient algorithm was used than in these prior calculations [6.35]. In it the summation over intermediate states is achieved by numerically solving a single inhomogeneous differential equation in an extension of a method developed by SHULMAN and DETRANO [6.36]. In this method the excited state energy $\hbar\omega_k$ is replaced by its Hamiltonian H' and a new function $|\psi\rangle$ is defined as

$$|\psi\rangle = \frac{1}{H' - \omega_L}|v\rangle \quad . \tag{6.19}$$

By writing out the Hamiltonian in a positional representation this results in a second-order differential equation that may be solved numerically. The resonance Raman polarizability then becomes

$$(\underline{\underline{\alpha}})_{uv} = |M_{ig}(\underline{R}_0)|^2 \frac{1}{\hbar} \langle u|\psi\rangle \quad . \tag{6.20}$$

Using this technique WILLIAMS et al. [6.35] have been able to make systematic comparisons between theory and experiment and thereby very accurately determine (±0.01 Å) the $^1\Pi_{1u}$ potential function of I_2. Such experiments clearly point out the utility of RRS as a means of obtaining excited electronic state information.

6.2.3 Perturbed Diatomic Molecules

In the gas phase at low pressure the bound potential functions of diatomic molecules consist of very sharp vibrational-rotational levels. For example, in the halogens the level widths are less than 1 MHz. In contrast when the molecule is placed either in a high pressure environment, in solution, or in a solid matrix the discrete levels become significantly broadened. The study of such systems is important both in order to learn the basic physical processes of RRS from perturbed diatomic molecules and to clarify the as yet unclear interpretation of experiments in gaseous, solution, and matrix phases.

The first clear set of experiments on the influences of pressure broadening on discrete RRS was reported by BERJOT et al. [6.37]. As shown in Fig.6.12, when exciting molecular I_2 at resonance with no buffer gas present they observed the characteristic discrete resonance Raman spectrum (resonance fluorescence). As they added buffer gas the spectrum continuously changed into one with features characteristic of continuum RRS.

These investigators [6.37] interpreted their data as resulting from the continuous passage from discrete RRS to continuum RRS. At low gas pressures the scattering is dominated by a single intermediate state as in (6.10). As the gas pressure is in-

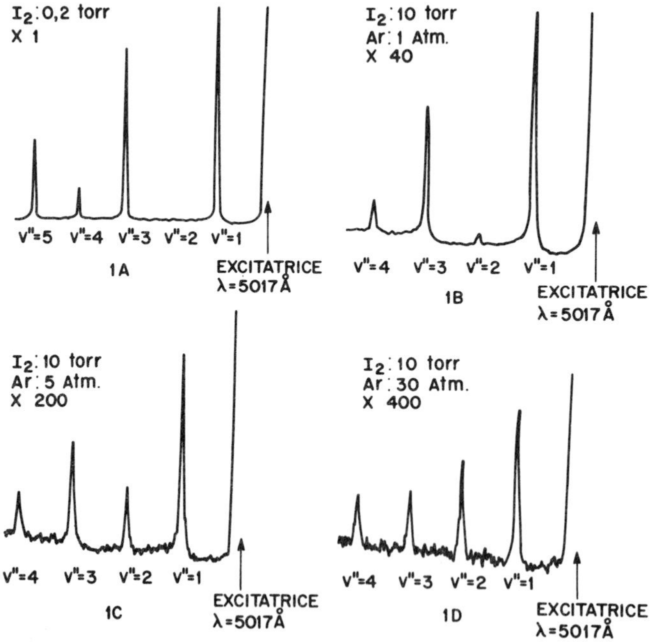

Fig. 6.12. Progressive change from discrete to continuum resonance Raman scattering in molecular iodine in the presence of argon. The position of the excitation frequency in each spectrum is indicated by the arrows. (From [6.37])

creased, however, the excited state damping increases until the energy denominator in (6.9) no longer discriminates against other nearby states and increased numbers of initial and intermediate states play important roles in the scattering. The increased damping drastically quenches the one exactly resonant state, but has little effect on other nearby states due to the decreased re-emission lifetime associated with those off-resonance transitions. Presumably, if the damping were increased still further overtone lines would disappear relative to the fundamental and scattering would take on many of the characteristics of ordinary non-resonance Raman scattering as discussed in the introduction to this chapter.

SILVERSTEIN and St. PETERS [6.38] questioned this interpretation and argued instead that the data could be explained by quenching out of the discrete RRS to allow experimental observation of the much weaker, but always present, continuum RRS produced by excitation of thermally populated excited vibrational levels of the ground electronic state to continuum levels of the B state. Due to the very short re-emission time of this latter process it is always present independent of pressure (in the experimental region used) but only becomes experimentally observable when the discrete scattering is quenched out.

Questions concerning the interpretation of RRS of small molecules dissolved in solvents have also been raised. I_2 is particularly useful for such studies because it forms charge transfer complexes with organic solvents. The strengths of these complexes depends on the particular solvent chosen and can be quantified by a shift in the I_2 resonance frequency. Because in solutions the interaction with the solvent causes substantial broadening of the discrete levels no distinction between excitation in the "discrete" or "continuum" regions may be made.

By utilizing their rotating cell, KIEFER and BERNSTEIN [6.39] were able to observe an extensive overtone progression in the Raman scattering from I_2 in various solvents using 5145 Å excitation. Typical of their data are the three spectra shown in Fig.6.13. Analysis of these data yielded spectroscopic constants for the iodine molecules in a variety of solvents.

In later studies by several groups the variation of the Raman intensity as a function of the incident laser frequency (excitation profile) were obtained for such solution data. As may be seen in Fig.6.14, for I_2 in hexane the peak of the Raman excitation profile is shifted to the red of the absorption profile [6.40]. The overtones also show a slight shift but not as great as that of the fundamental mode shown in Fig.6.14. Red shifted excitation profiles have been observed in other charge transfer complexes as well such as those observed by MICHAELIAN et al. [6.41] in complexes between tetracyanoethylene and various aromatic groups.

In an effort to interpret the red shift in I_2 dissolved in chloroform FENSTERMACHER and CALLENDER [6.42] did some calculations of the frequency dependence of the Raman intensity. Although their calculational technique accurately reproduced the absorption spectrum, it could not reproduce the red shifted excitation profile. They con-

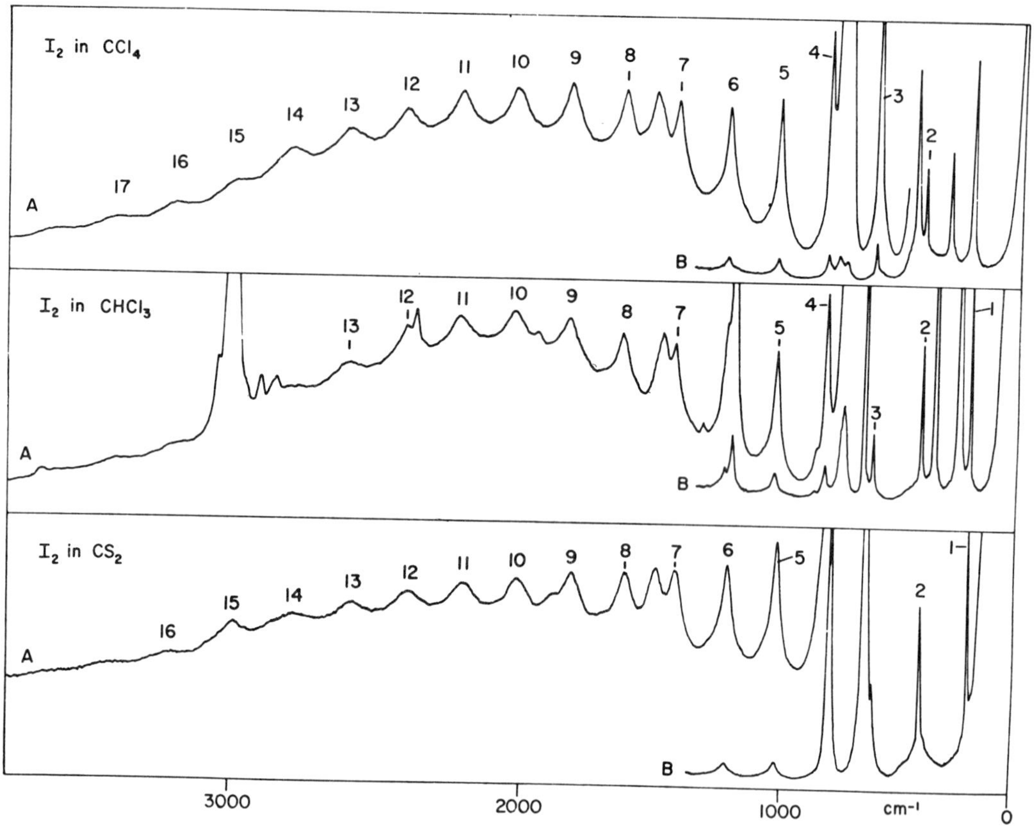

Fig. 6.13. Resonance Raman spectra of 10^{-2} molar iodine in carbon tetrachloride, chloroform and carbon disulfide. 5145 Å excitation was used and the spectral slit width for A was ~7.5 cm^{-1} and for B it was ~3 cm^{-1}. (From [6.39])

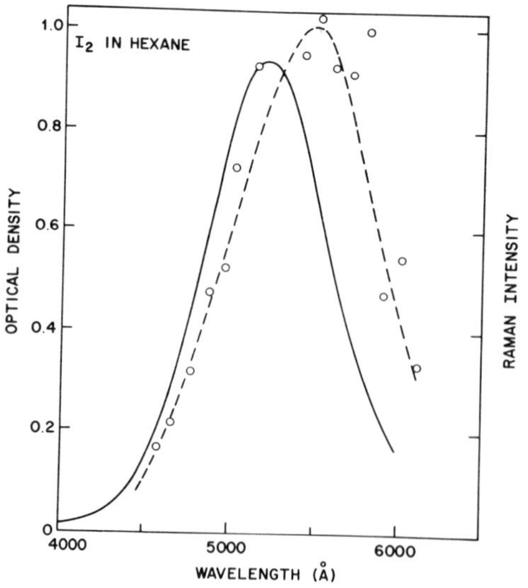

Fig. 6.14. Comparison between excitation profile and optical absorption of I_2 in hexane. The optical absorption is represented by the solid line and the points correspond to the intensity of the Raman fundamental (~211 cm^{-1}) for several excitation frequencies. The dashed line was drawn through the data points to help guide the eye. (From [6.40])

cluded that the most likely source of the discrepancy was their neglect of the $^1\Pi_{1u}$ state in the summation over intermediate states.

MATSUZAKI and MAEDA [6.43] measured the resonance Raman excitation profile of I_2 in chloroform, deuterated benzene and tetrahydrofuran. In addition to observing a red shift of the excitation profile, they observed an additional increase in the Raman intensity to the blue of the absorption peak in deuterated benzene. They interpreted this phenomenon by invoking a semi-classical treatment which suggested the possibility of an interference effect between the I_2 visible absorption band and the ultraviolet charge transfer band.

MICHAELIAN et al. [6.41] after observing a red shift of the excitation profile of the tetracyanoethylene complexes considered the origin of the red shift in the I_2 complexes. They argued that on the basis of the high energy of the charge transfer band and on the basis of their observation of independent resonances associated with charge transfer states, that frequency dependent damping of the discrete states is a more likely explanation of the red shifted excitation profiles than interference with the charge transfer state, as proposed by MATSUZAKI and MAEDA [6.43].

In experiments carried out at room temperature as these solution studies were, the thermal population of many vibrational-rotational states serve to complicate the basic phenomena. These problems may be overcome by matrix isolation studies in which the molecule of interest is placed in low concentration in a rare gas solid matrix held at about 4 K. ANDREWS and his co-workers [6.22,44,45] have reported several such experiments on matrix isolated halogens and interhalogens. They found that in I_2 and Br_2 isolated in rare gases intense overtone progressions could be observed. Qualitatively similar progressions were observed independent of whether the incident frequency was above or below the gas phase dissociation limit. In their work on matrix isolated Br_2 [6.22], in addition to the resonance Raman progression, a relaxed fluorescence series of lines was also observed. These data are shown in Fig.6.15. As indicated schematically on the bottom of Fig.6.15 in relaxed fluorescence the incident photon populates a discrete level of the excited state manifold from which it relaxes non radiatively to the $v = 0$ level. It then fluoresces to the ground electronic state vibration manifold resulting in a series of lines with the spacing of the ground electronic state [6.46].

In an effort to better understand the red shift of the resonance Raman excitation profile we have also studied matrix isolated Br_2 recently [6.47]. Br_2 was selected since the simultaneous observation of relaxed fluorescence and RRS permits direct monitoring of the degree of damping (via the widths) of discrete levels. To monitor these widths the spectrometer is set to the frequency of a relaxed emission line and the incident laser is tuned through the Br_2 absorption spectrum. The resulting excitation spectrum is shown in Fig.6.16. For the lowest vibrational levels of the upper state manifold the line widths are limited by the instrumental resolution. However, on going to higher and higher vibrational levels the lines become progressively

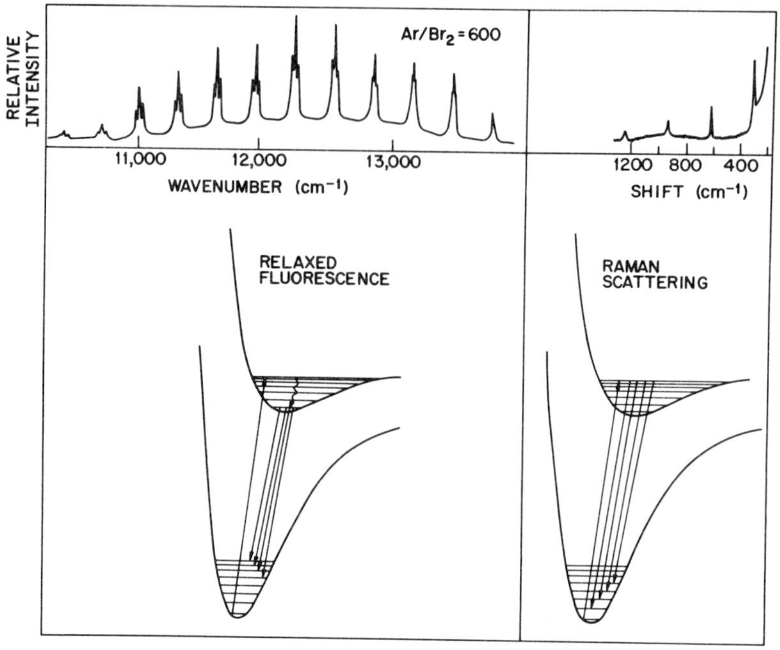

Fig. 6.15. The relaxed fluorescence and resonance Raman spectra of bromine isolated in an argon matrix at 15 K. The data on the top was taken from AULT et al. [6.22]. The diagrams on the bottom demonstrate the difference between the relaxed fluorescence process and the resonance Raman scattering process

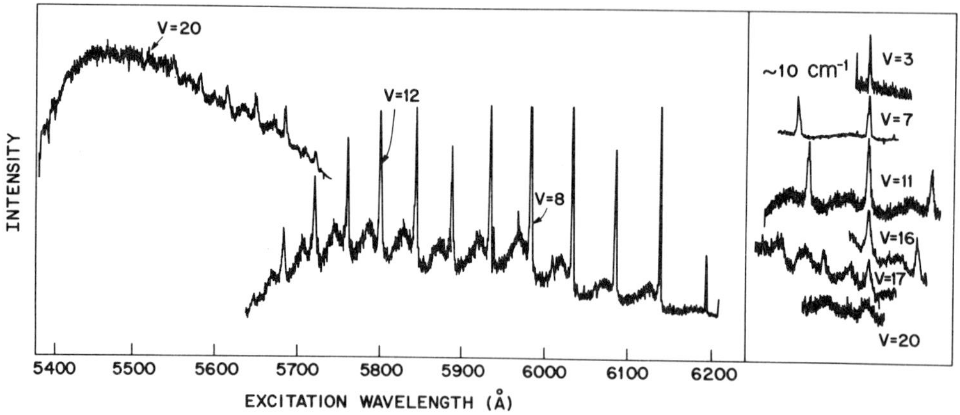

Fig. 6.16. Excitation profile of the relaxed fluorescence of Br_2 in argon showing the change in the excited state line widths as a function of vibrational quantum number. The data was not corrected for the variation in the dye laser intensity (rhodamine 110 and rhodamine 6G) accounting for the variation of intensity in the spectra on the left. On the right several vibrational levels are superimposed to better demonstrate the variation in line width as a function of vibrational level. (From [6.48])

broader until the width of the lines becomes comparable to the spacing and consequently the discrete features wash out.

To determine the magnitude of the contribution of one of these single discrete transitions to the Raman intensity for this system, one needs only compare the Raman intensity on and off resonance with the transition. Such measurements [6.48] that we have carried out in matrix isolated Br_2 surprisingly have shown no change in the Raman intensity in going from exact resonance to well off resonance with any of the discrete transitions. We concluded from these data that the Raman intensity for matrix isolated Br_2 is derived from the real part of the transition amplitude, (i.e., the off-resonance contribution) not the imaginary part, at least in this region of the spectrum. This is in marked contrast to the gaseous results on I_2 and Br_2 in which the Raman scattering (discrete RRS) varies dramatically as the incident frequency is tuned through the discrete lines of the absorption spectrum. The data are analogous to the high pressure gas phase studies [6.37] in which the discrete RRS was quenched out and a continuum type RRS spectrum became evident. In the case of matrix isolated Br_2 it is clear that there is an underlying contribution to the Raman intensity due to the off-resonance (real) part of the transition amplitude. This contribution is very much stronger than that due to resonance with discrete levels to the extent that their contribution is not observable. Presumably it is damped out by matrix induced vibrational relaxation.

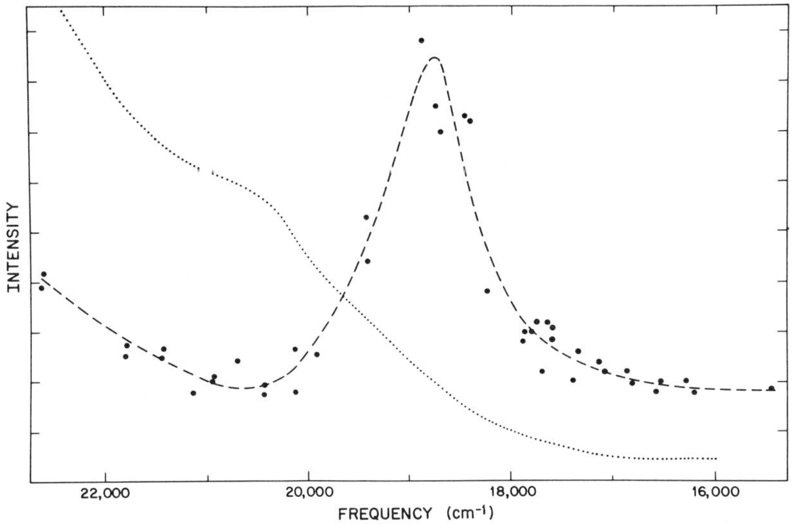

Fig. 6.17. Excitation profile of the fundamental vibrational mode of Br_2 isolated in a rare-gas matrix. The points are the intensity data of the Raman mode obtained in several experiments. The dashed line is a curve drawn through the data points to guide the eye. The dotted curve is the absorption spectrum. (From [6.47])

In the frequency region where the discrete lines may be seen in the absorption spectrum, 16,000-18,000 cm^{-1}, only the fundamental vibrational mode is very strong. In tuning toward the blue the intensity of the fundamental increases as does the relaxed fluorescence, and the overtones become prominent. Studies by BONDYBEY and co-workers [6.46] have shown that the excitation profile of the relaxed fluorescence accurately follows the absorption shape of the B state and they observed no contribution to the relaxed fluorescence from the $^1\Pi_{1u}$ state. In contrast to this behavior one would expect that the Raman scattering excitation profile would follow the profile of the absorption, reflecting both the presence of the $B(^3\Pi_{0^+u})$ and $^1\Pi_{1u}$ states. However, just as in the case of liquid solutions of halogens the excitation profile [6.47] peaks far to the red of the absorption maximum as shown in Fig.6.17. Although qualitative agreement between the experiment and a calculation was obtained using an oversimplified model, we have recently carried out more accurate calculations [6.49] which indicate that the proper qualitative interpretation of these data requires explicit inclusion in the Raman amplitude expression of frequency dependent damping factors. A quantitative explanation of these data has not yet been reached. However, we expect that from accurate calculations of excitation profiles in this matrix isolated system it will be possible to reach a quantitative understanding of the origin of red shifted excitation profiles and the role damping factors play in influencing intensities of Raman scattering data. It is then hoped that arriving at such understanding in relatively simple systems will further the understanding of the behavior in more complex systems.

6.3 Polyatomic Molecules

A vast number of resonance Raman investigations have been carried out on polyatomic molecules and interesting interference, polarization, and enhancement phenomena have been observed. Several recent reviews [6.8,10-15] have discussed these various phenomena. In this present section we wish to only cite a few examples of scattering experiments from polyatomics which typify the type of information that may be obtained and which exemplify the general physical processes that may occur. After this qualitative introduction to polyatomics in the following section on porphyrin we present a detailed account of RRS from polyatomic systems with well characterized energy levels.

Raman scattering from diatomic molecules, discussed in the prior section, can be understood in a quantitative way because the precise forms of the excited state potential functions are so accurately known. Such detail is lacking for most polyatomic molecules that have been subjected to resonance Raman studies thus far. This, in addition to the greater complexity of polyatomic molecules, makes understanding of the scattering mechanisms on a quantitative level far more difficult than that obtainable for diatomics.

For the diatomic halogen molecules the excited state potential functions are very different in shape and equilibrium internuclear separation from the ground state. In many polyatomic molecules this is not the case since many excited electronic states have the same shape and internuclear equilibrium position as the ground state. Consequently for such systems the Franck-Condon term makes only a minor contribution to the Raman intensity and the Herzberg-Teller term dominates the scattering. Some polyatomic molecules do display overtone progressions characteristic of Franck-Condon scattering but most tend to show a rather uniform enhancement of several fundamental vibrational modes with only little enhancement of overtones. This behavior is characteristic of scattering via a Herzberg-Teller mechanism.

In general when interpreting data from polyatomic molecules most investigations have treated the nuclear dependence of the electronic matrix elements by the Herzberg-Teller perturbation. However, recently WARSHEL [6.15] has been using the more general Taylor series expansion (6.7) of the electronic moment in order to interpret resonance Raman data from complex systems. By making some approximations about the general form of the electronic potentials he has been able to rewrite the Raman intensities so as to reveal geometric changes upon electronic excitation. His calculational techniques offer promise of being able to gain a more quantitative understanding of Raman scattering from complex molecules. In addition application of the technique to molecules of biological interest may be of use in studying molecular conformations.

Just as for diatomic molecules Franck-Condon scattering with its overtone progressions may be observed in polyatomics as well. Due to the symmetry selection rules only the totally symmetric modes are allowed by this mechanism. The symmetric modes that get maximally enhanced are those modes that correspond to the internuclear separation change that occurs in the excited state.

Typical of the type of data [6.50] observed for Franck-Condon type scattering is that shown in Fig.6.18 for $(NH_4)_5Mo_2Cl_9 \cdot H_2O$, and $Cs_4Mo_2Cl_9$. ν_1 corresponds to an A_{1g} mode resulting from the Mo-Mo stretch and ν_4 is thought to result from an A_{1g} stretching vibration of the Mo-Cl bond. The ν_1 progression is analogous to the previously discussed results from diatomic molecules. The $\nu_4 + n\nu_1$ progression is also characteristic of Franck-Condon type scattering as may be seen by writing the vibrational matrix element as $\langle u_1;u_2|k\rangle\langle k|v_1;v_2\rangle$ where v_1 and v_2 represent two different initial vibrational modes and u_1 and u_2 represent their respective final states. Since the normal coordinates are orthogonal these independent progressions may occur. In the recorded Raman spectrum this appears as a progression of the combination. By analyzing data such as this CLARK and FRANKS [6.50] have been able to draw conclusions regarding the symmetry of the excited state from the requirement that a fully allowed transition is needed for the observation of Franck-Condon term scattering. In addition analysis of the frequencies of the overtones has allowed calculation of the anharmonicity of the ground electronic state potential.

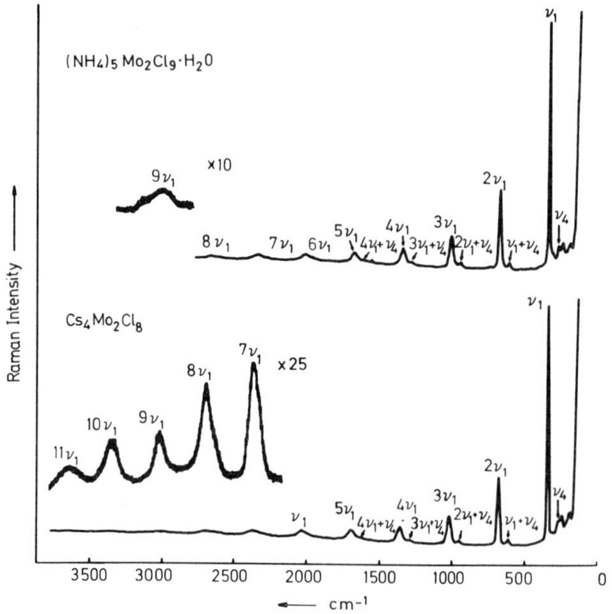

Fig. 6.18. Resonance Raman spectra of $(NH_4)_5Mo_2Cl_9 \cdot H_2O$ and $Cs_4Mo_2Cl_9$ obtained with 5145 Å excitation. (From [6.50])

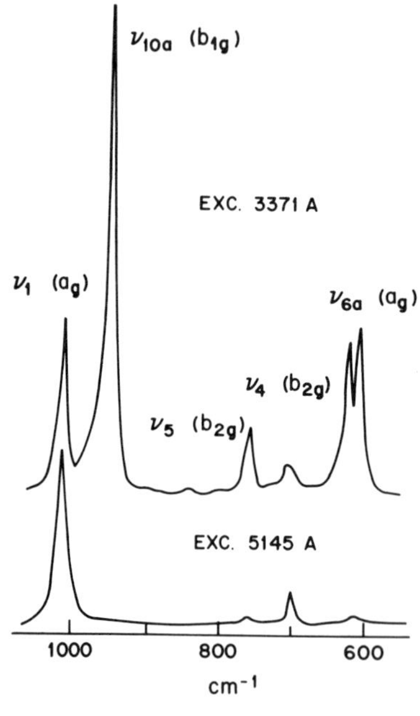

Fig. 6.19. Resonance Raman spectrum of pyrazine crystal obtained at two different excitation frequencies. The fine structure of the ν_{6a} band is due to Davydov splitting. (From [6.54])

Several molecules are known to have bands that get intensity from Herzberg-Teller coupling and Raman scattering in the region of these absorption bands exhibit behavior expected for a vibronic interaction. Early observations of such effects were seen in heme proteins [6.51] and N-heterocyclic aromatic molecules [6.52,53]. In the latter molecules (pyridine, pyrazine, and quinoxaline) very substantial frequency reduction of the mode responsible for the coupling has been observed in the excited state [6.52-54]. Raman scattering from such systems exhibit very strong enhancement of these asymmetric modes as may be seen in Fig.6.19 for the pyrazine crystal [6.54].

The strong generalized enhancement of a vast number of modes via the Herzberg-Teller type of mechanism such as that of the above mentioned heterocycles stands as perhaps the single most important characteristic of RRS. Because of this general enhancement, information regarding many of the vibrational degrees of freedom of molecules may be obtained in resonance Raman experiments. Low concentration solutes may be studied without detectable solvent interference. Similarly specific chromophoric sites may be selectively investigated in large complex molecules, or even in exceedingly complex biological systems. As more and more laser frequencies become available one might hope to be able to completely map out a complex molecules or protein by studying each component independently.

6.4 Porphyrin Compounds

Without question the clearest and most beautiful example of scattering by the Herzberg-Teller mechanism occurs in porphyrin compounds. An enormous number of experiments have been done on porphyrin compunds studied both as isolated species [6.55-58] and as the visible chromophore in heme proteins [6.10]. At first one might think that such compounds are too complex to be able to make any headway in the physical understanding of the scattering processes. However, this is indeed not the case. Instead porphyrins are a near ideal class of molecules with which to study RRS phenomena. Because of the many subtle variations between the various porphyrins one can systematically alter the electronic state properties by choosing the appropriate compound. Parameters such as excited state lifetime, symmetry, electronic degeneracy, and vibronic coupling display nearly the complete gamut of values within this class of homologous molecules. Furthermore by using hemeproteins one has the effect of a matrix isolated porphyrin even at elevated temperatures.

The basic porphyrin ring is a conjugated heterocycle such as that shown in Fig.6.20. We show it here with an iron atom at its center but there are porphyrins with other metals as well. Additional ligands above and below the plane may be made to the metal atoms and a variety of groups may be substituted for the protons at the numbered and lettered positions. The numbered positions are referred to as the β positions and

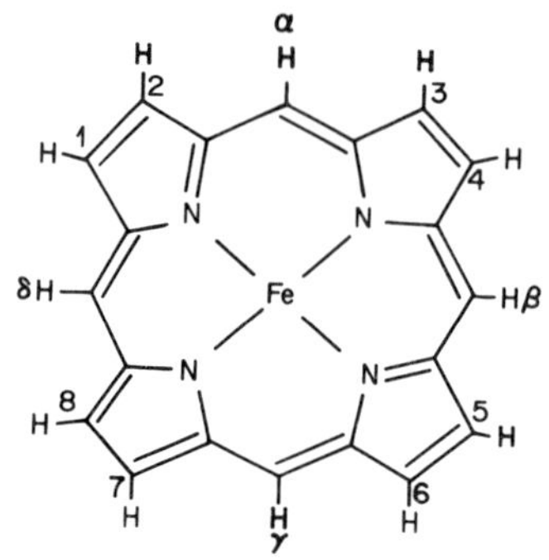

Fig. 6.20. The structure of the porphyrin macrocycle

the lettered as the *meso* positions. Proteins are generally bound through the β positions as well as through the metal atom.

The understanding of RRS from complex systems such as porphyrins depends on the understanding of the resonant state. This is especially true of systems with many degrees of freedom where the excited state potential surfaces may be complicated functions of the molecular nuclear coordinates. There have been a number [6.2-6,23, 59-64] of theoretical treatments of RRS using different descriptions of the resonant excited state and this has led to a certain amount of confusion about the appropriate form for the resonant Raman amplitude. Of course, all representations must be mathematically equivalent as long as a complete basis set is used in the description. However, if one wants to designate a single term in the Raman amplitude as the resonance term and be able to relate that resonance in a simple and straight forward manner to the state involved in the resonant absorption then care must be exercised in the choice of the basis set. Ideally we want a choice of basis states such that the summation over I in (6.2) contains as few terms as possible for a given resonance process and we want to be able to relate the energies $\hbar\omega_{GI}$ of the resonant states I to those of the states that are traditionally used in describing the relevant absorption process. If the excited electronic states are narrow and well separated it is generally possible to treat the scattering process as one in which the initial ground vibronic state is radiatively coupled to the excited intermediate state and it in turn is radiatively coupled to the final vibronic state. If, however, there exist close lying states that couple non-radiatively then the process cannot be viewed in such a simplified fashion. To make clear precisely what the appropriate states are to use in discussing RRS from porphyrins, in the next subsection we present a rather detailed discussion of the electronic structure as determined from an analysis of the absorption spectrum.

6.4.1 Porphyrin Absorption Spectrum

Although porphyrins have a fairly complex electronic structure, their absorption spectrum (Fig.6.21) is relatively simple and the properties are rather straight for-

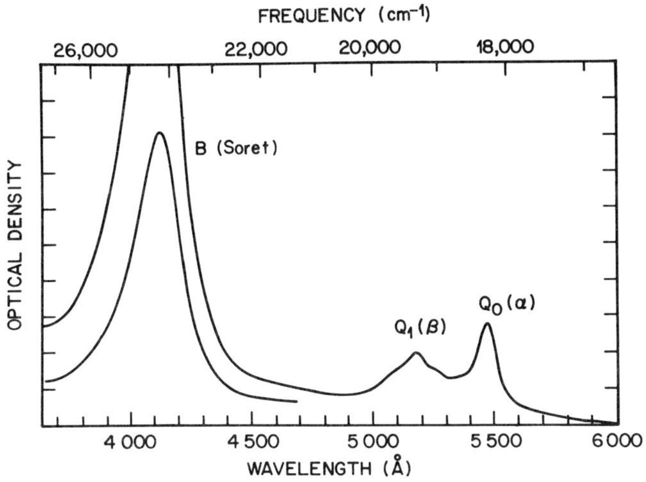

Fig. 6.21. Absorption spectrum of cytochrome c. Top spectrum 160 μmolar. Bottom 60 μmolar. (From [6.68])

ward to analyze because the aromatic system conforms very well to a simple π orbital description of quantized levels of an electron with circular boundary conditions [6.65-67]. The electronic transitions that give rise to typical porphyrin visible absorption spectra involve the promotion of a π electron to an unfilled π* orbital. These π-π* excited electronic states are modified in various metalloporphyrins by interaction between the porphyrin π orbitals and the metal orbitals.

There are two π-π* excited electronic states in porphyrin molecules in the visible and near uv. These are nearly degenerate configurations which through configuration interaction via the electron repulsion, mix and give rise to two bands in the absorption spectrum. Since the porphyrin molecule has D_{4h} symmetry these in plane transitions are degenerate in the X,Y direction and both belong to the E_u representation. The two zeroth order configurations add to give rise to the very strong B state (also called the Soret band) and subtract to give rise to the much weaker Q state. An absorption spectrum of cytochrome c which illustrates these features is shown in Fig. 6.21.

As may be seen in Fig.6.21, the B band typically gives rise to a strong absorption peak ($\varepsilon \simeq 10^4$) between 4000 and 4400 Å. The low energy component of the Q band occurs at about 5500 Å and results from a transition connecting the v = 0 levels of the ground and excited states. The higher energy peak (~5200 Å) is a vibrational side band consisting of an envelope of many singly excited vibrational levels of Q. The zero vibrational band and the vibrational envelope are also referred to as the α and β bands, respectively.

To understand these electronic transitions it is instructive to show that the absorption spectrum may be treated in a fashion parallel [6.18,69] to the general treatment given in Sect.6.1. We may apply the Born Oppenheimer approximation and obtain adiabatic electronic wavefunctions $Q(\underline{r},\underline{R})$ as well as the corresponding vibrational wavefunctions $q(\underline{R})$ just as was done in (6.3). Similarly the adiabatic wavefunctions for the B state may be written as $B(\underline{r},\underline{R})b(\underline{R})$. In order to be able to treat the nuclear coordinate dependence of those states we expand them in terms of fixed nuclear coordinate wavefunctions but in this case for convenience using the Herzberg-Teller perturbation rather than the more general Taylor series expansion that was used in (6.6). We obtain

$$Q_\rho(\underline{r},\underline{R}) = Q_\rho(\underline{r},\underline{R}_0) + \sum_{\xi,\rho} h^\xi_{QB} R_\xi B_\sigma(\underline{r},\underline{R}_0) + \dots$$

$$B_\rho(\underline{r},\underline{R}) = B_\rho(\underline{r},\underline{R}_0) + \sum_{\xi,\rho} h^\xi_{QB} R_\xi Q_\sigma(\underline{r},\underline{R}_0) + \dots \quad . \tag{6.21}$$

In these expressions the indices σ and ρ both may take on the values X and Y and they distinguish between the doubly degenerate members of each electronic state. Since these are π states, they also correspond to the polarization direction of the associated dipole moment. h^ξ_{QB} is the Herzberg-Teller perturbation expression, i.e.,

$$h^\xi_{QB} = \frac{1}{\hbar} \frac{\langle Q_\rho(\underline{r},\underline{R}_0)|\partial H/\partial R_\xi|B_\sigma(\underline{r},\underline{R}_0)\rangle}{\omega_{Q_\rho(\underline{r},\underline{R}_0)} - \omega_{B_\sigma(\underline{r},\underline{R}_0)}} \tag{6.22}$$

and R_ξ is the coordinate operator of the ξth vibration mode. Here $Q_\rho(\underline{r},\underline{R})$ and $B_\rho(\underline{r},\underline{R})$ are adiabatic Born-Oppenheimer electronic states and $Q_\rho(\underline{r},\underline{R}_0)$ and $B_\sigma(\underline{r},\underline{R}_0)$ are crude Born-Oppenheimer states. The reader is referred to the articles by AZUMI and MATSUZAKI [6.70] and by BALLHAUSEN and HANSEN [6.71] for excellent discussions of vibronic coupling and definitions of the various types of Born-Oppenheimer approximations. It should be noted that because the energies in the denominator in (6.22) are the crude Born-Oppenheimer energies they may not correspond very well to the observed energies of the molecular eigenstates from absorption measurements. In writing out (6.21) we have included only the lowest-order terms in the coordinate expansion. The next order correction terms were not included because experimentally they make very little contribution. In addition, we have neglected at this juncture non-adiabatic interactions. Such interactions have been found to be important when studying the intensity behavior. In the treatment presented here we will reserve inclusion of non-adiabatic effects for a later subsection.

To determine the symmetries of the vibrational modes that may be Herzberg-Teller active it is necessary to examine the matrix element in (6.22). For the matrix element to be non-zero it must contain the totally symmetric representation. Therefore the symmetry of the modes that can vibronically couple the two configurations is determined by the group theoretical direct product of the symmetries of the electronic states Q and B [6.72]. For a D_{4h} molecule we obtain

$$E_u \times E_u = A_{1g} + A_{2g} + B_{1g} + B_{2g} \quad . \tag{6.23}$$

Therefore a coordinate operator with any of the above symmetries will result in a totally symmetric matrix element that will be active in vibronic coupling.

The absorption spectrum of porphyrins exhibits the relatively simple structure shown in Fig.6.21. They do not exhibit any vibronic progression in either absorption or fluorescence, characteristics of Franck-Condon interactions. Instead in the red-yellow region of the absorption spectrum the two observed bands may be characterized by one transition to a zero vibrational state and one to $v = 1$ states of many differing vibrational modes. Since the amplitude A for an absorption transition from G to I is proportional to the matrix element $\langle I|r|G\rangle$ we may write out the absorption in the region of the state Q by using the Born-Oppenheimer wavefunctions $Q(\underline{r},\underline{R})q(\underline{R})$ and the coordinate expansion for Q given in (6.20). We obtain

$$A \propto \langle Q(\underline{r},\underline{R}),k|r|r,v\rangle = \langle Q(\underline{r},\underline{R}_0)|r|g\rangle\langle k|v\rangle + \sum_{\xi} \langle B(\underline{r},\underline{R}_0)|r|g\rangle h^{\xi}_{QB}\langle k_{\xi}|R_{\xi}|v_{\xi}\rangle + \dots \tag{6.24}$$

In (6.24) just as in our previous definitions k corresponds to a vibrational level of an excited state Q, and v to a vibrational level of the ground state g.

From the fact that only one vibrational side band is observed in the Q band absorption it appears that for the porphyrins the ground and excited vibrational potential functions are nearly identical. Due to this and also due to the large oscillator strength of the B transition, Herzberg-Teller vibronic coupling plays an important role in the Q state optical properties. With these features in mind we may reexamine (6.24). If we assume the potentials are nearly identical and nearly harmonic at the bottom of the wells, then $\langle k|v\rangle = \delta_{v,k}$ and $\langle k_{\xi}|R_{\xi}|v_{\xi}\rangle = \delta_{v,k\pm1}$. In other words, if the temperature is sufficiently low that only the lowest vibrational levels $v = 0$ are populated then the first term in (6.24) must result in a transition to $v = 0$ of the Q state. Similarly the second term will be a highly structured band containing transitions to the $v = 1$ states of the modes that are vibronically active. Therefore the first term corresponds to the α-band and the second term to the β-band. It will be shown in a later subsection that these conclusions are born out by RRS experiments. Before discussing the Raman experiments though we present some diagrams that we have found useful in cataloguing the many types of interaction processes that may take place in resonance Raman scattering from such complex molecules.

6.4.2 Helpful Diagrams

In metalloporphyrin compounds there are many terms that may contribute to the resonance Raman intensity. The formalism, while accurate in presenting the origin of each contribution often becomes difficult to follow due to different terminology used by different groups as well as due to complexity resulting from an often exces-

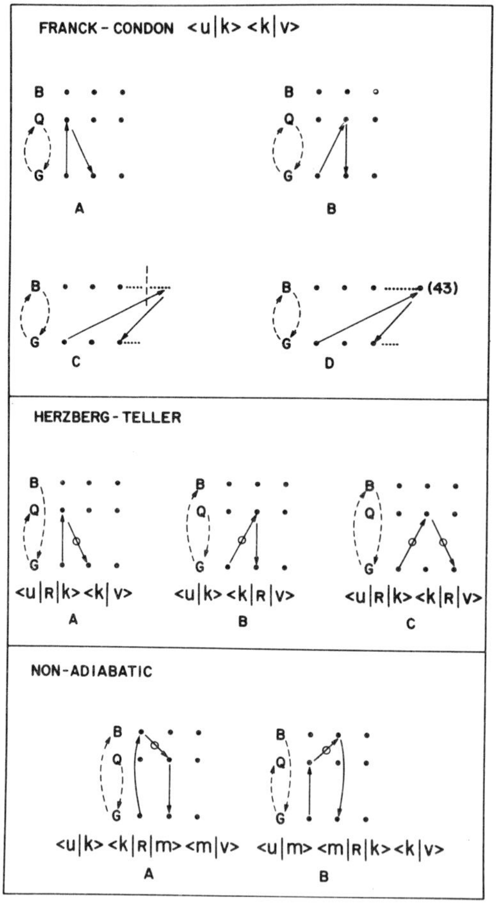

Fig. 6.22. Diagrams to aid in the interpretation of Raman scattering from complex systems. For each diagram the appropriate vibrational matrix elements are indicated. The Franck-Condon type vibrational matrix elements are indicated by the solid lines and the vibrational matrix elements involving a coordinate operator by a solid line intersecting a circle. The appropriate electronic matrix element for each transition is indicated by a dashed line. The ordinate indicates electronic states and the dots along the abscissa indicate vibrational levels. The G,Q,B labelling is appropriate for the porphyrin class of compunds. Franck-Condon diagram C represents the first overtone scattering from continuum levels (the dissociation limit is indicated by the dashed line); and diagram D represents first overtone discrete resonance Raman scattering from the B state level $v = 43$

sive number of subscripts and superscripts needed to completely define a state. To overcome these difficulties we find it very useful to utilize diagrams which with a few definitions rather easily define the scattering process. The diagrams appropriate for the basic mechanisms discussed in this chapter are presented in Fig.6.22.

In these diagrams we schematically plot the energy of the electronic states on the ordinate and we indicate the vibrational levels of each state on the abscissa. The solid lines indicate transitions between the indicated vibrational levels. The solid line with a circles in it represents a vibrational transition through a coordinate operator, i.e., $\langle k|R_\xi|v\rangle$; and the plain solid line corresponds to a vibrational overlap matrix element, i.e., $\langle k|v\rangle$. The dotted lines indicate the appropriate electronic matrix elements needed for writing out the appropriate term in the Raman amplitude. Along with each diagram we write out the matrix elements represented by the diagram. Throughout the remainder of this chapter we will refer to the diagram appropriate for the mechanism being discussed.

6.4.3 Scattering from Q_0, Q_1, and B

The analysis of the absorption spectrum, presented in a prior subsection, indicated that the ground state potential and the potential for the Q state have nearly the identical shape. As such we would expect strong scattering from the Herzberg-Teller mechanism. The resonance Raman spectrum of heme proteins bears out this conclusion. In Fig.6.23 the Raman spectra of cytochrome c obtained at about 6 K are presented. Inset in the figure is the absorption spectrum and the arrows indicate the excitation frequencies of each of the lettered spectra. The line indicated with an asterisk re-

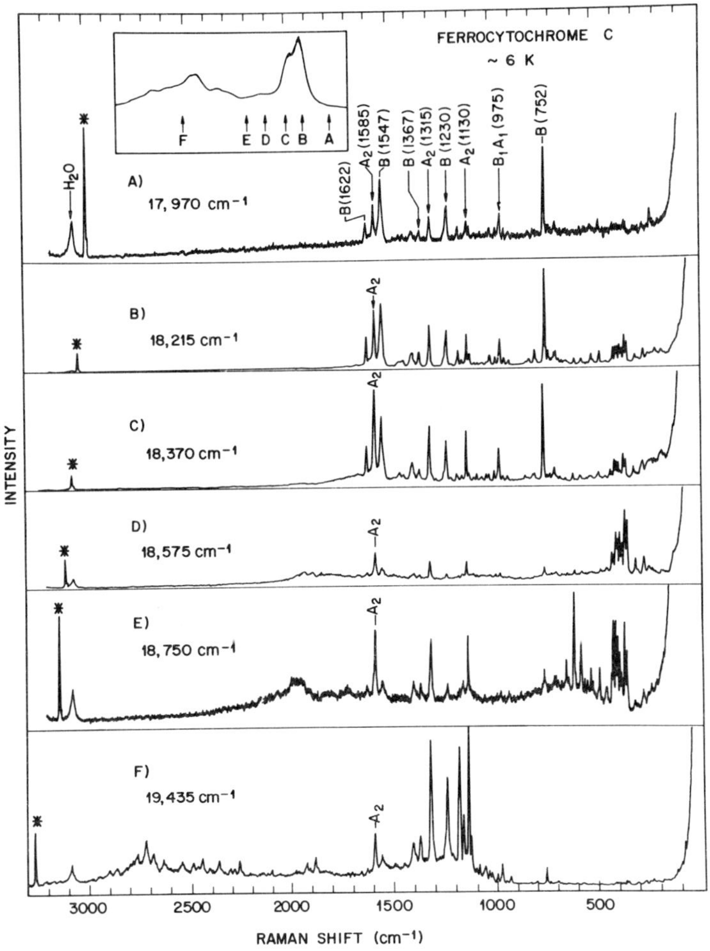

Fig. 6.23. Resonance Raman scattering from cytochrome c at several excitation frequencies. In the top spectrum the symmetries and the frequencies of the prominent lines are labeled. For ease of identification the A_2 line at 1585 cm^{-1} is labeled in all spectra. The line labeled by an asterisk is a grating ghost. To compare the intensity of the spectra the ice line near 3100 cm^{-1} may be used. The inset indicates the position of each excitation frequency on the absorption spectrum

sults from a spurious grating artifact. The H_2O band near 3100 cm^{-1} may be used as an intensity reference to gauge the enhancement of the porphyrin Raman modes. In the upper spectra we have labeled the symmetries of some of the most prominent vibrational modes.

In going from an excitation frequency of 17,970 cm^{-1} (below resonance with the Q_0 band) to 18,215 and 18,370 cm^{-1} (in resonance with the X and Y split components of the Q_0 band) we observe a general enhancement of the Raman modes. The absence of overtones rules out Franck-Condon scattering and the general enhancement of A_1, A_2, B_1, and B_2 modes is the expected result from a Herzberg-Teller mechanism. The resonance Raman polarizability for the scattering may be written

$$(\underline{\underline{\alpha}})_\xi \propto M_{Qg}(\underline{R}_0)M_{Bg}(\underline{R}_0)\frac{1}{\hbar}\frac{h^{\xi}_{QB}\langle u|R_\xi|k\rangle\langle k|v\rangle}{\omega_{vk}-\omega_L-i\Gamma_k} \quad . \tag{6.25}$$

This expression gives the polarizability $(\underline{\underline{\alpha}})_\xi$ for the ξ mode. $M_{Qg}(\underline{R}_0)$ and $M_{Bg}(\underline{R}_0)$ correspond to the matrix elements $\langle Q(\underline{r},\underline{R}_0)|r|g\rangle$ and $\langle B(\underline{r},\underline{R}_0)|r|g\rangle$, respectively. This mechanism is represented by Herzberg-Teller diagram A in Fig.6.22.

The change in relative intensities of the modes as a function of the excitation frequency seen in Fig.6.23 result from constructive and destructive interference effects [6.73] and will be discussed in greater detail in Sect.6.4.5. Because we are considering resonance with the Q_0 state then in (6.25), assuming relatively low temperature, the matrix element $\langle k|v\rangle \simeq 1$ and corresponds to the overlap of $v=0$ wavefunctions of identical oscillators. Additionally within the harmonic oscillator approximation for a $\Delta v=1$ transition the matrix element $\langle u|R_\xi|k\rangle \simeq \hbar/2\omega_\xi$ where ω_ξ is the frequency of the ξth normal mode. Within the approximation of identical oscillators in the ground and the excited Q states, the orthogonality of the vibrational wavefunctions prevents any contribution to the scattering intensity from the Franck-Condon mechanism.

As the incident laser is shifted out of resonance with Q_0, the intensity decreases as expected (spectrum D in Fig.6.23). In addition, as may be seen in spectra D, E, and F, as the incident frequency is tuned into resonance with Q_1 (the β band), completely different sets of modes become enhanced. The origin of each of the sets of enhanced lines is evident when remembering the origin of the absorption in this region which depends on matrix elements of the type $\langle k_\xi|R_\xi|v_\xi\rangle$. For a given shift from Q_0 in the Q_1 band those modes ξ that are responsible for that Q_1 absorption are those that have the same frequency as that shift. Consequently in spectrum D in which the incident frequency is only slightly shifted from Q_0 only low frequency modes are enhanced; in the spectrum E which is further shifted from Q_0 higher frequency modes become enhanced; and finally in spectrum E which is substantially shifted from Q_0 still higher frequency modes become enhanced.

The diagram appropriate to behavior of these data in the Q_1 band is the Herzberg-Teller B diagram and from it for a mode ξ we may write the polarizability $\underline{\underline{\alpha}}$ as

$$(\underline{\underline{\alpha}})_\xi \propto M_{Qg}(\underline{R}_0)M_{Bg}(\underline{R}_0)\frac{1}{\hbar}\frac{h_{QB}^{\xi}\langle u|k\rangle\langle k|R_\xi|v\rangle}{\omega_{vk}-\omega_L-i\Gamma_k} \quad . \tag{6.26}$$

From this we see that the incident photon interacts with the system via the coordinate operator $\langle k|R_\xi|v\rangle$ just as it does in absorption and the re-emission occurs via an overlap factor $\langle u|k\rangle$ in which the final state has the same vibrational quantum number as the intermediate state.

Combinations and overtones may also be active under the Herzberg-Teller mechanism through the product of two Herzberg-Teller interactions, i.e., Herzberg-Teller C diagram in Fig.6.22. We may write out the polarizability for such a process as

$$(\underline{\underline{\alpha}})_{\xi_1\xi_2} \propto |M_{Bg}(\underline{R}_0)|^2\frac{1}{\hbar}\frac{h_{QB}^{\xi_1}h_{QB}^{\xi_2}\langle u|R_{\xi_1}|k\rangle\langle k|R_{\xi_2}|v\rangle}{\omega_{vk}-\omega_L-i\Gamma_k} \quad . \tag{6.27}$$

Here $|u\rangle$ and $|v\rangle$ differ by one quantum in the ξ_1 and ξ_2 modes. In spectrum F of Fig.6.23 the overtones and combination in the 2000-3000 cm^{-1} region may be seen. As expected from a mechanism that can create only modes with a total quantum number changes of two, higher overtones and combinations are not observed. By comparing (6.24) to (6.26) and (6.27) it may be seen that the ratio of a given overtone to its fundamental equals twice the Q_0/Q_1 absorption intensity ratio for that mode. In principle by obtaining very complete Raman intensity data, this information could be used to completely deconvolute the Q_1 band to determine whether or not there is any underlying structure.

Raman excitation profiles in the region of the Q transitions have been shown [6.74] to be useful to determine accurate lineshapes without the hindrance of vibrational side bands present in absorption spectra. In cases where the linewidth is homogeneously broadened this allows direct determination of the rate of decay, i.e., lifetime of the excited state on the subpicosecond time scale. FRIEDMAN et al. [6.75] have used such data to measure lifetimes in cytochrome c and cytochrome b_5 where lifetimes of 20-50 fs were measured. It is interesting to note that the study of repopulation of the ground state following picosecond pulse excitation is a complementary technique to this in that it monitors the rate of repopulation of the ground state in contrast to the rate of depopulation of the excited state measured by the Raman measurements. Application of both techniques to the same system offers the possibility of precisely locating the rate limiting steps in the relaxation process. cess.

The intense Soret transition originates from a strongly allowed electric dipole transition. Consequently it is expected that RRS will be dominated by the Franck-Condon mechanism and the totally symmetric modes will be enhanced. Due to the unavailability of convenient laser frequencies very little data have been obtained in this region however. It is a particularly interesting region to study since the damping factor in heme proteins for example is about an order of magnitude larger than it is in the region of the Q transition. Detailed Raman investigations in this region should help elucidate the precise origin of the very large frequency width of the Soret band.

6.4.4 Anomalous Polarization

In general Raman scattering from liquids results in modes that are either fully polarized or depolarized. Fully polarized modes are characterized by a depolarization ratio (intensity of light scattered perpendicular to the incident polarization divided by that scattered parallel to it) that is very small (zero for isotropic molecules) and depolarized modes are characterized by a depolarization ratio of three fourths. Porphyrin compounds, however, display anomalously polarized lines (also called inverse polarization) in which the depolarization ratio is greater than 3/4 and may approach infinity. This effect was first predicted many years ago by PLACZEK [6.76] but was only first observed in 1972 in experiments of cytochrome c and hemoglobin by SPIRO and STREKAS [6.51] (see Fig.6.24). Since this time it has been observed in many other porphyrin compounds and in octahedral iridium ions [6.77].

Anomalous polarization reflects asymmetry in the scattering tensor and because of destructive interference effects can only be observed near resonance. Those tensor components that display the effect are those that transform as the rotation about one of the axes. For example, for D_4 symmetry the rotation about the Z axis transforms as an A_2 representation and this asymmetric polarizability tensor has the form XY-YX. Consider an experiment in such a molecule in which the incident light is polarized in the XY plane and the angle between the incident and scattered polarization directions is ϕ. We then let θ be the angle between the incident polarization direction and the molecule fixed X-axis and we assume that the X-like and Y-like electronic states are degenerate. We may then write the intensity of the anomalously polarized A_2 modes in terms of direction cosines using standard expressions [6.78] for a fixed orientation molecule

$$I_{A_2} \propto \left(E_X^L E_Y^S - E_Y^L E_X^S\right)^2 \propto [\cos\theta \sin(\theta+\phi) - \sin\theta \cos(\theta+\phi)]^2 = \sin^2\phi \quad . \tag{6.28}$$

In this expression E^L and E^S represent the electric field of the incident laser and the scattered photons, respectively. From (6.28) it should be noted that the A_2 mode intensity is *independent* of molecular orientation and has a maximum when the incident and scattered polarization directions are at 90° from each other. For scattering from modes of other symmetry a plus sign appears in place of the minus sign in (6.28) so the scattered polarization properties are dependent on molecular orientation, and averaging over all molecular orientations results in the usual depolarization ratios.

In the above discussion we have assumed that there is a four fold axis of symmetry so that the X and Y components of all the levels in the Q and B manifold are degenerate. If the X and Y components are slightly nondegenerate then the anomalous po-

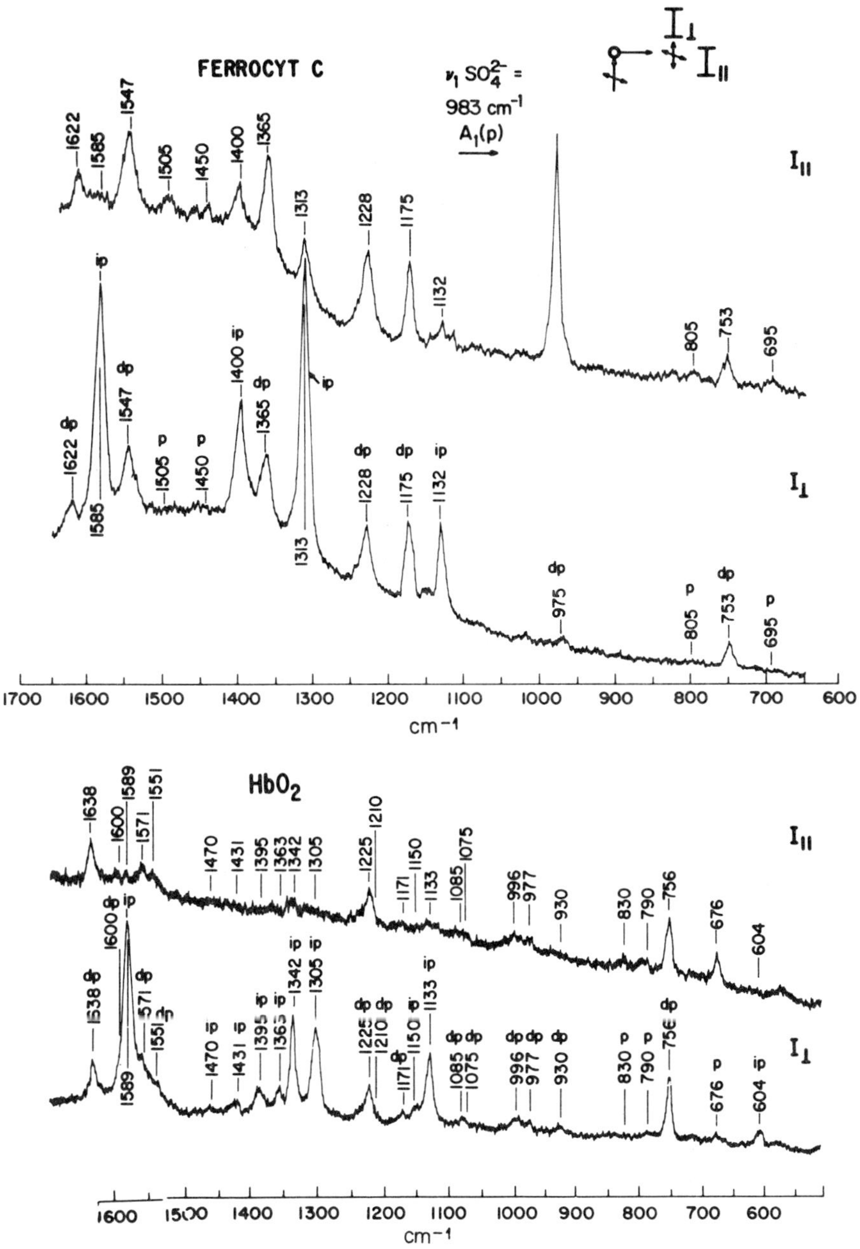

Fig. 6.24. Resonance Raman spectra of oxyhemoglobin (bottom pair of curves) and ferrocytochrome c (top pair). The scattering geometry is shown schematically in the diagram at the top. Both the direction and the polarization vector of the incident laser radiation are perpendicular to the scattering direction. The scattered radiation is analyzed into components perpendicular ($I_\perp$) and parallel ($I_\parallel$) to the incident polarization vector. The exciting wavelength was 568.2 nm (5682 Å) for HbO_2 and 514.5 nm (5145 Å) for cytochrome c. The slit width was about 10 cm^{-1}. Concentrations about 0.5 mM for HbO_2 and about 0.5 mM for cytochrome c. The anomalously polarized lines are indicated by ip. (From [6.51])

larization ratio will be less than infinity due to the addition of a symmetric component to the scattering amplitude. As pointed out by COLLINS et al. [6.79] advantage of this may be taken to study the degree of X,Y degeneracy in complex systems. The most sensitive region to detect the break down of perfect anomalous depolarization is the excitation region between the X and Y components of a near-degenerate absorption. Here, a clear dip in the anomalous depolarization ratio may be detected. The origin of this dip is readily explained by considering the A_2 mode scattering from the X,Y split component of the Q_0 band. If the frequency splitting between the X,Y components is ω_0, and $\Delta\omega$ is the shift in frequency from the center point between these two components, then again for a fixed orientation the scattering polarizability α may be written as

$$\alpha \propto \frac{E_X^L E_Y^S}{\Delta\omega + \frac{\omega_0}{2} + i\Gamma} - \frac{E_Y^L E_X^S}{\Delta\omega - \frac{\omega_0}{2} + i\Gamma} \quad . \tag{6.29}$$

Here the first term represents the incident photon interacting with the lower energy component, arbitrarily taken as the X component and the second term represents interaction with the higher energy component (Y). (6.29) may be rewritten as

$$\alpha \propto \frac{(\Delta\omega + i\Gamma)\left(E_X^L E_Y^S - E_Y^L E_X^S\right) - \frac{\omega_0}{2}\left(E_X^L E_Y^S + E_Y^L E_X^S\right)}{(\Delta\omega + i\Gamma)^2 - \frac{\omega_0^2}{4}} \quad . \tag{6.30}$$

From (6.30) it may be seen that the term in ω_0 makes a symmetric contribution to the scattering polarizability, thereby decreasing the degree of anomalous polarization in the region between the X,Y components giving a minimum exactly between them at $\Delta\omega = 0$. In addition, the greater the breakdown of the degeneracy (the greater the magnitude of ω_0), the lesser the degree of anomalous polarization.

6.4.5 Interference Effects

Just as interference effects were observed in diatomic molecules as discussed in Sect.6.2.2 interference effects also are frequently observed in the porphyrins. Depending on the phase of the terms involved in the scattering amplitude constructive or destructive interference may occur. A detailed discussion of interference effects in such systems was recently reported by FRIEDMAN and HOCHSTRASSER [6.73].

In the previous subsections of this section we treated each resonance independently. However, the full expression for the Raman amplitude must include all the terms that may make contributions. Nevertheless on resonance it is often possible to consider only the resonant contribution. If, however, an understanding of the Raman intensity over an extended region of the absorption is sought after it becomes necessary to consider contributions from all nearby states and to appropriately treat their relative phases in the Raman amplitude expression.

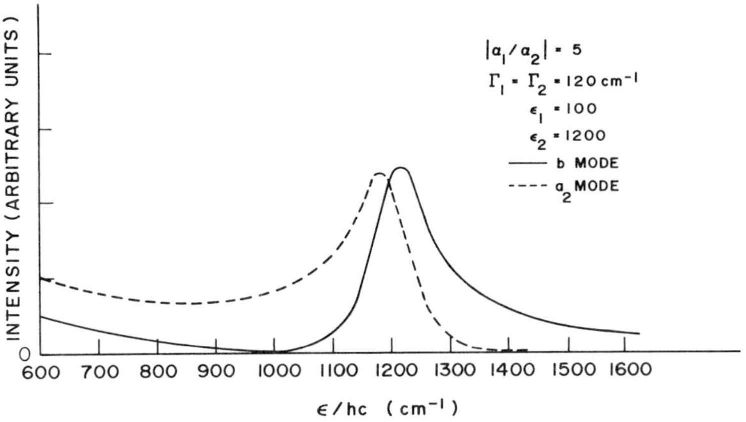

Fig. 6.25. The calculated resonance Raman scattering excitation profile for fundamentals of A_2 modes (dashed line) and B modes (unbrocken line) in the region of Q_1 band. Subscripts 1 and 2 refer to the Q_0 and Q_1 bands, respectively. α corresponds to the value of the product of matrix elements in the Raman amplitude expression and Γ and ϵ correspond to the line width and energy, respectively. (From [6.73])

By considering a given incident polarization and a fixed molecular framework the phases of the Q_0 and Q_1 contributions to the Raman amplitude may be determined by inspection for each of the modes of interest. By doing this (for Herzberg-Teller scattering) it may be seen that the A_1, B_1 and B_2 modes all add constructively while the A_2 modes add destructively. Of course, due to the sign of the denominator in the amplitude expression the opposite interference effect will occur for excitation frequencies lying in between Q_0 and Q_1. A calculation by FRIEDMAN and HOCHSTRASSER [6.73] showing this effect is represented in Fig.6.25. As expected from the above considerations in examining Fig.6.23 the A_2 modes, in comparison to the B modes, continue to show substantial enhancement in the region between Q_0 and Q_1 excitation. This is particularly evident in spectra C and D where the bands at 1130, 1315 and 1585 cm^{-1} are among the most prominent.

In moving the excitation frequency away from the Q_1, Q_0 resonances the intensity of the A_2 modes go to zero since the contribution from Q_1 destructively interferes with that from Q_0. Similarly it should be noted that if the damping factors of Q_0 and Q_1 become very large the resonance denominators of Q_0 and Q_1 become more nearly identical and the two contributions will cancel.

6.4.6 Non-Adiabatic Effects

In the discussion presented thus far we have not included non-adiabatic contributions to the scattering intensity. Such contributions become significant when two excited states have nearly overlapping energy levels such that transitions between these levels may occur. This represents a breakdown in the Born-Oppenheimer approximation and therefore these terms are referred to as being non-adiabatic. These effects influence both the absorption spectrum and the resonance Raman spectrum [6.58,61,69, 80-82], and therefore must be included whenever the origin of the Raman intensity is being studied.

Incorporation of the non-adiabatic effects into the Raman amplitude may be made by introducing additional terms into the coordinate expansion of (6.7). To relate to the ALBRECHT notation the lowest order such term is referred to [6.58] as the "D" term, and using the same notation as (6.7) and (6.8) it may be written as

$$(\underline{\underline{\alpha}})_\xi = M_{ig}(\underline{R}_0)M_{jg}(\underline{R}_0)\frac{1}{\hbar}\sum_k\sum_m \frac{h_{ij}}{(\omega_i-\omega_j)(\omega_{vk}-\omega_L-i\Gamma_k)} \times\left\{\frac{\omega_\xi^i\langle u|k\rangle\langle k|R_\xi|m\rangle\langle m|v\rangle(\delta_{k,m-1}-\delta_{k,m+1})}{\omega_k-\omega_m}-\frac{\omega_\xi^i\langle u|m\rangle\langle m|R_\xi|k\rangle\langle k|v\rangle(\delta_{m,k-1}-\delta_{m,k+1})}{\omega_k-\omega_m}\right\}. \quad (6.31)$$

For simplicity in (6.31) we have considered only one normal mode ξ and two intermediate states i and j. ω_ξ^i is the frequency of a normal mode ξ of electronic state i. All other notation is the same as that used earlier and listed in Table 6.1. More complete treatments of the development of non-adiabatic effects may be found in [6.18, 58,61,80,82].

By assuming that the incident frequency is far from the vibrational levels m of electronic state j so that completeness may be applied and by assuming states i and j may be represented by identical harmonic oscillators with the same internuclear separation, we may combine (6.31) with the Herzberg-Teller terms (6.25) and (6.26). We identify state i with the Q state and state j with the B state. Doing so leads to the expression

$$(\underline{\underline{\alpha}})_\xi \propto M_{Qg}(\underline{R}_0)M_{Bg}(\underline{R}_0)\frac{1}{\hbar}\frac{h_{QB}^\xi\langle u|R_\xi|k\rangle\langle k|v\rangle}{\omega_{vk}-\omega_L-i\Gamma_k}\left(1-\frac{\omega_\xi}{\omega_B-\omega_Q}\right) \quad (6.32)$$

for resonance with Q_0 and

$$(\underline{\underline{\alpha}})_\xi \propto M_{Qg}(\underline{R}_0)M_{Bg}(\underline{R}_0)\frac{1}{\hbar}\frac{h_{QB}^\xi\langle u|k\rangle\langle k|R_\xi|v\rangle}{\omega_{vk}-\omega_L-i\Gamma_k}\left(1+\frac{\omega_\xi}{\omega_B-\omega_Q}\right) \quad (6.33)$$

for resonance with Q_1. Note that (6.32) and (6.33) are identical with (6.25) and (6.26), respectively, except for the inclusion of the factor $\omega_\xi/(\omega_B-\omega_Q)$ in each equation. The diagram for the non-adiabatic contribution to the scattering is included in Fig.6.22.

Fig. 6.26. Excitation profiles of nickel etioprophyrin in chloroform. The top two modes (1205 and 1118 cm^{-1}) are anomalously polarized; the next three (1654, 1159, and 752 cm^{-1}) are depolarized and the bottom three (1518, 1139 and 342 cm^{-1}) are polarized. -h_4 denotes meso-protio nickel etioporphyrin and -d_4 is meso-deutero nickel etioporphyrin. The frequency of the exciting light is given in units of kk (1kk =1000 cm^{-1}). The deviation from equal strength of the two components of each of the anomalously polarized lines results from non-adiabatic effects and the variation in relative strength of the remaining lines results from additional Jahn-Teller distortion effects. (From [6.81])

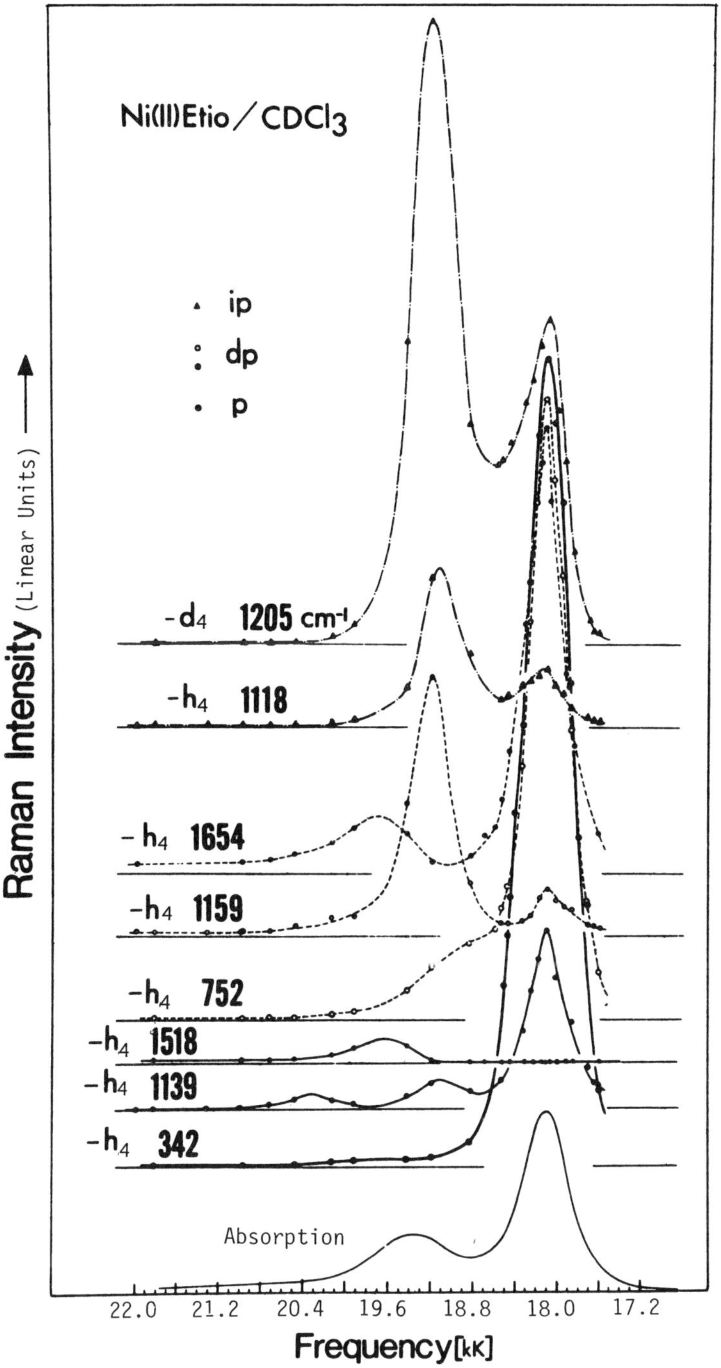

Fig.2.26. Figure caption see oposite page

From (6.32) and (6.33) we see that the non-adiabatic contribution adds amplitude to the Q_1 scattering and subtracts it from the Q_0 scattering relative to what is expected for just a linear vibronic coupling model within the adiabatic approximation. In that approximation the Raman intensity for a fundamental vibrational mode should be the same for resonance with either Q_0 or Q_1 (assuming identical damping constants). Deviation from equal intensities for a Q_0 and a Q_1 transition is most strikingly seen in the recently reported data on nickel etioporphyrin by SHELNUTT et al. [6.81]. In Fig.6.26 we have reproduced the data from that report. Note that the top two modes of A_2 symmetry at 1205 and 1118 cm^{-1} which within the linear vibronic coupling approximation should have approximately equal intensities, have much greater intensity in the Q_1 band than they do in the Q_0 band due to this non-adiabatic effect. By taking the ratio of the Q_1 and Q_0 intensity determined by (6.32) and (6.33) we obtain

$$\frac{I^{Q_1}}{I^{Q_0}} = \left(\frac{\Delta\omega_{QB}+\omega_\xi}{\Delta\omega_{QB}-\omega_\xi}\right)^2 \quad . \tag{6.34}$$

Using a value of 6500 cm^{-1} for the Q-B energy spacing and 1200 cm^{-1} for ω_ξ we obtain a value of ~2 for this ratio. This is consistent with the data reported by SHELNUTT et al. [6.81] for the A_2 modes and reproduced in Fig.6.26.

In the region to the red of the B transition (in the foothills of the Soret) there is an absorption band in cytochrome c that mimics the shape of the β band (Q_1). FRIEDMAN and HOCHSTRASSER [6.18] have assigned it as a transition involving one quantum of a vibronically active asymmetric mode in addition to some number of quanta of totally symmetric modes. It is expected that such a transition may have a very large non-adiabatic contribution to the Raman amplitude for excitation frequencies in this region because of the proximity of this level to the B state. This should cause the energy denominator in the non-adiabatic term expression to be so small that the non-adiabatic contribution to the Raman amplitude may outweigh the preresonance enhancement of Franck-Condon scattering via the B state. Raman investigations in this region should illuminate this point and serve to further clarify the assignment of the weak bands to the red of the Soret.

The presence of non-adiabatic terms will also modify the previously discussed interference effects. The non-adiabatic terms will cause A_2 modes, which from considerations of the Herzberg-Teller term alone destructively interfere away from resonance, to remain finite, but small, away from resonance. Similarly B modes which destructively interfere for excitation between Q_0 and Q_1 will have increased intensity due to the inclusion of non-adiabatic terms. On the other hand for excitation between Q_0 and Q_1 the non-adiabatic contribution to the A_2 amplitude cancels. This difference between the behavior of A_2 modes and B modes in the region between Q_0 and Q_1 allows one to measure the magnitude of the non-adiabatic contribution by comparing experimental data for A_2 and B modes in this excitation frequency region.

6.4.7 Influence of Jahn-Teller Distortion

Throughout the above discussion we have assumed degeneracy of the Q_0 state. Although many porphyrins exhibit splitting of the X and Y components of Q_0 through static variations in the molecular structure, the splitting is small enough that the potentials remain harmonic and insignificantly displaced from their equilibrium positions. Recently, however, SHELNUTT et al. [6.81] have demonstrated that in some etioporphyrins and tetraphenylporphyrins significant distortion of the Q state may take place via a dynamic Jahn-Teller effect. The effect of the Jahn-Teller coupling is to shift the X and Y potential curves of the active modes in opposite directions. The X and Y vibrational levels remain degenerate, however.

The manifestation of this effect in the Raman spectrum is to cause enough distortion of the Q state that Franck-Condon type of scattering can occur. This additional contribution to the re-emission can interfere with scattering from the Herzberg-Teller term (modified to include non-adiabatic effects) and therefore distort the relative intensities of the Q_0 and Q_1 scattering. From symmetry considerations A_2 modes are not influenced by Jahn-Teller distortions but A_1, B_1 and B_2 modes are. Hence, in Fig.6.26 the top two spectra of inversely polarized modes demonstrate the expected Herzberg-Teller plus non-adiabatic intensity distribution. However, the remaining modes (both A_1 and B) have varying relative intensities of the Q_0 and Q_1 contributions due to the interference effects resulting from the Jahn-Teller distortion.

6.5 Concluding Remarks

In this chapter we have presented the basic mechanisms by which resonant Raman scattering may occur and illustrated them by treating a few examples in depth. We apologize to those readers who are interested in the applications of resonance Raman scattering to other systems but hope that the approach presented here is sufficiently general to be of use in a wide variety of molecules. The applications of resonant Raman scattering to various molecules has furthered the understanding of many important chemical and biological problems. Many of these are discussed in other reviews. Our failure to mention many of the important applications does not in any way indicate that we do not view them as significant. We felt it better to limit our examples to as few as possible for the sake of clarity of presentation.

While some aspects of the theories of the resonant Raman effect remain not well understood, for most systems a theoretical basis to treat the scattering phenomena is at hand. Further refinements both experimentally and theoretically of some of the more subtle aspects are however still needed. In the next few years we may look forward to the clarification of some of these as well as more quantitative calculations of Raman spectra and of excitation profiles from complex systems. From such studies detailed information about molecular and electronic structure will be learned. As more and more dye lasers become available studies over a wider range of frequencies will make the technique even more powerful. Indeed recently PETICOLAS and his co-workers [6.83] have developed a laser Raman spectrograph capable of obtaining high quality spectra as far into the ultraviolet as ~2700 Å. As more such laboratories become built up chemists and biologists will have at their disposal a flexible tool with which molecular and electronic properties of any molecule may be studied at very low concentrations.

Acknowledgments. We wish to thank Dr. J.A. Shelnutt for critically reading this manuscript.

References

6.1 Martin, R.M., Falicov, L.M.: In *Light Scattering in Solids*, ed. by M. Cardona, Topics in Applied Physics, Vol.8 (Springer, Berlin, Heidelberg, New York 1975) Chap.3, pp.79-145
6.2 Tang, J., Albrecht, A.C.: "Developments in the Theory of Vibrational Raman Intensities", in *Raman Spectroscopy Theory and Practice*, ed. by H.A. Szymanski, Vol.2 (Plenum Press, New York 1970) pp.33-67
6.3 Jacon, M.: "Theoretical Study of Resonance Raman Scattering by the Methods of Quantum Electronics", in *Advances in Raman Spectroscopy*, ed. by J.-P. Mathieu, Vol.1 (Heyden and Son, London 1973) p.325
6.4 Behringer, J.: "Theories of Resonance Raman Scattering", in *Molecular Spectroscopy*, ed. by Barrow,R.F., Long,D.A., Millen,D.J., Vol.2 (The Chemical Society, Burlington House, London 1974) pp.100-172;
"Experimental Resonance Raman Spectroscopy", in *Molecular Spectroscopy*, ed. by Barrow,R.F., Long,D.A., Millen,D.J., Vol.3 (The Chemical Society, Burlington House, London 1975) pp.163-280
6.5 Van Labeke, D., Jacon, M., Berjot, M., Bernard, L.: J. Raman Spectrosc. *2*, 219 (1974)
6.6 Novak, F., Friedman, J.M., Hochstrasser, R.M.: "Resonant Scattering of Light by Molecules: Time-Dependent and Coherence Effects", in *Lasers and Coherence Spectroscopy*, ed. by J.I. Steinfeld (Plenum Press, New York 1978) pp.451-518
6.7 Kiefer, W.: "Recent Techniques in Raman Spectroscopy", in *Advances in Infrared and Raman Spectroscopy*, ed. by Clark,R.J.H., Hester,R.E., Vol.3 (Heyden and Son, London 1977) Chap.1, pp.1-42
6.8 Lewis, A., Spoonhower, J.: "Tunable Laser Raman Spectroscopy in Biology", in *Neutron, X-Ray and Laser Spectroscopy in Biophysics and Chemistry*, ed. by Yip,S., Chen,S. (Academic Press, New York 1974) pp.347-375
6.9 Cooney, R.P., Curthoys, G., Tam, N.T.: Adv. Catal. *24*, 293 (1975)
6.10 Spiro, T.G.: Proc. R. Soc. London A*345*, 89 (1975)
6.11 Thomas, G.J. Jr.: "Raman Spectroscopy of Biopolymers", in *Vibrational Spectra and Structure*, ed. by J. Durig, Vol.3 (Marcel Dekker, New York 1975) pp.239-315
6.12 Johnson, B.B., Peticolas, W.L.: Ann. Rev. Phys. Chem. *27*, 465 (1976)
6.13 Lord, R.C.: Appl. Spectrosc. *31*, 187 (1977)
6.14 Spiro, T.G., Stein, P.: Ann. Rev. Phys. Chem. *28*, 501 (1977)
6.15 Warshel, A.: Ann. Rev. Biophys. and Bioengin. *6*, 273 (1977)
6.16 Compaan, A.: "Resonance Raman Spectroscopy With Tunable Lasers", in *Applied Spectroscopy Reviews* (in press)

6.17 Rousseau, D.L., Williams, P.F.: J. Chem. Phys. *64*, 3519 (1976)
6.18 Friedman, J.M., Hochstrasser, R.M.: Chem. Phys. *1*, 457 (1973)
6.19 Rousseau, D.L., Patterson, G.D., Williams, P.F.: Phys. Rev. Lett. *34*, 1306 (1975); *37*, 1441 (1975)
6.20 Gross, E.F., Permogorov, S.A., Travnikov, V.V., Sel'kin, A.V.: Fiz. Tverd. Tela *13*, 699 (1971) [English transl.: Sov. Phys. Solid State *13*, 578 (1971)]
6.21 Klein, M.V.: Phys. Rev. B*8*, 919 (1973)
6.22 Ault, B.S., Howard, W.F. Jr., Andrews, L.: J. Mol. Spectrosc. *55*, 217 (1975)
6.23 Albrecht, A.C.: J. Chem. Phys. *34*, 1476 (1961)
6.24 Shorygin, P.P.: Sov. Phys. - Usp. *16*, 99 (1973)
6.25 Silverstein, S.D.: General Electric Technical Report #73CRD195 (June 1973); Silverstein, S.D., St. Peters, R.L.: Chem. Phys. Lett. *23*, 140 (1973); Phys. Rev. A*9*, 2720 (1974)
6.26 Jacon, M., Van Labeke, D.: Mol. Phys. *29*, 1241 (1975)
6.27 Williams, P.F., Fernandez, A.: To be published
6.28 For example, see Kurzel, R.B., Steinfeld, J.I., Hatzenbuhler, D.A., Leroi, G.E.: J. Chem. Phys. *55*, 4822 (1971) and references therein
6.29 Williams, P.F., Rousseau, D.L., Dworetsky, S.H.: Phys. Rev. Lett. *32*, 196 (1974)
6.30 We present here an outline of the calculations initially reported in [6.29] and given in detail in [6.17]. Several other calculations of lifetime responses using a variety of incident pulse shapes have been reported. These include Silverstein, S.D.: General Electric Technical Report 73CRD196 (June 1973); Friedman, J.M., Hochstrasser, R.M.: Chem. Phys. *6*, 155 (1974); Berg, J.O., Langhoff, C.A., Robinson, G.W.: Chem. Phys. Lett. *29*, 305 (1974); Hilborn, R.C.: Chem. Phys. Lett. *32*, 76 (1975); Mukamel, S., Jortner, J.: J. Chem. Phys. *62*, 3609 (1975); Kimble, H.J., Mandel, L.: Opt. Commun. *14*, 167 (1975); Metiu, H., Ross, J., Nitzan, A.: J. Chem. Phys. *63*, 1289 (1975); Wilson, A.D., Friedman, H., Ahiman, V.: Chem. Phys. Lett. *43*, 539 (1976); Friedman, H., Wilson, A.D.: Chem. Phys. Lett. *46*, 307 (1977); Ron, A., Ron, A.: Chem. Phys. Lett., in press
6.31 Williams, P.F., Rousseau, D.L.: Phys. Rev. Lett. *30*, 961 (1973)
6.32 Holzer, W., Murphy, W.F., Bernstein, H.J.: J. Chem. Phys. *52*, 399 (1970)
6.33 Kiefer, W., Bernstein, H.J.: J. Mol. Spectrosc. *43*, 366 (1972)
6.34 Baierl, P., Kiefer, W., Williams, P.F., Rousseau, D.L.: Chem. Phys. Lett. *50*, 57 (1977)
6.35 Williams, P.F., Fernandez, A., Rousseau, D.L.: Chem. Phys. Lett. *47*, 150 (1977)
6.36 Schulman, J.M., Detrano, R.: Phys. Rev. A*5*, 1125 (1972)
6.37 Berjot, M., Jacon, M., Bernard, L.: Can. J. Spectrosc. *17*, 60 (1972)
6.38 Silverstein, S.D., St. Peters, R.L.: Phys. Rev. A*9*, 2720 (1974)
6.39 Kiefer, W., Bernstein, H.J.: J. Raman Spectrosc.*1*, 417 (1973)
6.40 Fortuno, G., Rousseau, D.L., Friedman, J.M., Williams, P.F.: To be published
6.41 Michaelian, K.H., Rieckhoff, K.E., Voigt, E.-M.: Proc. Nat. Acad. Sci. *72*, 4196 (1975)
6.42 Fenstermacher, P.R., Callender, R.H.: Opt. Commun. *10*, 181 (1974)
6.43 Matsuzaki, S., Maeda, S.: Chem. Phys. Lett. *28*, 27 (1974)
6.44 Howard, W.F. Jr., Andrews, L.: J. Raman Spectrosc. *2*, 447 (1974); Wight, C.A., Ault, B.S., Andrews, L.: J. Mol. Spectrosc. *56*, 239 (1975)
6.45 Grzybowksi, J.M., Andrews, L.: J. Raman Spectrosc. *4*, 99 (1975)
6.46 Bondybey, V.E., Bearder, S.S., Fletcher, C.: J. Chem. Phys. *64*, 5243 (1976)
6.47 Friedman, J.M., Rousseau, D.L., Bondybey, V.E.: Phys. Rev. Lett. *37*, 1910 (1976)
6.48 Friedman, J.M., Rousseau, D.L., Bondybey, V.E.: To be published
6.49 Williams, P.F., Biesle, R.A., Rousseau, D.L., Friedman, J.M.: To be published
6.50 Clark, R.J.H., Franks, M.L.: J. Am. Chem. Soc. *97*, 2691 (1975)
6.51 Spiro, T.G., Strekas, T.C.: Proc. Nat. Acad. Sci. *69*, 2622 (1972)
6.52 Ito, M., Suzuka, I., Udagawa, Y., Kaya, K., Mikami, N.: Chem. Phys. Lett. *16*, 211 (1972); Suzuka, I., Mikami, N., Udagawa, Y., Kaya, K., Ito, M.: J. Chem. Phys. *57*, 4500 (1972)
6.53 Kalantar, A.H., Franzosa, E.S., Innes, K.K.: Chem. Phys. Lett. *17*, 335 (1972)
6.54 Ito, M.: "Vibronic Coupling as Revealed by Pre-resonance Raman Effect", in *Proc. 5th Intern. Conf. on Raman Spectroscopy*, ed. by Schmid, E.D., Brandmüller, J., Kiefer, W., Schrader, B., Schrötter, H.W. (Schulz Verlag, Freiburg/Br. 1976) pp.267-276

6.55 Verma, A.L., Asselin, M., Sunder, S., Bernstein, H.J.: J. Raman Spectrosc. *4*, 295 (1976)
6.56 Kitagawa, T., Abe, M., Kyogoku, Y., Ogoshi, H., Watanabe, E., Yoshida, Z.: J. Phys. Chem. *80*, 1181 (1976)
6.57 Asher, S., Sauer, K.: J. Chem. Phys. *64*, 4115 (1976)
6.58 Shelnutt, J.A., O'Shea, D.C., Yu, N.-T., Cheng, L.D., Felton, R.H.: J. Chem. Phys. *64*, 1156 (1976)
6.59 Nafie, L.A., Pezolet, M., Peticolas, W.L.: Chem. Phys. Lett. *20*, 563 (1973)
6.60 Koningstein, J.A.: Opt. Spektrosk. *35*, 260 (1973) [English transl.: Opt. Spectrosc. *35*, 152 (1973)]
6.61 Mingardi, M., Siebrand, W.: J. Chem. Phys. *62*, 1074 (1975)
6.62 Hong, H.-K.: Chem. Phys. *9*, 1 (1975)
6.63 Yeung, E.S., Heiling, M., Small, G.J.: Spectrochim. Acta *31*A, 1921 (1975)
6.64 Tsuboi, M., Hirakawa, A.Y.: J. Raman Spectrosc. *5*, 75 (1976)
6.65 Gouterman, M.: J. Mol. Spectrosc. *6*, 138 (1961)
6.66 Perrin, M.H., Gouterman, M., Perrin, C.L.: J. Chem. Phys. *50*, 4137 (1969)
6.67 Adar, F.: "Electronic Absorption Spectra of Hemes and Hemeproteins", to appear in *The Porphyrins*, ed. by David Dolphin (□□□, □□□ 19□□)
6.68 Woodbridge, P.K.: Unpublished results
6.69 Shelnutt, J.A.: Ph.D. Thesis, Georgia Institute of Technology (1975)
6.70 Azumi, T., Matsuzaki, K.: Photochem. Photobiol. *25*, 315 (1977)
6.71 Ballhausen, C.J., Hansen, A.E.: Ann. Rev. Phys. Chem. *23*, 15 (1972)
6.72 Friedman, J.M., Hochstrasser, R.M.: J. Am. Chem. Soc. *98*, 4043 (1976)
6.73 Friedman, J.M., Hochstrasser, R.M.: Chem. Phys. Lett. *32*, 414 (1975)
6.74 Adar, F., Gouterman, M., Aronowitz, S.: J. Chem. Phys. *80*, 2184 (1976)
6.75 Friedman, J.M., Rousseau, D.L., Adar, F.: Proc. Nat. Acad. Sci. *74*, 2607 (1977)
6.76 Placzek, G.: "Rayleigh-Streuung und Raman Effekt", in *Handbuch der Radiologie*, ed. by E. Marx, Vol.VI, Pt.2 (Akademische Verlagsgesellschaft, Leipzig 1934) pp.205-374. English translation by A. Werbin: "Rayleigh and Raman Scattering", UCRL Transl. No. 526L, available from National Technical Information Service, U.S. Department of Commerce, Springfield, Va.
6.77 Hamaguchi, H., Harada, I., Shimanouchi, T.: Chem. Phys. Lett. *32*, 103 (1975); Hamaguchi, H., Shimanouchi, T.: Chem. Phys. Lett. *38*, 370 (1976)
6.78 Loudon, R.: Adv. Phys. *13*, 423 (1964)
6.79 Collins, D.W., Champion, P.M., Fitchen, D.B.: Chem. Phys. Lett. *40*, 416 (1976)
6.80 Small, G.S., Yeung, E.S.: Chem. Phys. *9*, 379 (1975)
6.81 Shelnutt, J.A., Cheung, L.D., Chang, R.C.C., Yu, N.-T., Felton, R.H.: J. Chem. Phys. *66*, 3387 (1977)
6.82 Zgierski, M.Z.: Chem. Phys. Lett. *36*, 390 (1975)
6.83 Peticolas, W.L.: "New Approaches to Raman Spectroscopy of Biomolecules", in *Proc. 5th Intern. Conf. on Raman Spectroscopy*, ed. by Schmid,E.D., Brandmüller, J., Kiefer,W., Schrader,B., Schrötter,H.W. (Schulz Verlag, Freiburg/Br. 1976) pp.163-175

7. Coherent Anti-Stokes Raman Spectroscopy

J. W. Nibler and G. V. Knighten

With 21 Figures

The development and application of Raman spectroscopy has followed closely the development of excitation sources of increased brightness and power. The laser provided a quantum jump in improvement over mercury arc sources, and cw visible lasers with average powers of 1-10 W are now in common use. Despite the fact that Raman scattering cross sections are quite small (Chap.4), the technique has proven an extremely valuable tool in molecular spectroscopy. Many of the more important applications are reviewed in the preceding chapters and, in Chap.6, ways to increase Raman cross sections via resonance enhancement have been discussed.

With pulsed lasers, peak powers in the megawatt and even gigawatt range are readily available and effects due to the nonlinear response of molecules to the high optical fields become observable. Much interest has centered on using such interactions to generate coherent beams at harmonic and at sum and difference frequencies. SHEN has recently reviewed such applications [7.1] and references to other developments in nonlinear optics are provided as an epilogue in the 1977 reprinting of BLOEMBERGEN's classic text [7.2]. At high pump powers, stimulated emission of Stokes and anti-Stokes Raman beams have been observed for many liquids and high pressure gases, in many cases with more than 20% conversion of the pumping radiation [7.3-6]. Despite this enormous gain in Raman signal strength, stimulated Raman spectroscopy has been of little use in molecular spectroscopy because its high threshold requirement limits emission to only the strongest Raman lines of materials of high number density.

Recently, however, a nonlinear optical technique known as coherent anti-Stokes Raman scattering (CARS)[1] has been used to obtain the coherence and signal strength advantages of stimulated Raman spectroscopy while retaining the general applicability of spontaneous Raman spectroscopy. Although the CARS process was demonstrated and understood over a decade ago [7.7,8], it is only in the last few years that the availability of tunable pulsed lasers of high peak power has enabled the advantages of the technique to be realized.

The CARS experiment is an example of a four-wave parametric process in which three waves, two at a pump frequency ω_1 and one at a "Stokes" frequency ω_2, are mixed in a sample to produce a new coherent beam at the anti-Stokes frequency $\omega_3 = 2\omega_1 - \omega_2$.

[1]This acronym is also used for coherent anti-Stokes Raman spectroscopy.

Table 7.1. Comparison of CARS and spontaneous Raman spectroscopy

Advantages of CARS

1. Very high signal levels (~10^5 × Raman signals).
2. Signal beam divergence ~10^{-4} × Raman values of 4π sr with consequently greater collection efficiencies, excellent spatial descrimination against luminous backgrounds such as occur in flames, discharges, chemiluminescent mixtures, etc.
3. Elimination of fluorescence problems since $\omega_3 > \omega_1$, ω_2.
4. Sample quantities very small, microliter volumes, millitorr pressures.
5. High resolution spectroscopy (0.03 to 0.5 cm^{-1}) without a monochromator is routine. Doppler broadening is reduced for forward scattering (CARS) versus 90° scattering (Raman). Very high resolution (0.001 cm^{-1}) possible with cw and pulsed sources.
6. CARS signals can be resonance enhanced when ω_3, ω_1 approach an electronic absorption.

Disadvantages of CARS

1. χ^{nr} limits detection of species of low *relative* concentration (~1% in solution, ~10^{-3}% in gases, if no resonance enhancement or χ^{nr} cancellation).
2. Samples must be optically transparent, not easily damaged by high power focused beams.
3. Equipment is expensive, not easily tuned over 0-4000 cm^{-1} spectral shift range.
4. Resonant frequencies and lineshapes are affected by χ^{nr} and by neighboring resonances.
5. Signals strongly dependent on laser amplitudes, wavelengths and alignment stability.
6. CARS signals can saturate for high laser powers, leading to anomalous relative intensities.

This mixing occurs for all samples but is greatly enhanced if the frequency difference $\omega_1 - \omega_2$ approaches a Raman active vibrational resonance in the medium. By varying ω_2, one can scan across the vibrational frequency and thereby generate a spectrum with all of the information normally contained in a spontaneous Raman spectrum.

The main advantages and disadvantages of CARS relative to spontaneous Raman spectroscopy are listed in Table 7.1 and most of these will be considered in more detail in subsequent sections. Foremost among the advantages is the large gain in Raman signal in a beam of small divergence. This permits the high-resolution Raman study of low pressure gases and luminous samples such as flames and plasmas. These and many other applications will be surveyed in this review of experimental work which employs the CARS technique. First, however, we discuss the key theoretical elements necessary to describe the CARS process and to provide a link with the more familiar features of Raman spectroscopy.

7.1 Theory

The theoretical basis for CARS spectroscopy has been reviewed recently by ANDERSON and HUDSON [7.9] and by DeWITT et al. [7.10]. These descriptions are in turn based upon earlier developments of the theory of nonlinear optics by BLOEMBERGEN and co-workers [7.2,11], MAKER and TERHUNE [7.8,12], and BUTCHER [7.13]. FLYTZANIS has also reviewed this work in his detailed discussion of nonlinear optical susceptibilities [7.14] and, in addition, others [7.15-23] have described the theory.

In this section we offer an outline of the procedure used in most of the above treatments but with particular emphasis on the application to CARS and Raman spectroscopy. Maxwell's equations are used to relate the amplitude of the anti-Stokes wave at frequency ω_3 to the third-order susceptibility term $\chi^{(3)}$ in the expansion of the polarization produced by incident waves of frequency ω_1 and ω_2. This is a macroscopic development of the theory and the resultant expression depends upon the bulk susceptibility of the sample. To connect $\chi^{(3)}$ to the microscopic properties of the molecules, the results of quantum-mechanical perturbation theory are used to express the molecular $\chi^{(3)}$ in terms of molecular wave functions and appropriate ensemble averages are then taken.

A similar approach is used to relate the amplitude of a stimulated Raman wave to the third-order susceptibility. In this case it is possible to relate $\chi^{(3)}$ to the cross section measured by spontaneous Raman spectroscopy and, in the absence of electronic resonance enhancement, to show that the Raman cross section is proportional to the imaginary part of the CARS susceptibility. This relation is compared with that commonly obtained from a phenomenological approach based on a damped harmonic oscillator model with a driving force produced by the applied fields. These results reveal the similarities and differences between CARS and Raman spectroscopy and these are discussed in detail in Sect.7.3.

7.1.1 Maxwell's Equations for Macroscopic Samples

To determine the magnitude of the anti-Stokes field produced in a sample as a result of the incident fields at radial frequencies ω_1 and ω_2, one starts with Maxwell's equations for macroscopic media containing no magnetization ($\underline{M}=0$) and no current due to free charges ($\underline{J}=0$). In Gaussian units, these are

$$\underline{\nabla} \times \underline{E} = -(1/c)\partial\underline{H}/\partial t \qquad \underline{\nabla} \times \underline{H} = (1/c)\partial\underline{D}/\partial t \tag{7.1}$$

$$\underline{\nabla} \cdot \underline{D} = 0 \qquad \underline{\nabla} \cdot \underline{H} = 0 \quad . \tag{7.2}$$

The electric displacement field $\underline{D}$ differs from the electric field $\underline{E}$ by the electric polarization $\underline{P}$ present in the sample according to the usual relation

$$\underline{D} = \underline{E} + 4\pi\underline{P} \quad . \tag{7.3}$$

$\underline{P}$ (the electric dipole moment per unit volume) is a function of time and position and is generally assumed to be related to $\underline{E}$ through a *constitutive relation* which we take below to be a power series expansion in the applied fields.

Equation (7.3) can be substituted into (7.1) and $\underline{H}$ eliminated to give the classical wave equation with a source term due to the polarization

$$\underline{\nabla} \times \underline{\nabla} \times \underline{E} + (1/c)\partial^2\underline{E}/\partial t^2 = -(4\pi/c^2)\partial^2\underline{P}/\partial t^2 \quad . \tag{7.4}$$

We assume that $\underline{E}(\underline{r},t)$ and $\underline{P}(\underline{r},t)$ will be sums of functions with frequencies of the applied fields, of the sample resonances and of the various sum and difference frequencies

$$\underline{E}(\underline{r},t) = \frac{1}{2}\sum_{i=-n}^{n} \underline{E}(\underline{r},i)\exp(-i\omega_i t) \tag{7.5}$$

$$\underline{P}(\underline{r},t) = \frac{1}{2}\sum_{i=-n}^{n} \underline{P}(\underline{r},i)\exp(-i\omega_i t) \quad . \tag{7.6}$$

For these definitions $\omega_{-i} = -\omega_i$ and, since $\underline{E}(\underline{r},t)$ and $\underline{P}(\underline{r},t)$ must be real quantities, $\underline{E}(\underline{r},-i) = \underline{E}(\underline{r},i)^*$ and $\underline{P}(\underline{r},i) = \underline{P}(\underline{r},-i)^*$ [7.8]. With (7.5,6) the wave equation separates into equations for each frequency component

$$\underline{\nabla} \times \underline{\nabla} \times \underline{E}(\underline{r},i) - (\omega_i/c)^2\underline{E}(\underline{r},i) = 4\pi(\omega_i/c)^2\underline{P}(\underline{r},i) \quad . \tag{7.7}$$

Next $\underline{P}(\underline{r},i)$ is written to show the linear and nonlinear contributions explicitly, i.e.,

$$\underline{P}(\underline{r},i) = \underline{P}^{(1)}(\underline{r},i) + \underline{P}^{(2)}(\underline{r},i) + \underline{P}^{(3)}(\underline{r},i) + \dots \tag{7.8}$$

where

$$\underline{P}^{(n)}(\underline{r},i) = \underline{\underline{\chi}}^{(n)}(\omega_1,\omega_2,\dots,\omega_n)\underline{E}(\underline{r},1)\underline{E}(\underline{r},2)\dots\underline{E}(\underline{r},n) \tag{7.9}$$

and $\underline{\underline{\chi}}^{(n)}$ is the susceptibility tensor of rank n+1.

For isotropic media such as gases or liquids and for any centrosymmetric crystal, $\underline{P}^{(2)}(\underline{r},i)$ (and all other even order terms) are zero because of inversion symmetry [7.14]. We neglect nonlinear terms higher than cubic and let

$$\underline{P}^{(1)}(\underline{r},i) = \chi^{(1)}(\omega_i)\underline{E}(\underline{r},i) = (\varepsilon_i - 1)\underline{E}(\underline{r},i)/4\pi \tag{7.10}$$

where the conventional definition of the dielectric constant is employed

$$\varepsilon_i = 1 + 4\pi\chi^{(1)} \quad . \tag{7.11}$$

Substitution of (7.5) and (7.10) into (7.7) yields the desired wave equation which isolates the nonlinear source term on the right-hand side

$$\underline{\nabla} \times \underline{\nabla} \times \underline{E}(\underline{r},i) - (\omega_i/c)^2 \, \varepsilon_i \, \underline{E}(\underline{r},i) = 4\pi(\omega_i/c)^2 \underline{P}^{(3)}(\underline{r},i) \quad . \tag{7.12}$$

Also, by similar substitutions in the divergence equation (7.2), we get

$$\underline{\nabla} \cdot \underline{D} = \underline{\nabla} \cdot \left[\varepsilon_i \, \underline{E}(\underline{r},i) + 4\pi \underline{P}^{(3)}(\underline{r},i)\right] = 0 \quad . \tag{7.13}$$

If $\underline{P}^{(3)}(\underline{r},i)$ is neglected, the solution for $\underline{E}(\underline{r},i)$ takes the usual form for a wave traveling through a non-absorbing dielectric medium

$$\underline{E}(\underline{r},i) = \underline{E}(i) \exp(i\underline{k}_i \cdot \underline{r}) \tag{7.14}$$

where

$$k_i = \varepsilon_i^{1/2} \omega_i/c = n_i \omega_i/c \tag{7.15}$$

and, from (7.13)

$$\underline{E}(i) \cdot \underline{k}_i = 0 \quad . \tag{7.16}$$

For real values of ε_i, the wave is unattenuated and ε_i is equal to the square of the index of refraction n_i. Of course, to obtain the contribution to the real field $\underline{E}(\underline{r},t)$ by a particular ω_i, the two complex fields $\underline{E}(\underline{r},i)$ and $\underline{E}(\underline{r},-i)$ must be combined according to (7.5)

$$\underline{E}(\underline{r},t) = \frac{1}{2}\left\{\underline{E}(i) \exp[i(\underline{k}_i \cdot \underline{r} - \omega_i t)] + \underline{E}(i)^* \exp[-i(\underline{k}_i \cdot \underline{r} - \omega_i t)]\right\} \quad . \tag{7.17}$$

$\underline{E}(\underline{r},t)$ is a plane wave with its polarization direction orthogonal to the propagation direction $\underline{k}_i$. The time averaged flux of energy carried by this wave (energy per unit area per unit time) is given by the Poynting vector

$$\underline{S}_i = (cn_i/8\pi) |\underline{E}(i)|^2 \, \hat{\underline{k}}_i = I_i \, \hat{\underline{k}}_i \tag{7.18}$$

where $\hat{\underline{k}}_i$ is the unit k vector and I_i is the intensity of the wave. These results will be used below to describe the unattenuated propagation of the ω_1 and ω_2 beams in the CARS process. First, however, it is necessary to consider some of the symmetry properties of the $\chi^{(3)}$ tensor.

7.1.2 Symmetry Properties of $\chi^{(3)}$

In component form (7.9) for $\underline{P}^{(3)}$ can be written as [7.8,9]

$$P_\alpha^{(3)}(\underline{r},i) = \chi_{\alpha\beta\gamma\delta}^{(3)}(\omega_1,\omega_2,\omega_3)\ E_\beta(\underline{r},1)\ E_\gamma(\underline{r},2)\ E_\delta(\underline{r},3) \qquad (7.19)$$

where we use the convention that a summation is to be carried out over all repeated Cartesian indices ($\beta\gamma\delta$ in this case). There will be 27 terms in this summation for each component of $P^{(3)}$ but fortunately symmetry restrictions greatly reduce this number [7.8,13,14]. First one may note that $\chi^{(3)}$ is invariant to the six permutations of the pairs $(\beta\omega_1)$, $(\gamma\omega_2)$, $(\delta\omega_3)$. Thus there will be six equivalent terms for any chosen order of the fields. If n of the frequencies are identical, n! of these permutations are indistinguishable and one obtains

$$P_\alpha^{(3)}(\underline{r},i) = (6/n!)\ \chi_{\alpha\beta\gamma\delta}^{(3)}(\omega_1,\omega_2,\omega_3)\ E_\beta(\underline{r},1)\ E_\gamma(\underline{r},2)\ E_\delta(\underline{r},3) \quad . \qquad (7.20)$$

In addition, $\chi^{(3)}$ must transform according to the macroscopic symmetry properties of the medium. For a crystal, this means that $\chi^{(3)}$ must be unchanged by any of the symmetry operations appropriate to the crystal. BUTCHER [7.13] has examined these restrictions and FLYTZANIS [7.14] tabulated the results for different crystal classes and for isotropic media such as liquids or gases. In the latter case there are 21 nonzero elements of which only 3 are independent. They are

$$\begin{aligned}
\chi_{1111} &= xxxx = yyyy = zzzz\\
\chi_{1122} &= xxyy = yyxx = xxzz = zzxx = yyzz = zzyy\\
\chi_{1212} &= xyxy = yxyx = xzxz = zxzx = yzyz = zyzy\\
\chi_{1221} &= xyyx = yxxy = xzzx = zxxz = yzzy = zyyz \quad \text{with}\\
\chi_{1111} &= \chi_{1122} + \chi_{1212} + \chi_{1221} \quad . \qquad (7.21)
\end{aligned}$$

The combination of the permutation and isotropy restrictions greatly simplifies the summations in (7.19) and, in the general case of four wave mixing, we obtain

$$\begin{aligned}
P^{(3)}(\underline{r},4) = {} & 6\chi_{1122}^{(3)}(\omega_1,\omega_2,\omega_3)\ E_\alpha(\underline{r},1)\ E_\beta(\underline{r},2)\ E_\beta(\underline{r},3)\\
& + 6\chi_{1212}^{(3)}(\omega_1,\omega_2,\omega_3)\ E_\beta(\underline{r},1)\ E_\alpha(\underline{r},2)\ E_\beta(\underline{r},3)\\
& + 6\chi_{1221}^{(3)}(\omega_1,\omega_2,\omega_3)\ E_\beta(\underline{r},1)\ E_\beta(\underline{r},2)\ E_\alpha(\underline{r},3) \quad . \qquad (7.22)
\end{aligned}$$

If we use (7.14) for the incident fields, we can write (7.22) as

$$\begin{aligned}
P^{(3)}(\underline{r},4) = {} & [6\chi_{1122}^{(3)}(\omega_1,\omega_2,\omega_3)\hat{\underline{e}}_1(\hat{\underline{e}}_2\cdot\hat{\underline{e}}_3)+6\chi_{1212}^{(3)}(\omega_1,\omega_2,\omega_3)\hat{\underline{e}}_2(\hat{\underline{e}}_1\cdot\hat{\underline{e}}_3)\\
& + 6\chi_{1221}^{(3)}(\omega_1,\omega_2,\omega_3)\hat{\underline{e}}_3(\hat{\underline{e}}_1\cdot\hat{\underline{e}}_2)]\ E(1)\ E(2)\ E(3)\ \exp[i(\underline{k}_1+\underline{k}_2+\underline{k}_2)\cdot\underline{r}] \quad , \qquad (7.23)
\end{aligned}$$

where $\hat{\underline{e}}_i$ is the unit polarization[2] vector of E(i).

For CARS, two of the frequencies are identical so we divide by 2! and also equate χ_{1122} to χ_{1212} because of permutation symmetry. This gives

$$
\begin{aligned}
P^{(3)}(\underline{r},3) &= \left[6\chi^{(3)}_{1122}(\omega_1,\omega_1,-\omega_2)\hat{\underline{e}}_1(\hat{\underline{e}}_1\cdot\hat{\underline{e}}_2)+3\chi^{(3)}_{1221}(\omega_1,\omega_1,-\omega_2)\hat{\underline{e}}_2\right] \\
&\quad + E(1)^2\ E(2)^*\ \exp[i(2\underline{k}_1-\underline{k}_2)\cdot\underline{r}] \\
&= \chi_{CARS}\hat{\underline{e}}_{CARS}\ E(1)^2\ E(2)^*\ \exp[i(2\underline{k}_1-\underline{k}_2)\cdot\underline{r}] \quad ,
\end{aligned}
\tag{7.24}
$$

where

$$
\chi_{CARS}\hat{\underline{e}}_{CARS} = 6\chi^{(3)}_{1122}(\omega_1,\omega_1,-\omega_2)\hat{\underline{e}}_1(\hat{\underline{e}}_1\cdot\hat{\underline{e}}_2) + 3\chi^{(3)}_{1221}(\omega_1,\omega_1,-\omega_2)\hat{\underline{e}}_2 \quad . \tag{7.25}
$$

If $\hat{\underline{e}}_1 = \hat{\underline{e}}_2$, we see that $\hat{\underline{e}}_{CARS} = \hat{\underline{e}}_1$ and

$$
\chi_{CARS} = 6\chi^{(3)}_{1122} + 3\chi^{(3)}_{1221} = 3\chi^{(3)}_{1111}(\omega_1,\omega_1,-\omega_2) \quad . \tag{7.26}
$$

For perpendicular polarizations ($\hat{\underline{e}}_1 \perp \hat{\underline{e}}_2$) we have $\hat{\underline{e}}_{CARS} = \hat{\underline{e}}_2$ and

$$
\chi_{CARS} = 3\chi^{(3)}_{1221}(\omega_1,\omega_1,-\omega_2) \quad . \tag{7.27}
$$

Thus the parallel and perpendicular polarization measurements suffice to completely determine all elements of the macroscopic susceptibility tensor.

Expression (7.23) can be applied to stimulated Raman scattering which is also known to depend upon the third-order susceptibility [7.2-5]. The third-order polarization in this case is written as

$$
\begin{aligned}
P^{(3)}(\underline{r},2) &= \left[6\chi^{(3)}_{1122}(\omega_2,\omega_1,-\omega_1)\hat{\underline{e}}_2+6\chi^{(3)}_{1212}(\omega_2,\omega_1,-\omega_1)\hat{\underline{e}}_1(\hat{\underline{e}}_1\cdot\hat{\underline{e}}_2)\right. \\
&\quad \left. + 6\chi^{(3)}_{1221}(\omega_2,\omega_1,-\omega_1)\hat{\underline{e}}_1(\hat{\underline{e}}_1\cdot\hat{\underline{e}}_2)\right]\ E(1)\ E(1)^*\ E(2)\ \exp(i\underline{k}_2\cdot\underline{r}) \\
&= \chi_{RAM}\ \hat{\underline{e}}_{RAM}|E(1)|^2\ E(2)\ \exp(i\underline{k}_2\cdot\underline{r}) \quad ,
\end{aligned}
\tag{7.28}
$$

where

$$
\chi_{RAM}\hat{\underline{e}}_{RAM} = 6\left[\chi^{(3)}_{1122}\ \hat{\underline{e}}_2 + \left(\chi^{(3)}_{1212}+\chi^{(3)}_{1221}\right)\ \hat{\underline{e}}_1(\hat{\underline{e}}_1\cdot\hat{\underline{e}}_2)\right] \tag{7.29}
$$

and we have used the fact that $\underline{k}_{-i} = -\underline{k}_i$ and $\hat{\underline{e}}_{-i} = \underline{e}_i$.

---

[2]Polarization is used in two senses here and elsewhere in the discussion but the meaning should be clear from the context.

For parallel polarization $\hat{\underline{e}}_1 = \hat{\underline{e}}_2$, $\hat{\underline{e}}_{RAM} = \hat{\underline{e}}_1$ and

$$\chi_{RAM} = 6\left(\chi^{(3)}_{1122}+\chi^{(3)}_{1212}+\chi^{(3)}_{1221}\right) = 6\chi^{(3)}_{1111}(\omega_2,\omega_1,-\omega_1) \tag{7.30}$$

while if $\hat{\underline{e}}_1 \perp \hat{\underline{e}}_2$ the polarization direction is $\hat{\underline{e}}_2$ and

$$\chi_{RAM} = 6\chi^{(3)}_{1122}(\omega_2,\omega_1,-\omega_1) \quad . \tag{7.31}$$

Equations (7.24) and (7.28) are the desired constitutive relations between the CARS and stimulated Raman polarizations and the applied fields. They show which susceptibility terms are important for various incident E field orientations and they enable us to predict the orientation of the CARS and Raman waves. The magnitude of these waves can now be obtained by using these relations in the wave equation (7.12).

7.1.3 CARS Amplitudes

The wave equation in this case is

$$\begin{aligned} &\underline{\nabla} \times \underline{\nabla} \times \underline{E}(\underline{r},3) - (\omega_3/c)^2 \varepsilon_3 \underline{E}(\underline{r},3) \\ &= 4\pi(\omega_3/c)^2 \chi_{CARS}\hat{\underline{e}}_{CARS} E(1)^2 E(2)^* \exp[i(2\underline{k}_1-\underline{k}_2)\cdot\underline{r}] \end{aligned} \tag{7.32}$$

and the divergence equation (7.13) is

$$\underline{\nabla} \cdot \left[\varepsilon_3 \underline{E}(\underline{r},3)+4\pi\chi_{CARS}\hat{\underline{e}}_{CARS} E(1)^2 E(2)^* \exp[i(2\underline{k}_1-\underline{k}_2)\cdot\underline{r}]\right] = 0 \quad . \tag{7.33}$$

From the latter we see that

$$\varepsilon_3 \underline{\nabla} \cdot \underline{E}(\underline{r},3) = 4\pi\chi_{CARS} E(1)^2 E(2)^* [\hat{\underline{e}}_{CARS}\cdot(2\underline{k}_1-\underline{k}_2)] \exp[i(2\underline{k}_1-\underline{k}_2)\cdot\underline{r}] \quad . \tag{7.34}$$

We expect $\underline{E}(\underline{r},3)$ to be amplified significantly only in the direction $2\underline{k}_1-\underline{k}_2$ which we will take to be the z-axis. The polarization vector $\hat{\underline{e}}_{CARS}$ will be orthogonal to $2\underline{k}_1-\underline{k}_2$ so that $\hat{\underline{e}}_{CARS}\cdot(2\underline{k}_1-\underline{k}_2) = 0$ and $\underline{\nabla}\cdot\underline{E}(\underline{r},3) = 0$. With this result and the vector equality $\underline{\nabla}\times\underline{\nabla}\times\underline{E} = \underline{\nabla}(\underline{\nabla}\cdot\underline{E}) - \nabla^2\underline{E}$ (7.32) reduces to

$$\begin{aligned} \partial^2\underline{E}(\underline{r},3)/\partial z^2 - (\omega_3/c)^2 \varepsilon_3 \underline{E}(\underline{r},3) &= 4\pi(\omega_3/c)^2 \chi_{CARS}\hat{\underline{e}}_{CARS} E(1)^2 E(2)^* \\ &\quad \times \exp[i(2k_1-k_2)z] \quad . \end{aligned} \tag{7.35}$$

We assume a solution of the form

$$\underline{E}(\underline{r},3) = \hat{\underline{e}}_{CARS} E^0(3) \exp(ik_3z) \tag{7.36}$$

where $k_3 = n_3\omega_3/c$ and $E^0(3)$ can be a function of z. On substitution, (7.35) becomes

$$\partial^2 E^0(3)/\partial z^2 + 2ik_3 \partial E^0(3)/\partial z = -4\pi(\omega_3/c)^2 \chi_{CARS} E(1)^2 E(2)^* \exp(i\Delta kz) \qquad (7.37)$$

with

$$\Delta k = 2k_1 - k_2 - k_3 \quad . \qquad (7.38)$$

If Δk is small the spatial variation of $E^0(3)$ is also small and the second derivative can be neglected. Integration of the first-order equation then gives the field at $z = \ell$ [we assume $E^0(3)$ is zero at z=0]

$$E^0(3) = (2\pi\omega_3/n_3 c)\, \chi_{CARS} E(1)^2 E(2)^* [\exp(i\Delta k\ell)-1]/\Delta k \quad . \qquad (7.39)$$

Substitution of this solution into (7.37) shows that neglect of the second derivative term is a good approximation as long as $\Delta k \ll 2k_3$, a condition usually met in all CARS experiments. The time averaged intensity of the anti-Stokes wave at z is therefore

$$I_3 = (n_3 c/8\pi)|\underline{E}(\underline{r},3)|^2 = \frac{256\pi^4\omega_3^2}{n_1^2 n_2 n_3 c^4}|\chi_{CARS}|^2 I_1^2 I_2 \ell^2 \left(\frac{\sin(\Delta k\ell/2)}{\Delta k\ell/2}\right)^2 \quad . \qquad (7.40)$$

This plane wave result for the intensity of the CARS beam reveals many of the characteristic features of the CARS process. Since the signal is proportional to I_1^2 and is linear in I_2 there are large gains in using pulsed lasers with high peak powers. I_3 depends upon the square of the bulk susceptibility and hence is quadratic in sample density. For exact phase matching (Δk=0) the term in parentheses [called sinc($\Delta k\ell/2$)] is unity and the signal increases as ℓ^2. For $\Delta k \neq 0$, the intensity becomes periodic and reaches its first maximum when $\ell = \pi/\Delta k$. The intensity at this *coherence length* ℓ_c is reduced by a factor of $(2/\pi)^2$ compared to the phase matched case. By expanding the index of refraction as a power series in frequency, ℓ_c is seen to be [7.18]

$$\ell_c = (\pi c/\omega_v^2)(2\partial n/\partial\omega + \omega^2\partial^2 n/\partial\omega^2)^{-1} \quad . \qquad (7.41)$$

For most gases ℓ_c is of the order of 100 cm at atmospheric pressure and this value varies inversely with pressure. For liquids, dispersion effects are much larger and ℓ_c is of the order of millimeters. To increase ℓ_c, $\Delta k = 2k_1 - k_2 - k_3$ is usually reduced in liquids by crossing the beams at a phase matching angle θ (Fig.7.1). For most liquids θ is 1-3°, hence the CARS signal is eventually limited by beam walk off.

The expression (7.40) for the CARS intensity was obtained by assuming plane waves for ω_1 and ω_2 whereas a more realistic picture involves the use of Gaussian beams.

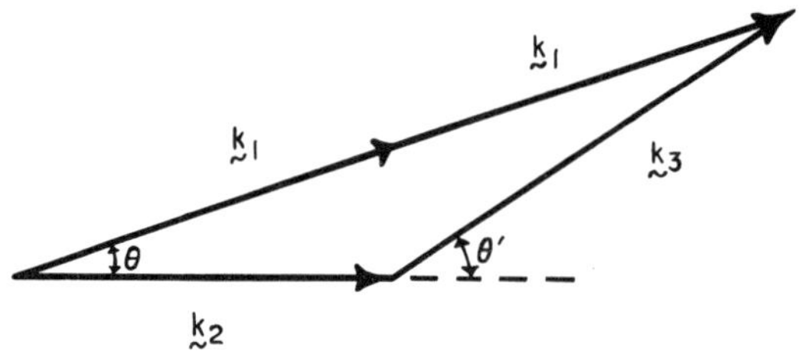

Fig. 7.1. Wave vector diagram for phase matching ($\Delta k = 2k_1 - k_2 - k_3 = 0$)

For such beams, the focal region is sometimes approximated by a cylinder of plane waves of diameter $w_0 = 4\lambda f/\pi w$ and length equal to the confocal parameter $b = \pi w_0^2/2\lambda$ [7.23,24]. λ is the wavelength of the beam, f the focal length of the lens and w the diameter at which the intensity has dropped to e^{-2} of the peak value. Taking $\ell = b$ and converting from intensity to power by the relation $I_i = P_i/\pi(w_0/2)^2$, (7.40) becomes

$$P_3 \simeq \left(\frac{2}{\lambda}\right)^2 \frac{256\pi^4\omega_3^2}{n_1^2 n_2 n_3 c^4} |\chi_{CARS}|^2 \, P_1^2 P_2 \quad . \tag{7.42}$$

In obtaining this result, an average $\lambda = \lambda_1$ is used and it is assumed that $\Delta k = 0$ so that $\mathrm{sinc}(\Delta k\ell/2) = 1$. This expression is identical to that given by REGNIER et al. [7.18] if one drops the n_i factors and accounts for the fact that their $\chi^{(3)}$ value is four times larger than ours due to their definition of $P^{(3)}(\underline{r},3)$[3].

It will be noted that (7.42) shows no dependence on the focal length of the lens or on the diameter of the beam. More accurate calculations involving numerical integration of (7.32) [7.26-29] show that about 75% of the CARS signal is generated within a length of 6b and also that there is a threefold gain in going from a focal length of 50 to 3 cm [7.29]. This result is consistent with experiments we have performed which show about a factor of two variation in going from f = 7 to 30 cm. Most of the CARS signal is thus generated in the focal zone of ~1 cm and (7.42) is probably good to within an order of magnitude. For solids and liquids there is a further loss since one trades off the gain in phase matching by beam crossing with the loss due to eventual walk off of the beams. The optimum signal is thus obtained at an angle smaller than the calculated phase matched angle.

At this point we have obtained a result (7.40) for the CARS intensity in terms of the unknown constitutive parameter χ_{CARS}. Our next task is to obtain a similar

---

[3] The reader is cautioned that such factors of 1/4 and/or the quantity 6/n! of (7.20) are sometimes absorbed into the definition of $\chi^{(3)}$ by various authors. In addition, the definition of $\chi^{(3)}$ may vary by a factor of $(4\pi\varepsilon_0)^2$ depending on the choice of Gaussian or MKSA units [7.25] and hence these factors must be considered in comparing expressions such as (7.40) or (7.42).

relation for the Raman intensity in terms of χ_{RAM} and to establish a connection with the cross section which can be measured by spontaneous Raman spectroscopy. Following this we will consider the theoretical form of $\chi^{(3)}$ and establish the desired link between χ_{CARS} and χ_{RAM}.

7.1.4 Raman Amplitudes

As in the CARS case, we can solve the classical wave equation for the change in amplitude of a field at the Stokes frequency ω_2 when the system is subjected to incident beams of frequency ω_1 and ω_2. Substituting (7.28) into (7.12) and following the same steps as before, we obtain the wave equation

$$-\nabla^2\underline{E}(\underline{r},2) - (\omega_2/c)^2\ \varepsilon_2\ \underline{E}(\underline{r},2) = 4\pi(\omega_2/c)^2\ \chi_{RAM}|\underline{E}(\underline{r},1)|^2\ \underline{E}(\underline{r},2) \quad . \tag{7.43}$$

If the nonlinear polarization is small, the solution for $\underline{E}(\underline{r},2)$ approaches that of a traveling wave (7.14) and hence we assume a solution with this form, i.e.,

$$\underline{E}(\underline{r},2) = \hat{\underline{e}}_{RAM}\ E(2)\ \exp(i\underline{k}\cdot\underline{r}) \quad . \tag{7.44}$$

Here $\underline{k}$ is parallel to the initial wave vector $\underline{k}_2$ and $\hat{\underline{e}}_{RAM}$ is given by (7.29). On insertion in (7.43), we see that this expression is suitable if the magnitude of k is

$$\begin{aligned} k &= (\omega_2/c)\left[\varepsilon_2+4\pi\chi_{RAM}|\underline{E}(\underline{r},1)|^2\right]^{1/2} \\ &= k_2\left(1+32\pi^2\chi_{RAM}\ I_1/n_1n_2^2c\right)^{1/2} = k_2\left(1+16\pi^2\chi_{RAM}\ I_1/n_1n_2^2c\right) \quad . \end{aligned} \tag{7.45}$$

In obtaining the latter result, we have used the definitions $k_2 = n_2\omega_2/c$, $I_1 = cn_1|\underline{E}(\underline{r},1)|^2/8\pi$, $\varepsilon_2 = n_2^2$ and have expanded the square root expression, keeping only the first two terms under the assumption that the nonlinearity is small. Since χ_{RAM} is complex ($\chi_{RAM}=\chi'_{RAM}+i\chi''_{RAM}$), k is complex and

$$\underline{E}(\underline{r},2) = \hat{\underline{e}}_{RAM}\ E(2)\ \exp(i\underline{k}'_2\cdot\underline{r})\ \exp(\underline{k}''_2\cdot\underline{r}) \tag{7.46}$$

where

$$\underline{k}'_2 = \underline{k}_2\left(1+16\pi^2\ I_1\ \chi'_{RAM}/n_1n_2^2c\right) \tag{7.47}$$

$$\underline{k}''_2 = -\underline{k}_2\left(16\pi^2\ I_1\ \chi''_{RAM}/n_1n_2^2c\right) \quad . \tag{7.48}$$

From this we find that the real part of the susceptibility produces a change in the index of refraction so that the wave length is altered slightly. If I_1 is very large the index change due to χ'_{RAM} can produce self-focusing [7.5]. According to (7.46), for positive values of $\underline{k}_2''$, one gets an increase in the Stokes amplitude as the wave propagates and hence $\underline{k}_2''$ is a measure of Raman gain. At high I_1 powers, significant conversions to I_2 can occur, a feature exploited in the Raman laser where ω_2 feedback is provided by placing the sample in a laser cavity [7.5].

It is important to note that (7.46) requires that a field at ω_2 be present initially at $r = 0$ in order for amplification to occur. In stimulated Raman spectroscopy, such a field is provided by spontaneous Raman scattering of ω_1 to produced ω_2. Of course, the production of a spontaneous photon cannot be obtained from a classical treatment but rather requires quantization of both matter and radiation fields. However, it follows from such quantization that the correct description of spontaneous scattering can be had from the classical approach if one introduces "noise" in the radiation field [7.9,16]. Each allowed mode of the radiation field is assumed to have random phase, energy $\hbar\omega$ and wave vector $k = n\omega/c$. The sample thus lies in a radiation field containing an "extra photon" for each mode and, depending upon the frequency dependence of the gain factor $\underline{k}_2''$, certain modes can be amplified according to (7.47).

Under the low gain conditions of spontaneous Raman spectroscopy, the $\underline{k}_2''$ exponential in (7.47) can be expanded to give

$$|\underline{E}(\underline{r},2)|^2 = |E(2)|^2 \, (1+2\underline{k}_2''\cdot\underline{r})$$

or

$$I_2 = I_2^0(1+2\underline{k}_2''\cdot\underline{r}) \quad . \tag{7.49}$$

The ω_2 intensity thus builds up linearly as the wave propagates a distance ℓ in a volume V in which the field E(1) is present. The initial frequency I_2^0 is equal to the energy density $\hbar\omega_2/V$ times the propagation velocity c/n_2 so that the net ω_2 energy gain per unit volume per second (power per unit volume) is

$$(2\underline{k}_2''\ell)(I_2^0A)/V = 2\underline{k}_2''\, I_2^0 = 2k_2''\hbar\omega_2 c/n_2 V \quad . \tag{7.50}$$

This result gives the power gain for a single mode; in a volume V there is a density of $2V/(2\pi)^3$ modes corresponding to traveling waves [7.16]. These correspond to the number of allowed k vectors in this volume, with two polarizations possible for each k vector. We expect that k_2'' will produce amplification for a range of frequencies $\omega_2 \to \omega_2 + d\omega_2$. This implies a range in k of $dk = n_2 d\omega_2/c$ and, for a solid angle $d\Omega$ in k space, we have a volume of

---

$\hbar = h/2\pi$ (normalized Planck's constant).

$$k^2 dk d\Omega = \left(n_2^3 \omega_2^2 / c^3\right) d\omega_2 d\Omega \quad . \tag{7.51}$$

If we consider only one polarization, the total number of modes in this volume is

$$V/(2\pi)^3 \cdot \left(n_2^3 \omega_2^2 / c^3\right) d\omega_2 d\Omega \quad . \tag{7.52}$$

Each mode will be amplified according to (7.50) so that the total power generated per unit volume in the frequency range ω_2 to $\omega_2 + d\omega_2$ and in the solid angle $d\Omega$ is

$$\frac{P_2}{V} = \frac{k_2'' \hbar n_2^2 \omega_2^3 d\omega_2 d\Omega}{4\pi^3 c^2} = \frac{-4\hbar n_2 \omega_2^4 \chi''_{RAM}}{n_1 \pi c^4} I_1 d\omega_2 d\Omega \quad . \tag{7.53}$$

Thus

$$P_2 = \left(-4\hbar n_2 \omega_2^4 \chi''_{RAM} / n_1 \pi c^4\right) I_1 V \, d\omega_2 d\Omega = \left(-4\hbar n_1 \omega_2^4 \chi''_{RAM} / n_1 \pi c^3 N\right) N \, P_1 \ell d\omega_2 d\Omega \tag{7.54}$$

where in the last step we have inserted N = molecules/cc and replaced $I_1 V$ by $P_1 \ell$. Now we can recognize the term in brackets as the usual definition of the differential scattering cross section per molecule, i.e.,

$$d^2\sigma / d\Omega d\omega_2 = -4\hbar n_2 \omega_2^4 \chi''_{RAM} / n_1 \pi c^4 N \quad . \tag{7.55}$$

When proper account is taken of units and definition of $\chi^{(3)}$, this equation is identical to that obtained by LALLEMAND [Ref.7.5, Eqs.(6,29)], and by PANTELL and PUTHOFF [Ref.7.16, Eqs.(88,90)].

With (7.55) we have a means of determining the macroscopic χ''_{RAM} from an experimental measurement of the Raman intensities. In the next section we consider the general form of the microscopic susceptibility that one obtains from perturbation theory and we see that an alternate form for χ_{RAM} and χ_{CARS} can be written in terms of molecular wavefunctions.

7.1.5 Microscopic Third-Order Susceptibilities

The quantum-mechanical treatment of the third-order polarization involves computing the average molecular dipole moment arising from the third power of the macroscopic fields. For ideal gases the macroscopic and microscopic fields are the same but for liquids and solids, the local fields will be higher by a factor L which is of the order of $[(\varepsilon+2)/3]^4$ [7.2,8,9]. The bulk dipole $P^{(3)}$ is an average over all orientations of the molecules times the number density and $\chi^{(3)}$ is thereby obtained as the coefficient of the various third-order field products.

Using time-dependent perturbation theory ARMSTRONG et al. [7.11] and others [7.9, 10,13,14] have derived a general expression for the molecular $\chi^{(3)}$. The result con-

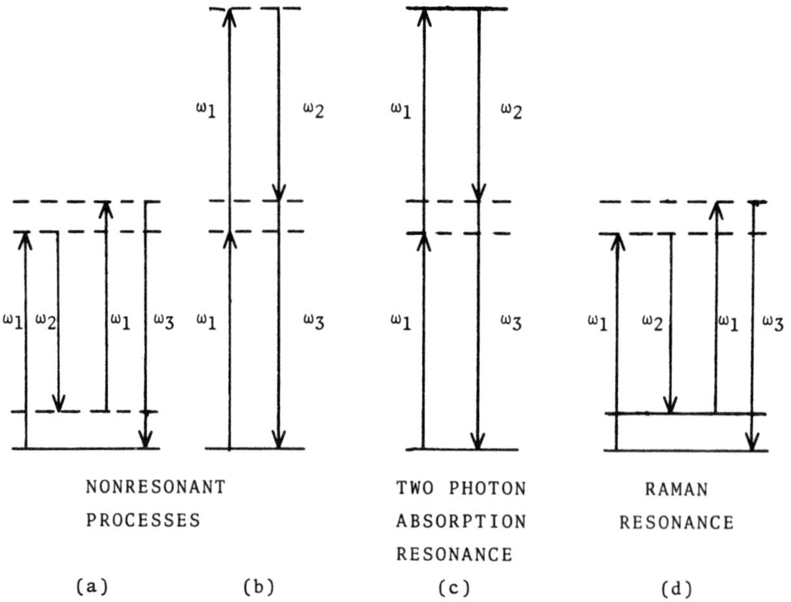

Fig. 7.2a-d. Processes contributing to CARS signals at ω_3. Dashed and solid lines stand for virtual and real states, respectively. Processes a and b contribute to the background signal, process c gives two photon absorption information while process d probes Raman active vibrational resonances

tains 24 terms, each having four fold sums over all molecular states with each term in these sums involving a product of four transition moment elements divided by three frequency factors. Two of these factors involve only one of the frequencies of the applied fields while the third factor involves one of the possible sums or differences of the incident fields. The former two can produce large increases in $\chi^{(3)}$ when an incident frequency lies near a single photon absorption for a molecule while the latter produces resonances in χ_{RAM}, χ_{CARS} and in other cases such as two photon susceptibilities. Diagrammatic perturbation theory [7.30-32] is a useful device in keeping track of the terms which are important for a particular frequency choice. For CARS, four types of processes can contribute to the susceptibility resonance at $\omega_3 = 2\omega_1 - \omega_2$. Figure 7.2 shows these cases and we see that the two processes a and b contribute a slowly varying nonresonant term χ_{NR} which is usually small. Process c corresponds to two photon absorption to a real level and has proved to be a useful way to do two photon absorption spectroscopy [7.33-36]. Process d is the one of principal interest for CARS.

In the absence of one or two photon resonances, the molecular CARS susceptibility for a transition "sequence" $n \to k \to n$ can be written [7.2,8,9,37]

$$\chi^{(3,mol)}_{\alpha\beta\gamma\delta}(\omega_1,\omega_1,-\omega_2) = \chi_{NR} + N_n L[24\hbar(\omega_{kn}-\omega_1+\omega_2-i\Gamma_{kn})]^{-1} \times \left[<n|\alpha^A_{\alpha\beta}|k><k|\alpha^S_{\gamma\delta}|n> + <n|\alpha^A_{\alpha\gamma}|k><k|\alpha^S_{\beta\delta}|n>\right] \quad . \tag{7.56}$$

Here N_n is the number density of state n, Γ_{kn} is the damping constant, $\omega_{kn} = (E_k - E_n)/\hbar$, and the Stokes (S) and anti-Stokes (A) polarizability elements are

$$<n|\alpha^A_{\alpha\beta}|k> = \frac{1}{\hbar}\sum_r \left(\frac{(M_\alpha)_{nr}(M_\beta)_{rk}}{\omega_r - \omega_3} + \frac{(M_\beta)_{nr}(M_\alpha)_{rk}}{\omega_r - \omega_1}\right)$$

$$<k|\alpha^S_{\gamma\delta}|n> = \frac{1}{\hbar}\sum_r \left(\frac{(M_\gamma)_{kr}(M_\delta)_{rn}}{\omega_r - \omega_2} + \frac{(M_\delta)_{kr}(M_\gamma)_{rn}}{\omega_r - \omega_1}\right) \tag{7.57}$$

where $(M_\alpha)_{kr} = <k|M_\alpha|r>$ is the transition moment element. It may be noted that the same polarizability elements occur in the sequence $k \to n \to k$ but with opposite sign and with a number density factor of N_k. Thus N_n in expression (7.56) should be written $N_n - N_k$ and we expect the CARS susceptibility to depend on this population density difference rather than N_n as in the Raman case.

The Raman susceptibility is written as [7.2,8,9,37]

$$\chi^{(3,mol)}_{\alpha\beta\gamma\delta}(\omega_2,\omega_1,-\omega_1) = \chi_{NR} + \frac{L}{24\hbar} N_n \left(\frac{<n|\alpha^S_{\alpha\beta}|k><k|\alpha^S_{\gamma\delta}|n>}{\omega_{kn}-\omega_1+\omega_2+i\Gamma_{kn}}\right) \quad . \tag{7.58}$$

We note that it is the imaginary part of this susceptibility which determines Raman intensities and, since χ_{NR} is real in the absence of one photon resonances, χ_{NR} does not influence the Raman spectrum. Both the CARS and Raman susceptibilities become resonant at $\omega_{kn} = \omega_1 - \omega_2$ and both involve the parameter Γ_{kn}. For each Raman transition, the imaginary part of $\chi^{(3)}$ has a Lorentzian form

$$\chi''^{(3,mol)} = \frac{-K\,\Gamma_{kn}}{(\omega_{kn}-\omega_1+\omega_2)^2+\Gamma^2_{kn}} = \frac{-K\,\Gamma_{kn}}{(\delta\omega)^2+\Gamma^2_{kn}} \quad . \tag{7.59}$$

Γ_{kn} is thus identified as the half width at half height for the spontaneous Raman line. This function has a peak value of $-K/\Gamma_{kn}$ when $\omega_1 - \omega_2 = \omega_{kn}$ and an integrated value of $-K\pi$ so that

$$\int_{-\infty}^{\infty} \chi''^{(3,mol)} d(\delta\omega) = \pi\Gamma_{kn}\,\chi''^{(3,mol)}(\text{peak}) \quad . \tag{7.60}$$

If all electronic absorptions are well removed from ω_1, ω_2 and ω_3, the frequency dependence of the α's is small and the elements α^A and α^S can be set equal. If we also neglect χ_{NR}, we find that at resonance (7.56) and (7.58) yield

$$\chi^{(3,\mathrm{mol})}_{\alpha\beta\gamma\delta}(\omega_1,\omega_1,-\omega_2)\mathrm{peak} = \frac{-2(N_n-N_k)}{N_n}\,\chi^{(3,\mathrm{mol})}_{\alpha\beta\gamma\delta}(\omega_2,\omega_1,-\omega_1)\mathrm{peak} \quad . \tag{7.61}$$

Equation (7.61) provides a useful connection between the microscopic CARS and Raman susceptibilities. There remains, however, the task of relating these to the macroscopic susceptibilities which are measured in a laboratory axis system. This involves an orientation average using direction cosines linking the laboratory and molecular frames and yields the macroscopic $\chi^{(3)}$ values as a linear combination of $\chi^{(3,\mathrm{mol})}$ terms. This is considered further later but for the present we simply note that identical combinations occur for $\chi^{(3)}_{1111}(\omega_1,\omega_1,-\omega_2)$ and $\chi^{(3)}_{1111}(\omega_2,\omega_1,-\omega_1)$ and hence relation (7.61) holds for the macroscopic susceptibilities also. Accordingly, when $\hat{e}_1 = \hat{e}_2$ (7.60) and (7.61) can be used to rewrite (7.55) as

$$(d\sigma/d\Omega) = (d^2\sigma/d\Omega d\omega_2)d\omega_2 = 2\hbar n_2\omega_2^4\,\Gamma_{kn}\,\chi_{\mathrm{CARS}}(\mathrm{peak})/(N_n-N_k)n_1c^4 \quad . \tag{7.62}$$

From the frequency dependence of χ_{CARS} (7.56) and from (7.62) we obtain finally

$$\chi_{\mathrm{CARS}} = 3\chi^{(3)}_{1111}(\omega_1,\omega_1,-\omega_2) = \frac{(N_n-N_k)n_1c^4}{2\hbar n_2\omega_2^4}\left(\frac{d\sigma}{d\Omega}\right)\frac{1}{\omega_{kn}-\omega_1+\omega_2-i\Gamma_{kn}} \quad . \tag{7.63}$$

A similar expression for χ_{CARS} results from a semi-classical model in which it is assumed that the field $E = E_1 + E_2$ produces a driving force $f = 1/2(d\alpha/dq)E^2$ on a harmonic oscillator with vibrational coordinate q. Solving the damped oscillator equation for q and relating $(d\alpha/dq)^2$ to $d\sigma/d\Omega$ yields [7.10,18,23]

$$\chi_{\mathrm{CARS}} = \frac{N_nc^4}{\hbar\omega_2^4}\left(\frac{d\sigma}{d\Omega}\right)\frac{\omega_{kn}}{\omega_{kn}^2-(\omega_1-\omega_2)^2-i2\Gamma_{kn}(\omega_1-\omega_2)} \quad . \tag{7.64}$$

This is basically the same as Equation 4 of REGNIER et al. [7.18] when one notes that their $\chi = 4\chi_{\mathrm{CARS}}$ and $\gamma = 2\Gamma_{kn}$. Equations (7.63) and (7.64) give nearly the same peak susceptibility at resonance and both yield essentially the same line shape.

7.2 Comparison of CARS and Raman Spectroscopy

7.2.1 Signal Strengths

From (7.42) and (7.63) the peak CARS power at resonance for a $0\rightarrow 1$ vibrational transition (neglecting n_i factors and N_1) is given by

$$P_{\mathrm{CARS}} \approx \left(\frac{8\pi c\omega_1\omega_3}{\hbar\omega_2^4}\right)^2\left[\frac{N_0}{\Gamma}\left(\frac{d\sigma}{d\Omega}\right)\right]^2 P_1^2P_2 \quad . \tag{7.65}$$

In contrast, the total spontaneous Stokes Raman power which would be scattered from the ω_1 beam into 4π steradians over a sample length of 1 cm is obtained by integrating (7.54) over $d\omega_2$ and $d\Omega$ to yield

$$p_{RAM} = \Omega\ell\left(\frac{d\sigma}{d\Omega}\right)N_0p_1 = 4\pi\left(\frac{d\sigma}{d\Omega}\right)N_0p_1 \quad . \tag{7.66}$$

For gas phase experiments, typically $(d\sigma/d\Omega) \sim 10^{-30}/cm^2sr$ and $\Gamma \sim 0.1\ cm^{-1}/2\pi c$ (assumed limited by laser linewidths) and, if we consider P[torr] of a gas with a vibrational shift of $\sim$2000 cm^{-1} from a 532 nm source, we calculate a power ratio

$$\frac{p_{CARS}}{p_{RAM}} \approx 10^{-5}\ p_1[W]\ p_2[W]\ P[torr] \quad .$$

With pulsed lasers, the p_1p_2 product can easily be greater than $10^{10}\ W^2$ so that, even at 1 torr, CARS enjoys a gain of 10^5 over Raman spectroscopy. Moreover this ratio *increases* if higher resolution laser sources are used since the natural Raman linewidths can be 0.01 cm^{-1} or less. In addition, the detection efficiency is much higher for the monochromatic CARS beam than for a spontaneous Raman signal which must be collected and dispersed with a spectrometer. The quantum efficiency of detectors for the blue ω_3 photons is higher than for red ω_2 light and fluorescence produced by ω_1 is not a problem. The small divergence of the CARS beam [mrad] also affords good discrimination against sample luminescence, an important advantage for combustion and plasma diagnostic applications.

In practice, we have observed gas phase CARS signals which are lower by factors of 10-100 than predicted by (7.65) [7.38]. This may be due in part to the plane wave approximation used in the derivation as well as to the relatively poor quality of the Nd-YAG beam used. In addition, in sample regions subjected to very high intensities, saturation effects can limit the CARS signal due to "pumping" of molecules from the 0 to 1 vibrational level [7.6,18,23,39-45]. HENESIAN et al. [7.46] also observed a discrepancy of $\sim$100 in their cw measurements on CH_4 but BARRETT [7.47,48] found good agreement in his work on H_2 and CH_4.

In liquids, the CARS signal is larger due to the $\sim 10^3$ increase in number density but this gain is reduced in part by the greater linewidths ($\sim$1 cm^{-1}) and shorter beam overlap regions ($\sim$1 mm vs $\sim$1 cm). Conversion efficiencies (p_3/p_2) of 0.1 to 1% have been observed in benzene for p_1 powers of 1-50 kW [7.49,50]. At higher powers, stimulated scattering by ω_1 begins to supply the ω_2 beam so that the ω_3 signal is no longer sensitive to scanning of the ω_2 laser. This power limitation exists for many liquids but is not a problem for low pressure gases.

7.2.2 Lineshapes and Maxima

Equation (7.63) for the CARS susceptibility can be rewritten for a single resonance as

$$\chi_{CARS} = A\,\frac{\delta\omega + i\Gamma}{\delta\omega^2+\Gamma^2} \tag{7.67}$$

and the real and imaginary part of this function are plotted in Fig.7.3. From (7.59) and (7.61) we see that χ''_{CARS} represents the lineshape of the Raman transition. The CARS lineshape is identical since it is proportional to

$$|\chi_{CARS}|^2 = \frac{A^2}{\delta\omega^2+\Gamma^2} = \frac{A}{\Gamma}\,\chi''_{CARS} \quad . \tag{7.68}$$

However, addition of χ_{NR} produces interference with χ'_{CARS} and distorts the peak as shown in Fig.7.4. The derivative of $|\chi_{CARS}|^2$ is zero at

$$\frac{\delta\omega}{\Gamma} = \frac{-\chi''_{MAX}}{2\chi_{NR}} \pm \left[\left(\frac{\chi''_{MAX}}{2\chi_{NR}}\right)^2 + 1\right]^{1/2} \tag{7.69}$$

so that the maximum shifts to higher $\delta\omega$ slightly (by less than Γ) while the minimum lies lower than the maximum by an amount of $\Gamma\chi''_{max}/\chi_{NR}$ ($\chi''_{max}=A/\Gamma$ is the maximum when $\chi_{NR}=0$). Since χ''_{max} and Γ can be obtained from Raman cross sections and linewidths, measurement of the frequency difference between the maximum and minimum can yield accurate values for the electronic nonresonant susceptibilities [7.33,34,51,52].

Because χ_{NR} includes contributions from solvent as well as sample molecules, the distortion shown in Fig.7.4 can become extreme for solute resonances at low concentrations. The bulk susceptibility is

$$|\chi^{(3)}|^2 = |\chi_{NR}+\chi'_{CARS}+i\chi''_{CARS}|^2 = \chi_{NR}^2 + 2\chi_{NR}\chi'_{CARS} + \chi'^2_{CARS} + \chi''^2_{CARS} \tag{7.70}$$

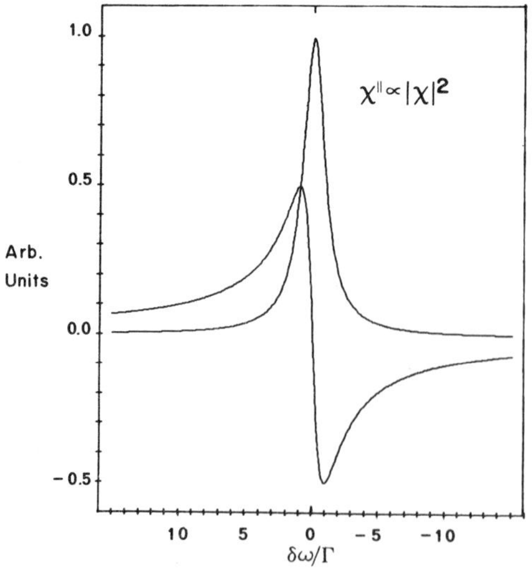

Fig. 7.3. Plot of the real (χ') and imaginary (χ'') parts of the third-order susceptibility in the region of a Raman resonance at $\delta\omega=0$. The horizontal scale is in units of Γ, the HWHM of the Raman transition. The total susceptibility $|\chi|^2$ has the same shape as χ'' (the upper trace) but with a different vertical scale, see (7.68)

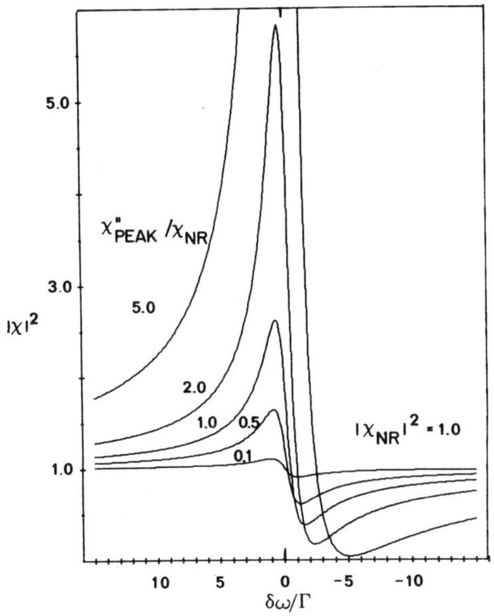

Fig. 7.4. Effect of χ_{NR} on a CARS resonance. Note that the maximum shifts no more than Γ from the true resonance frequency but that the minimum shifts significantly as χ_{RES}/χ_{NR} changes

so that, when χ_{NR} is large, the spectrum takes the dispersive form of χ'_{CARS} on a high nonresonant background. This feature of CARS spectroscopy is its principal limitation for analytical applications since it is found that solute resonances are lost in the solvent background for concentrations below about 1% [7.50]. Because of the narrower lines for gases, the detection limits are somewhat better, ~10 ppm [7.18,29,53-54]. TOLLES and TURNER [7.55] have recently considered the signal to noise characteristics for CARS signals in the presence of a strong χ_{NR} background and compare the analytical capabilities with those of Raman and absorption spectroscopy. Several methods have been proposed to reduce the contribution of χ_{NR} and these are discussed in a later section.

If several resonances lie in the same spectral region then additional interferences can occur in the real and in the imaginary parts of the sum

$$|\chi_{CARS}|^2 = |\chi_{NR} + \sum_j \chi'_j + i\sum_j \chi''_j|^2 \quad . \tag{7.71}$$

Figure 7.5 shows the real and imaginary parts of χ_{CARS} for the Q branch transitions of N_2. The latter portrays the spontaneous Raman spectrum while the CARS spectrum is expected to appear like $|\chi_{CARS}|^2$. The experimental trace shows good agreement with the calculated curve. However, examination of the calculated maxima of the CARS peaks shows small shifts ($< \pm 0.1\ cm^{-1}$) from the known Raman resonances. Corrections for such shifts may be necessary in high resolution studies aimed at accurately measuring vibrational-rotational Raman frequencies by CARS.

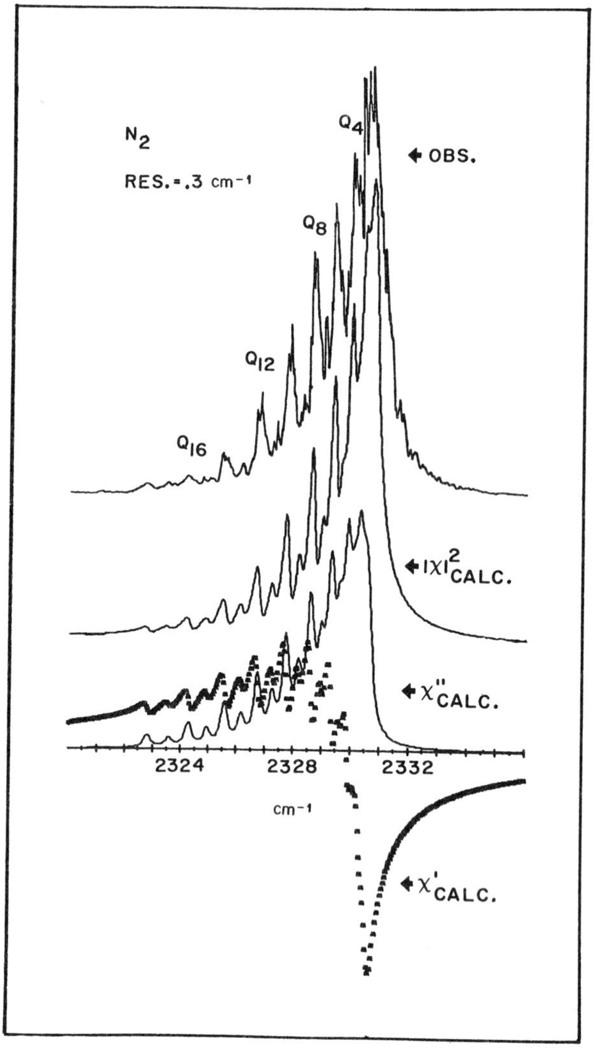

Fig. 7.5. Calculated and observed Q branch for N_2, $\Gamma = 0.15\ \mathrm{cm}^{-1}$, $\chi''_{\mathrm{Peak}}/\chi_{\mathrm{NR}} = 125$

7.2.3 Selection Rules

The selection rules for CARS are seen by examining the expectation value for a susceptibility element which is resonant at a frequency ω_J corresponding to a transition between vibration-rotation states 0, J and 1, J'. Neglecting χ_{NR} and using the population difference between the two states, (7.56) for the molecular susceptibility in the laboratory frame can be written

$$\langle 0J|\chi^{(3,\mathrm{mol})}_{\alpha\beta\gamma\delta}(\omega_1,\omega_1,-\omega_2)|0J\rangle = \frac{L(N_0-N_1)}{24\hbar}\,\frac{\langle 0J|\alpha^A_{\alpha\beta}|1J'\rangle\langle 1J'|\alpha^S_{\gamma\delta}|0J\rangle+(\beta\Leftrightarrow\gamma)}{\omega_J-\omega_1+\omega_2-i\Gamma_j} \quad . \tag{7.72}$$

To express these integrals in terms of the molecular axis system we make use of the transformation relation

$$\alpha_{\alpha\beta} = \sum_{\alpha'\beta'} \phi_{\alpha\alpha'} \phi_{\beta\beta'} \alpha_{\alpha'\beta'} \tag{7.73}$$

where the ϕ's are direction cosines. If the vibrational and rotational wave functions are assumed separable, (7.72) becomes

$$\langle 0J|\chi^{(3,mol)}_{\alpha\beta\gamma\delta}|0J\rangle = \frac{L(N_0-N_1)}{\omega_J-\omega_1+\omega_2-i\Gamma_J} \sum_{\alpha'\beta'\gamma'\delta'} \Big[\langle 0|\alpha^A_{\alpha'\beta'}|1\rangle\langle 1|\alpha^S_{\gamma'\delta'}|0\rangle \times \langle J|\phi_{\alpha\alpha'}\phi_{\beta\beta'}|J'\rangle\langle J'|\phi_{\gamma\gamma'}\phi_{\delta\delta'}|J\rangle + (\beta\leftrightarrow\gamma)\Big] \quad . \tag{7.74}$$

It is apparent that each integral for a given term in this summation must be invariant to any molecular symmetry operation for a non-zero contribution to the susceptibility. For the vibrational part, $\alpha^A_{\alpha'\beta'}$ and $\alpha^S_{\gamma'\delta'}$ must belong to the same symmetry species as the normal coordinate q since $\Psi_0(q)\Psi_1(q)$ transforms as q. Accordingly, the vibrational selection rules for CARS are identical to those for spontaneous Raman spectroscopy. By analogous reasoning, the two rotational integrals in (7.74) yield exactly the same selection rules as in the normal Raman case. Thus, unlike the hyper Raman [7.56] or $3\omega_1 - \omega_2$ [7.30] Raman processes, CARS does not permit the observation of any vibration-rotation transition forbidden to two photon Raman spectroscopy. YURATICH and HANNA [7.22] discuss the selection rules in more detail and reach the same conclusions. They point out, however, that some effective differences may occur between CARS and resonance Raman spectra because the former involves $|\alpha^A\alpha^S|$ while the latter depends on $|\alpha^S|^2$.

7.2.4 Orientation Averaging and Polarization Behavior

It is important to note that the CARS process is a parametric process in which the ω_3 field amplitude at a point is a constructive sum of contributions by all the molecules in the medium. For gases and liquids, the molecules are randomly oriented and this must be taken into account in evaluating the total amplitude. This means that the CARS intensity I_3 is proportional to $|\bar{E}_3|^2$ so that we need $|\bar{\chi}^{(3)}|^2$ (and not $|\overline{\chi^{(3)}|^2}$) where the bar indicates an orientation average. For liquids, where the rotational motion is assumed to be quenched, a classical average of the four fold product of direction cosines is appropriate in evaluating $\chi^{(3)}$. MONSON et al. [7.57] have evaluated these averages, and ANDERSON and HUDSON [7.9] used them in arriving at combinations of the molecular susceptibility elements for each macroscopic susceptibility term.

YURATICH and HANNA [7.22,58] have also considered the orientation averaging process for both liquids and gases using spherical tensor techniques. For forward scattering they show that the Raman intensity is proportional to

$$\overline{|\alpha^S|^2} = 3\alpha^2\theta^0_{RAM} + \frac{2}{9}\gamma_a^2\theta^1_{RAM} + \frac{2}{15}\gamma_s^2\theta^2_{RAM} \tag{7.75}$$

[α^S is the element defined in (7.57)] while the CARS intensity is proportional to

$$\overline{|\alpha^A\alpha^S|^2} = |3\bar{\alpha}^2\theta^0_{CARS} + \frac{2}{9}\bar{\gamma}_a^2\theta^1_{CARS} + \frac{2}{15}\bar{\gamma}_s^2\theta^2_{CARS}|^2 \quad . \tag{7.76}$$

In these expressions, α^2 is related as usual to the isotropic part of the molecular scattering tensor while γ_a^2 and γ_s^2 are associated with the antisymmetric and symmetric parts of the anisotropy. They are defined in Table 7.2 in terms of polarizability elements in a complex polar rather than Cartesian basis. Also shown are the angular θ^K factors which describe the intensity variation with different polarization angles for the forward scattering geometry.

Expressions (7.75) and (7.76) can be used with Table 7.2 to define depolarization ratios such as

$$\rho_{RAM} = \frac{I_{RAM}(\perp)}{I_{RAM}(||)} = \frac{5\gamma_a^2+3\gamma_s^2}{45\alpha^2+4\gamma_s^2} \tag{7.77}$$

and

$$\rho_{CARS} = \frac{I_{CARS}(\perp,\perp)}{I_{CARS}(||,||)} = \left|\frac{-5\bar{\gamma}_a^2+3\bar{\gamma}_s^2}{45\bar{\alpha}^2+4\bar{\gamma}_s^2}\right|^2 \tag{7.78}$$

where $I_{CARS}(\beta_{12},\beta_{13})$ indicates the polarization directions of ω_2 and ω_3 with respect to ω_1. We note that $I(\perp,||)$ and $I(||,\perp)$ are both zero, results also predicted by (7.25). In the absence of resonance enhancement, γ_a^2 is zero and ρ_{RAM} takes the customary form which permits distinction between totally symmetric vibrational modes ($\rho_{RAM} \leqq 3/4$) and others ($\rho_{RAM} = 3/4$). Similarly $\rho_{CARS} \leqq (3/4)^2$ and $= (3/4)^2$ in these two cases. From (7.26,27) we make the identification

$$\rho_{CARS} = \left|\frac{\chi^{(3)}_{1221}}{\chi^{(3)}_{1111}}\right|^2 = \left|\frac{3\bar{\gamma}_s^2}{45\bar{\alpha}^2+4\bar{\gamma}_s^2}\right|^2 \tag{7.79}$$

and it is clear that, when $\bar{\gamma}_a^2 = 0$, only two invariants of the isotropic $\chi^{(3)}$ tensor can be determined. CARS and Raman spectroscopy thus offer equivalent information regarding the symmetry of vibrational modes. However, depolarization measurements should be more accurate for CARS because of the squared relation and because the beams are more intense and are collimated. These advantages may be offset somewhat by the small but variable contribution of χ_{NR} which should be added to $\chi^{(3)}_{1221}$ and $\chi^{(3)}_{1111}$ in (7.79).

Table 7.2. Polarizability and angular factors for Raman and CARS spectroscopy (forward scattering, all beams linearly polarized) [7.22,58]

	K = 0 Isotropic Scattering	K = 1 Antisymmetric Anisotropic Scattering	K = 2 Symmetric Anisotropic Scattering
Raman	$\alpha^2 = \frac{1}{3}\,\lvert(\alpha^S)_0^0\rvert^2$	$\gamma_a^2 = \frac{3}{2}\sum_{Q=-1}^{1}\lvert(\alpha^S)_Q^1\rvert^2$	$\gamma_S^2 = \frac{3}{2}\sum_{Q=-2}^{2}\lvert(\alpha^S)_Q^2\rvert^2$
	$\Theta_{RAM}^0 = \frac{1}{3}\cos^2\beta_{12}$	$\Theta_{RAM}^1 = \frac{1}{2}\sin^2\beta_{12}$	$\Theta_{RAM}^2 = \frac{1}{6}\left(3+\cos^2\beta_{12}\right)$
	$\bar{\alpha}^2 = \frac{1}{3}\,\lvert(\alpha^A)_0^{0*}(\alpha^S)_0^0\rvert$	$\bar{\gamma}_a^2 = \frac{3}{2}\sum_{Q=-1}^{1}\lvert(\alpha^A)_Q^{1*}(\alpha^S)_Q^1\rvert$	$\bar{\gamma}_S^2 = \frac{3}{2}\sum_{Q=-2}^{2}\lvert(\alpha^A)_Q^{2*}(\alpha^S)_Q^2\rvert$
CARS	$\Theta_{CARS}^0 = \frac{1}{3}\cos\beta_{12}\cos\beta_{13}$	$\Theta_{CARS}^1 = \frac{-1}{2}\sin\beta_{12}\sin\beta_{13}$	$\Theta_{CARS}^2 = \frac{1}{6}\,(4\cos\beta_{12}\cos\beta_{13}+3\sin\beta_{12}\sin\beta_{13})$

β_{ij} is the angle between ω_i, ω_j polarization vectors.

7.3 Experimental Aspects of CARS

A variety of laser sources have been used for the ω_1, ω_2 beams necessary for CARS. The earliest experiments used stimulated Stokes Raman scattering of a ruby laser beam by various liquids to provide a "point tunable" ω_2 beam for mixing experiments [7.8,18,29,53,54,60-65]. By using part of the ruby output to pump a near infrared dye laser, continuous CARS spectra were subsequently recorded [7.66-79]. TARAN has discussed the requirements for high spectral purity and amplitude stability in the ruby and dye laser outputs [7.68,70,71]. High resolution spectra (0.01 cm^{-1} or less) can be obtained with such a system because it is possible to obtain MW powers from an amplifier following a ruby laser operated in a single longitudinal mode [7.68-70]. The principal disadvantages of this source are that the pulse repetition rates are low (generally less than 1 pps) and the invisible infrared dye output is less convenient to align than visible beams.

For liquid, solid and solution phase studies, power levels of 1-50 kW are sufficient and these are conveniently provided by two dye lasers pumped by a single N_2 laser source [7.49,72-74]. Commercial components are available which provide 6-8 ns pulses of 0.03 to 0.3 cm^{-1} bandwidth at repetition rates up to 50 pps. The poor beam quality of the N_2 laser precludes its use as ω_1 and necessitates transverse pumping of the two dye lasers. The dye output quality is consequently less than ideal but is good enough to permit doubling of one or both dye laser beams into the ultraviolet region. This feature, plus the tunability of both ω_1 and ω_2 make this system particularly useful in resonance CARS studies [7.73-78]. Figure 7.6 illustrates a typical experimental arrangement and improved procedures to optimize the beam overlap phase matching and sampling conditions are discussed in [7.73,79,80].

Another common arrangement utilizes a doubled Nd:YAG laser at 532 nm for the ω_1 beam. Part of this output (or a source at 354 nm produced by summing 532 and 1060 nm beams) is used to pump a dye laser which provides the ω_2 wave [7.48-51,81-84]. A description of such a system in use at the Naval Research laboratory was given by TOLLES et al. [7.23] and NIBLER et al. [7.84]. Figure 7.7 shows a similar experimental arrangement used by the authors for gas phase CARS at Oregon State University. A commercial Nd-YAG oscillator-amplifier combination provides 70 MW, 10 ns pulses of 1060 nm output at ~15 pps. The beam mode is donut shaped but is highly collimated and reproducible. A one inch KDP* type II doubler is used to provide more than 20 MW of 532 output of 0.1 cm^{-1} bandwidth. The leftover 1060 nm light is redoubled and used to pump a dye oscillator and amplifier in a semilongitudinal manner. A 20× telescope and a 316 1/mm echelle in the oscillator cavity provide a 0.25 cm^{-1} tunable beam which mirrors the donut mode of the pump laser and which, after amplification, reaches powers of 50 to 500 kW. The beams are made colinear with a dichroic mirror and brought it to a common focus in the sample with a 10 cm focal length lens. The beam is recollimated and the ω_3 signal is isolated with two prisms and dichroic mir-

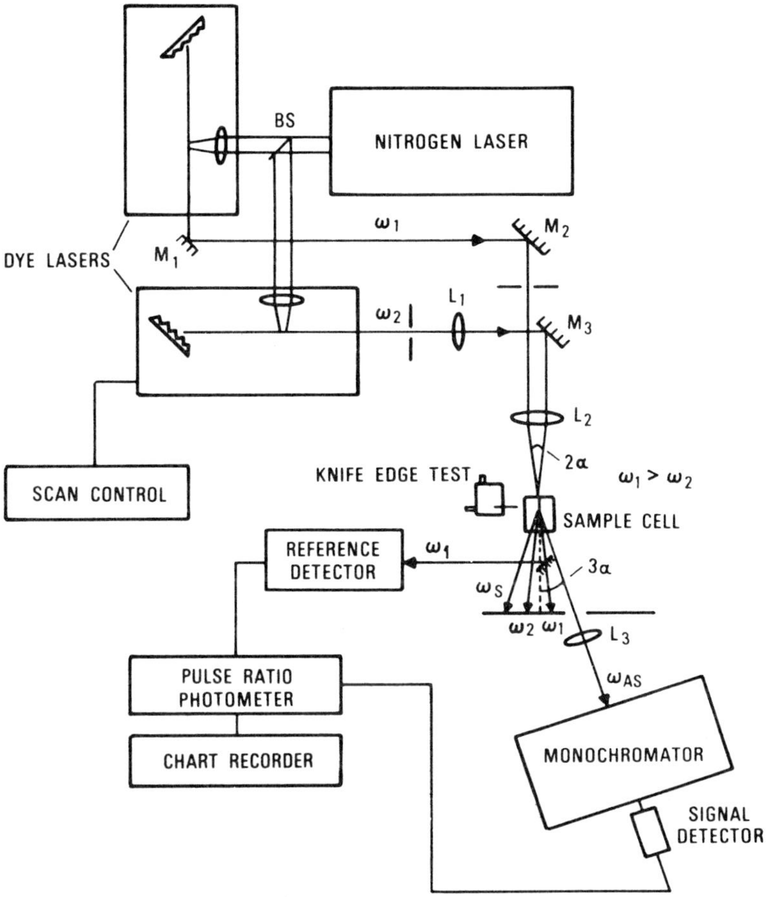

Fig. 7.6. Schematic diagram of the optics of CARS. M_1, M_2, and M_3 are mirrors; L_1, L_2, and L_3 are lenses; BS is a beam splitter which reflects 33% of the light. The focal length of L_1 is 500-800 mm, that of L_2 is 200 mm. The crossing angle is 2α. From [7.73] with permission

rors for detection with a photodiode or photomultiplier. A simple boxcar integrator is used to reduce the shot-to-shot signal fluctuations which are typically 10-50%. Further improvement can be had by passing the leftover ω_1, ω_2 beams through a non-resonant sample such as argon or NaCl. This CARS reference signal varies with the ω_1, ω_2 beam fluctuations in the same way as the sample and the S/N is increased by averaging the ratio of these signals [7.50-54,66-68, see also 7.73]. Very precise relative frequency measurements can be made by passing the ω_2 beam through an analyzing etalon and absolute frequency measurements are made easier by the high ω_1 and ω_2 intensities.

An alternative to the dye laser for tunable ω_2 radiation is an optical parametric oscillator (OPO) [7.19,85-88]. Here the 1064 nm output of a Nd-YAG laser pumps a

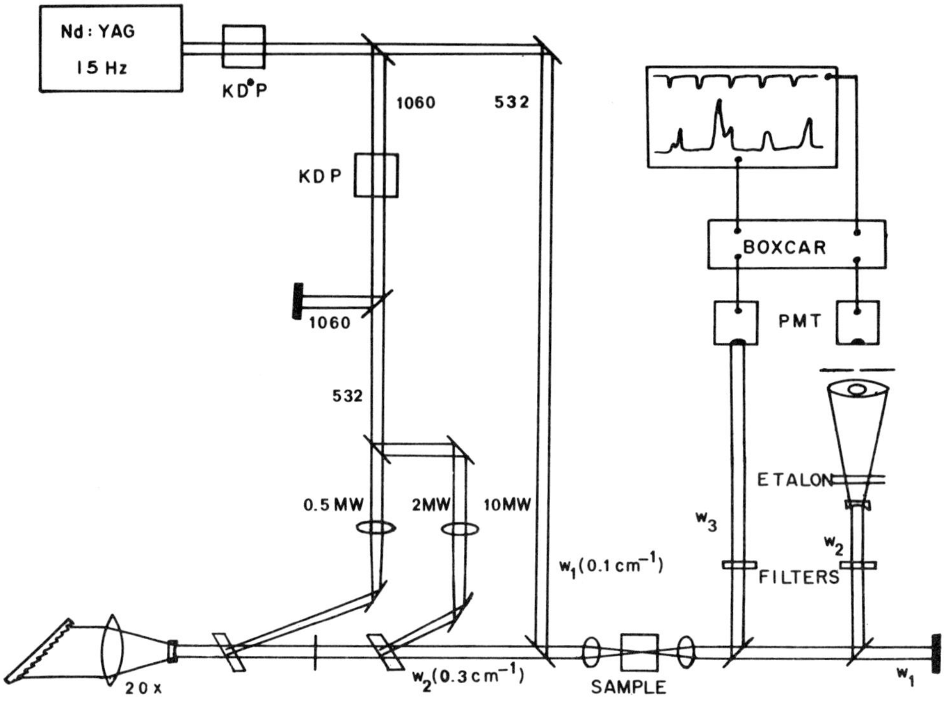

Fig. 7.7. Apparatus used at OSU to record CARS spectra of gases. The diffraction coupled YAG laser has a linewidth of 0.1 cm^{-1} in a donut mode. The dye laser provides a linewidth of ~0.3 cm^{-1} in a similar mode. Part of the ω_2 beam is passed through an etalon to provide an accurate relative frequency mark

temperature or angle tuned OPO crystal ($LiNbO_3$) to produce an ω_2 beam which can be scanned over a wide range in the near infrared (>4000 cm^{-1}). This output is then mixed in the sample with the 1064 nm beam to produce the CARS signal. In one modification, a 532 nm probe beam is also used with the other two to shift the output into the visible and thereby improve the sensitivity [7.19,85,86]. The principal disadvantage of the OPO approach is that it is difficult to get tunable narrow linewidths because the free running OPO has a very variable linewidth. The efficiencies are low compared to dye lasers and the detection efficiencies in the near infrared are not as good as in the visible region. Beam alignment is also not as convenient since the beams are not visible.

It is also possible to use a broad band source for ω_2 so that the sample can select those ω_2 components which match a Raman resonance [7.59,87,89-91]. The ω_3 beam then contains several frequencies and these must be dispersed by a monochromator. The latter can be scanned to yield a CARS spectrum under high resolution conditions because of the high signal intensities compared to Raman spectroscopy. Alternatively, the entire spectrum can be recorded photographically [7.91] or with an optical multi-

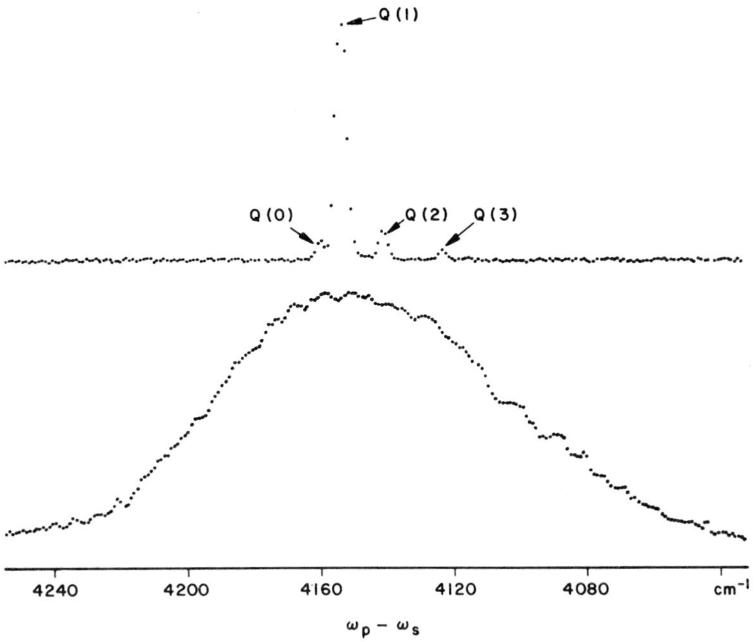

Fig. 7.8. Anti-Stokes spectra as recorded simultaneously by a dual-spectrum optical multichannel analyzer: Upper and lower traces are for H_2 sample and reference gases, respectively. The reference anti-Stokes linewidth is ~25 Å. From [7.89] with permission

channel analyzer (OMA) [7.38,59,79,89,91]. The advantages of such an approach are the simplicity of the dye laser and the ability to record "single shot" CARS spectra. Figure 7.8 shows the CARS spectrum of the Q-branch region of H_2 recorded by ROH et al. by this technique [7.89]. Also shown is the CARS spectrum of a nonresonant reference gas which was obtained on half of the same SIT OMA during the same 20 ns laser flash. The latter permits normalization of the resonant spectrum so that accurate intensity ratios can be obtained, an important feature for temperature measurements in flames and plasmas. With one or two stages of intensification, commercial OMA's rival photomultipliers in sensitivity. Figure 7.9 displays a spectrum of the ν_1 Q-branch of the symmetric stretch of CH_4 recorded with a PAR 1205I ISIT OMA in a single laser flash for a sample pressure of 0.6 torr [7.59,90]. Despite the low pressure and ω_1 and ω_2 powers, the signal to noise is quite good. The dynamic range is limited, however, (~750) and signal "blooming" between channels can be troublesome if one tries to detect a weak signal near a strong one.

Using cw sources, significant improvements in ω_1, ω_2 linewidths and amplitude stability are possible, albeit at a considerable sacrifice in the ω_3 CARS signal. BARRETT and BEGLEY were the first to report a cw CARS spectrum, of the ν_1 Q-branch of CH_4 [7.47]. This work was rapidly followed by similar experiments by others [7.46,

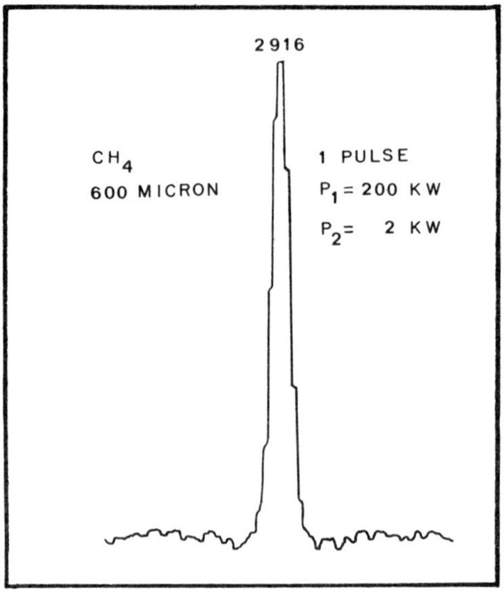

Fig. 7.9. A CARS spectrum of CH_4 on a single laser pulse taken at a sub torr pressure. A broad band dye laser was used and the signal was dispersed with a 0.6 m spectrometer and recorded with a PAR ISIT OMA tube

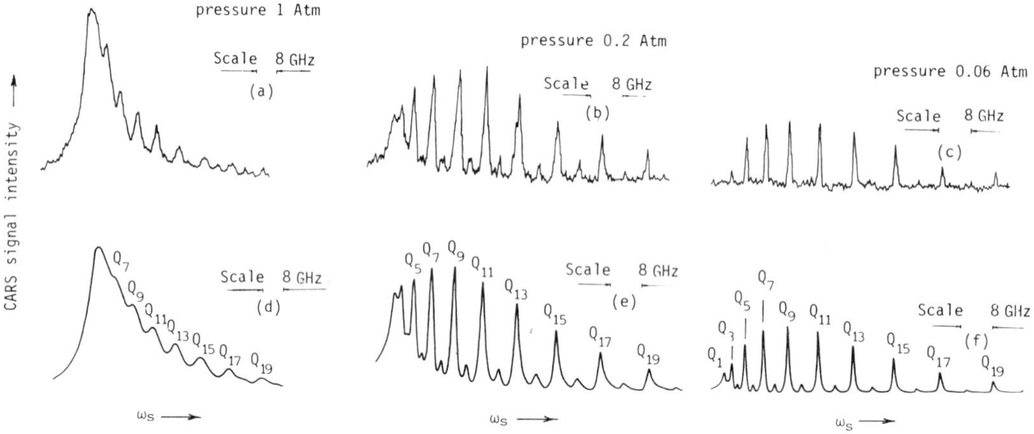

Fig. 7.10. Experimental (a,b,c) and calculated (d,e,f) CARS spectra of the Q-branch spectrum of the 1974 cm^{-1} ν_2 transition of C_2H_2 at different pressures. Single frequency cw lasers with 0.001 cm^{-1} resolution were used to obtain these spectra. From [7.94] with permission

47,86,87,92-95]. HIRTH and VOLLRATH [7.93] greatly increased the cw CARS signal by placing the sample at a common focal point within the Ar and dye laser cavities. Powers at ω_1 approached 500 W and N_2 vibrational signals of 3×10^7 photons/s were recorded. FABELINSKY et al. [7.94] have recently reported the use of single mode cw lasers with linewidths of 0.001 cm^{-1} in a study of the ν_2 mode of acetylene (Fig.7.10). Further improvements in cw laser linewidths (to the kHz range) are feasible, and KHOO and YEUNG have recently proposed that Doppler-free measurements of homogeneous linewidths are possible if one fixes ultranarrow ω_1, ω_2 beams at a resonance peak and analyzes the ω_3 output with an interferometer [7.96,97]. Pressure broadening may be limiting in such studies, however, since cw spectra below ~40 torr have not been reported and it will be very difficult to obtain cw CARS signals at pressures in the sub-torr range. The use of hollow waveguides may be of aid here since MILES et al. [7.87] have recently reported gains of 10^2-10^3 for such waveguide cells over cells with the usual tight focusing arrangement.

7.4 Survey of CARS Experimental Results

7.4.1 Solids

A number of solids have been studied by CARS, with much of the early interest centering on the measurement on nonresonant electronic susceptibilities. In their original investigation in 1964, MAKER and TERHUNE [7.8,12] used benzene and other aromatic compounds to generate ω_2 by the stimulated Raman process. These beams allowed the determination of χ^{nr} for calcite and a number of alkali halides and glasses.

In 1969 COFFINET and DeMARTINI [7.98] used CARS to investigate polaritons in GaP. Landau level excitations have been studied in the conduction band of InSb [7.99,100]. Using two synchronized CO_2 lasers, anisotropic interference in beams generated by $\chi^{(2)}$ and $\chi^{(3)}$ processes was observed and measured in GaAs [7.101]. Other solid state studies include the observation by WYNNE [7.72] of resonances in lithium niobate between 50 and 700 cm^{-1} using two dye lasers. LEVENSON et al. described CARS measurements in diamond, calcite, SrF_2, CaF_2, BaF_2, and CdF_2 [7.34]. Interference between the resonant and nonresonant susceptibilities as well as polarization experiments were employed to deduce different components of the χ tensor. In diamond an optical phonon resonance was observed at 1332 cm^{-1} in which the CARS signal varied by 10^8 across the resonance [7.34,51,102]. AKHMANOV and KOROTEEV [7.19,85,86] have also determined magnitudes and signs of resonant and nonresonant $\chi^{(3)}$ components for calcite using four wave mixing ("active Raman spectroscopy") in which all photon frequencies are different. BLOEMBERGEN and co-workers [7.102-106] have reported double resonance interference effects in the CARS spectra of CuCl. By tuning $2\omega_1$ into resonance with a longitudinal-exciton resonance of the crystal and then tuning ω_2 across a $\omega_1 - \omega_2 = 210\ cm^{-1}$ Raman resonance, striking interferences were noted.

CARS studies of crystals are of particular interest since polarization variations permit many of the $\chi^{(3)}$ tensor components to be determined. In addition, it is possible to probe $\Delta k \neq 0$ transitions in solids by varying $\Delta k = 2k_1 - k_2 - k_3$. This allows spectra of the excitation dispersion curve to be obtained, a new kind of "spectroscopy in momentum space" [7.72,98].

7.4.2 Liquids

CARS spectra have been recorded for benzene [7.8,12,19,33,34,37,50,79] and several substituted aromatic compounds; C_6H_5Br [7.8,12], C_6H_5Cl [7.33,34], $C_6H_5CH_3$ [7.8,12, 19,87] and $C_6H_5NO_2$ [7.33,34,60]. One of the objectives in these studies has been the determination of the relative contribution of the resonant and nonresonant susceptibilities from measurements of $\delta\omega/\Gamma$ in (7.69). From KATO and TAKUMA's measurements of the peak differential Raman cross section of the 992 cm^{-1} band of benzene [7.107], one obtains $\chi^{(3)}_{1111}(\omega_2,\omega_1,-\omega_1) = 15 \pm 1.3 \times 10^{-14}$ cm^3/erg for $\lambda_1 = 532$ nm. Very recently, OWYOUNG and PEERCY have reported a value of $15.9 \pm 1.1 \times 10^{-14}$ cm^3/erg for this same quantity as measured by a new technique of two-beam nonlinear inteferometry [7.108]. The latter method also yields a ratio of 10.2 ± 0.7 for $\chi^{(3)}_{1111}(\text{peak})/\chi_{NR}(\text{elec})$, a ratio measured by CARS to be 9.4 ± 0.5 [7.108], 10.9 ± 0.5 [7.34], $14,9 \pm 4.5$ [7.19, 108]. The relative CARS signals scale as the square of this ratio so that one sees why it is difficult to distinguish solute resonances from nonresonant solvent signals for concentration below ~1%.

In Fig.7.11 the CARS spectrum obtained by ITZKAN and LEONARD [7.109] for water is compared with the normal Stokes Raman spectrum. Because of the large linewidth the susceptibility ratio drops from ~10 for benzene to a value of ~2 for water. Considerable disparity exists in measurements of $\chi_{NR}(H_2O)/\chi_{NR}(C_6H_6)$; 0.25 [7.35] versus

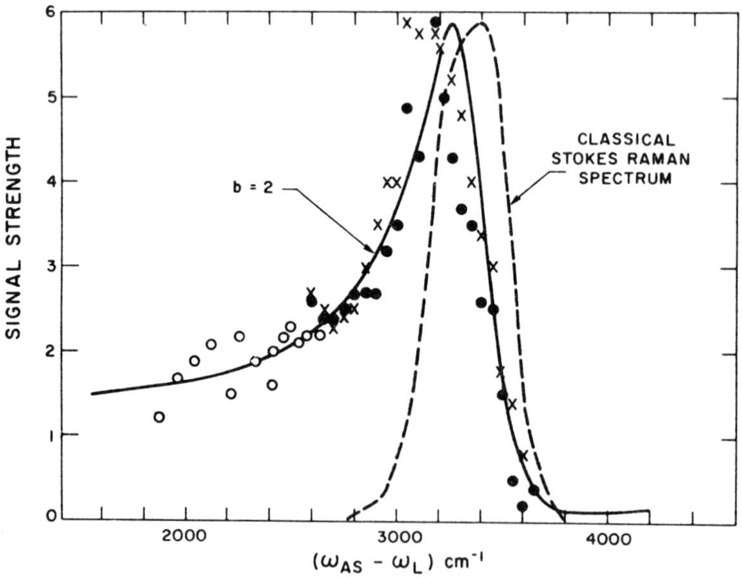

Fig. 7.11. CARS signal for liquid water vs Raman shift in cm^{-1}. The open circles were obtained with the dye Pilot chemical No.481 and the closed circles and crosses represent experimental data obtained at two different times with a coumarin 6/7D4MC mixture. The solid trace is a theoretical curve, the dashed trace a conventional Raman spectrum. From [7.109] with permission

11.0 [7.76]. A susceptibility ratio of 200 ± 40 has been measured for liquid N_2 in the only application of cw CARS to liquids reported to date [7.37,110]. The use of CARS for fluorescent samples has been demonstrated with β ionone, a yellow material of importance in the study of visual processes [7.81]. Other liquids which have been examined are CCl_4 [7.60,79], CH_3OH [7.78], CS_2 [7.35,85,111], $CHCl_3$ [7.35], C_6H_{12} [7.35-37] and naphthalene [7.35]. A number of these studies have been aimed at obtaining two photon absorption cross sections from interference effects between processes such as Fig.7.2c,d and it is concluded that the cross sections are more accurate than those obtained by classical two photon absorption measurements [7.35,36, 111].

7.4.3 Gases

The first report of CARS in gases was by RADO [7.65]. Using a stimulated Stokes signal in H_2 for ω_2, he determined χ_{NR} for a number of gases. By varying the pressure in the H_2 cell used to generate ω_2, DeMARTINI et al. [7.61-64,112] made linewidth measurements of the $J = 1$ Q-branch line of H_2 and observed Dicke narrowing as a function of pressure. Since these early experiments, tunable lasers have become more common in CARS experiments and the gas phase work has tended to exploit the advantages of CARS for the study of luminous samples and for high resolution spectroscopy.

a) Combustion and Gas Diagnostics

TARAN and co-workers at ONERA [7.18,29,53,54,66-68,70,71,89,111-116] have pioneered in the application of CARS to concentration and temperature measurements of combustion systems. Figure 7.12 shows a concentration profile of the H_2 produced by pyrolysis

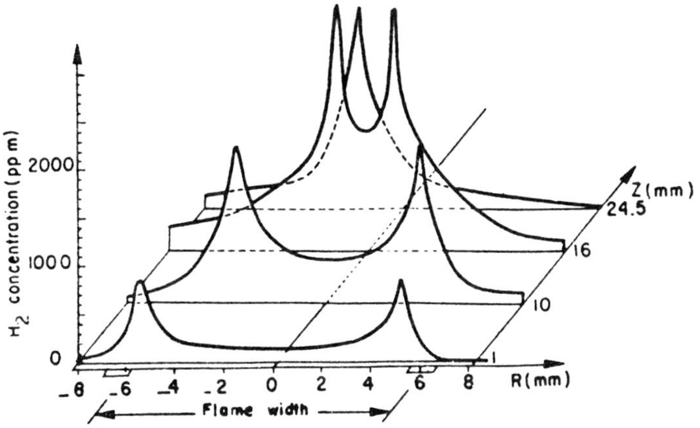

Fig. 7.12. H_2 distribution in a horizontal natural gas flame. R is the distance from the burner axis; Z is the distance along the axis; coordinate R is vertical, pointing downward. From [7.53] with permission

in an air-natural gas (75% methane) flame [7.18,53,54]. From relative intensities of the Q-branches for H_2 near the top of the flame, a rotational temperature of 1350 ± 30 K was deduced, a value in good agreement with a thermocouple measurement of 1340 ± 20 K [7.66]. Single-pulse measurements have also been demonstrated using a broad-band dye laser, Fig.7.8 [7.89]. A detection limit of ~10 ppm for H_2 in 1 atm of N_2 was determined [7.18,53,54]. It may be noted that H_2 is a particularly favorable candidate for detection by CARS since $\chi^{(3)}_{RES}(H_2)/\chi^{(3)}_{NR}(N_2) = 1.6 \times 10^4$ [7.65] and elimination of χ_{NR} and/or improvements in the S/N ratio will be necessary for comparable (or better) detectivity limits of other sample molecules.

In other work by the ONERA group, vibration-rotation spectra of H_2, N_2 and O_2 were obtained [7.66,68]. Figure 7.13 shows the results for atmospheric N_2 at room temperature; in the inset is a higher resolution (0.16 cm^{-1}) scan of the central Q-branch. A logarithmic plot of $(p_3/p_{3,ref})^{1/2}$ is used to provide a larger dynamic range than available in a plot such as Fig.7.5 and to more directly permit concentration and temperature measurements [7.68]. For example, from the slope of the Q-branch, a temperature of 1250 K was deduced in a probe volume 2 mm from the surface of a spherical porous burner fueled by ethylene glycol. A nitrogen mole fraction of 0.68 was determined at this position and temperature and N_2, CO concentration maps were made as a function of position. It is important to note that normal Raman scattering would offer poorer detectivities in these flames because of chemiluminescence and incandescence of a particulate matter excited by the laser. Similar work on combustion systems is underway by HARVEY et al. at NRL [7.117], SCHREIBER et al. at Wright-Patterson Air Force base [7.89,118], and ECKBRETH et al. at United Technologies Research Center [7.119].

b) Plasma Diagnostics

Discharge media and plasmas represent another area in which CARS has found diagnostic application. At NRL, spectra of D_2 and N_2 have been recorded at the center of a 10 KV 4ma dc discharge of about 7 cm length. The spectrum for 48 torr of D_2 is displayed in Fig.7.14 and, from the rotational intensities of the Q_J structure of the $v = 0$ level, a rotational temperature of ~400 K was calculated [7.120]. The extent of vibrational excitation was quite low; the $v = 1/v = 0$ intensity ratios yield a vibrational temperature of ~1050 K. In contrast, the spectrum in Fig.7.15 for 10 torr of N_2 under similar conditions shows extensive pumping of vibrational levels up to $v = 7$ [7.38, 59]. The energy distribution is non-Boltzmann but the relative populations of the vibrational levels can be calculated by computer iteration [7.122]. These population distributions explain why N_2 is an efficient energy transfer additive in CO_2 lasers whereas D_2 is not effective in a D_2/HCl transfer laser [7.122].

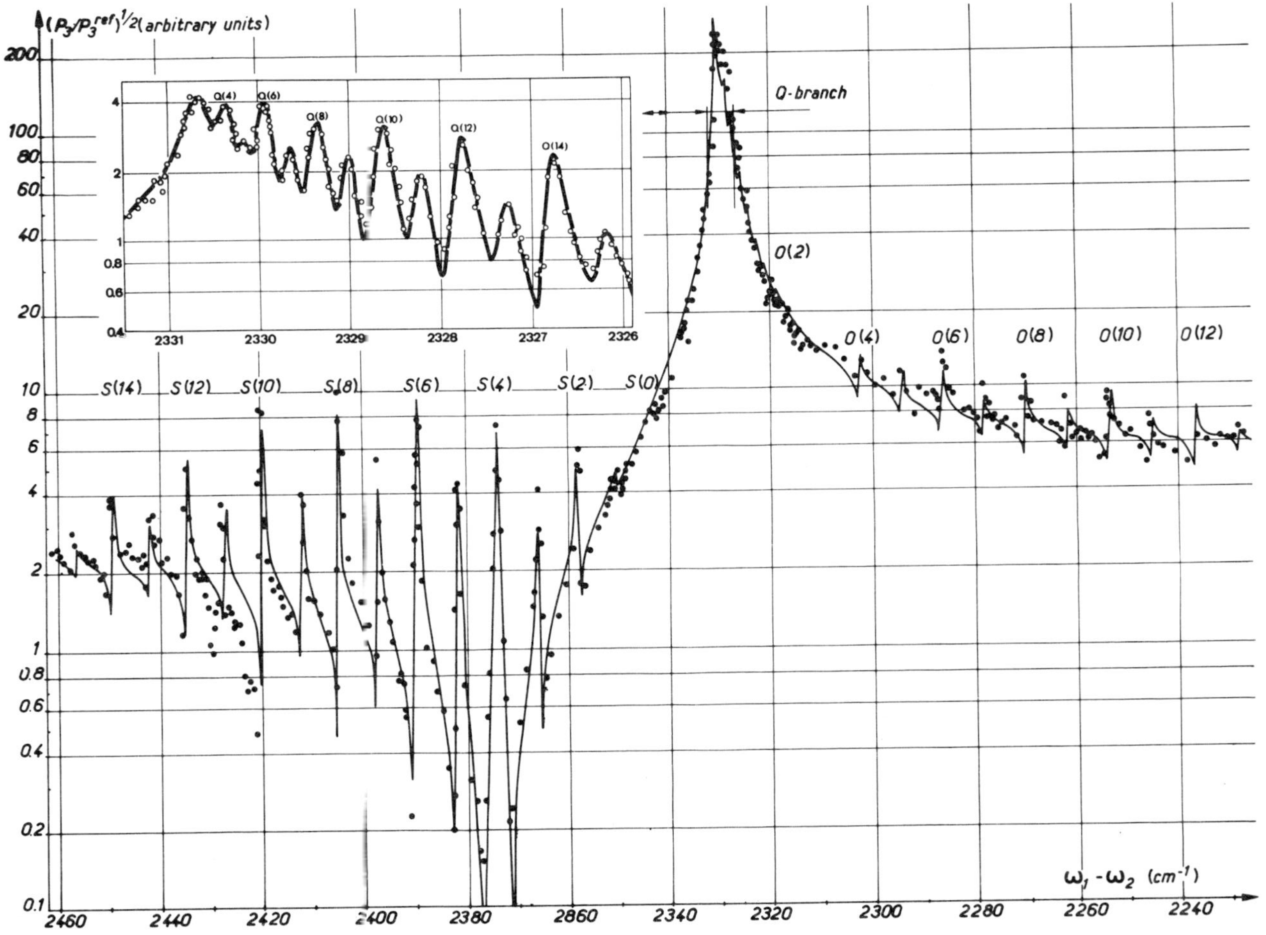

Fig. 7.13. Spectrum of atmospheric N_2. The curves are calculated for Boltzmann equilibrium assuming $(d\sigma/d\Omega)Q = 1.1 \times 10^{-31}$ cm^2. Fitting parameters were $(d\sigma/d\Omega)O,S = 0.55 \times 10^{-32}$ cm^2, $\chi^{nr} = 4 \times 10^{-18}$ cm^3/erg. The ordinate is roughly in units of 10^{-18} cm^3/erg. From [7.68] with permission

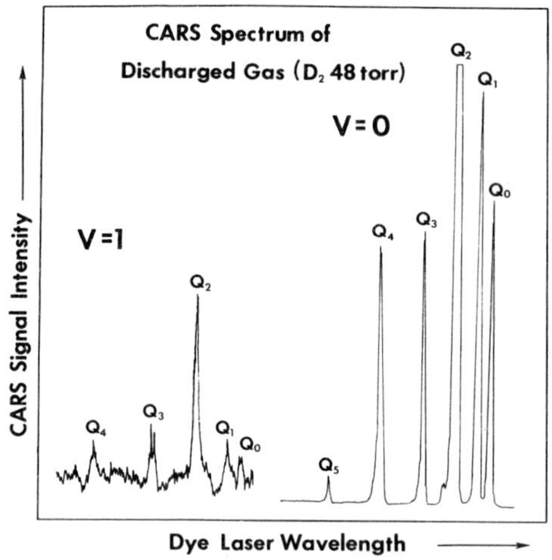

Fig. 7.14. The CARS spectrum of D_2 in an electrical discharge taken with the apparatus described in [7.120]. The vibrational temperature calculated from these spectra is ~1050 K, while the rotational temperature is ~400 K. From [7.120] with permission

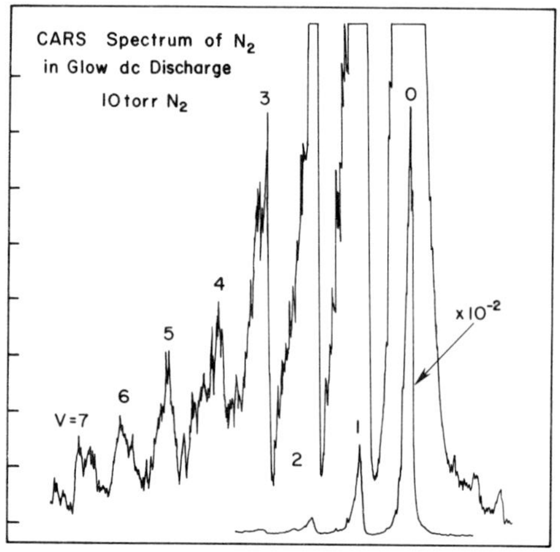

Fig. 7.15. Vibrational spectrum of N_2 (10 torr) in an electrical discharge

c) Photochemical, Kinetic and Excited State Studies

Although little work has yet been done in this area, the sensitivity and the short duration of the probing pulses make CARS promising in the study of photofragments, excited states and reaction products. For example, if a portion of a 532 nm beam is doubled to produce light at 266 nm, photochemical fragments produced by the ultraviolet beam can be followed by CARS [7.38,59]. In Fig.7.16 benzene vapor is shown to form a number of complex photochemical products as a result of 266 nm photolysis. Evidence for acetylenic, olefinic and saturated organic groups is suggested in the resulting spectrum. Although, no photochemistry of benzene has been reported to occur at 266 nm, it is known [7.123] that vacuum ultraviolet photolysis by normal single photon processes ($g \rightarrow u$) produces classes of compounds like the ones detected by CARS.

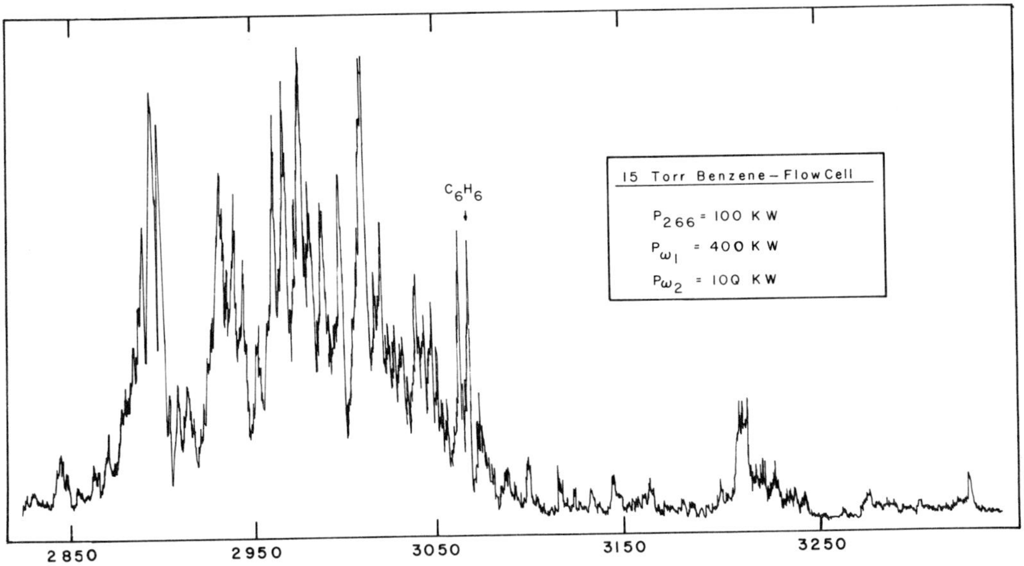

Fig. 7.16. A spectrum observed for benzene vapor (15 torr) when subjected to a 266 nm laser beam simultaneous in time and space with the CARS beams. In the absence of the uv beam, only a single strong CARS line is observed in the CH stretching region (ν_1 at 3073 cm^{-1}). On exposure to the 266 nm beam, the 3073 cm^{-1} feature is decreased tenfold and the complex spectrum shown in this figure results

The region in the VUV corresponds to twice the frequency of the 266 nm beam. The photolysis is therefore attributed to two photon $g \rightarrow g$ absorption in the VUV. It should be mentioned that for the pressures used in these experiments (3-15 torr), the 15 ns laser pulse width places photolysis and CARS detection in a nearly collision-free regime. Moreover, CARS can be carried out with picosecond lasers [7.124] and can be used, therefore, as a diagnostic tool for very fast relaxation processes such as occur in large molecules or in liquids.

d) High-Resolution Spectroscopy

CARS has definite advantages for high-resolution studies of molecular gases. In conventional Raman spectroscopy one loses signal by narrowing spectrometer slits so that a spectral resolution of 0.5 to 0.05 cm^{-1} is very difficult to obtain. For CARS, because of the inverse dependence on Raman linewidth, one actually *increases* the signal by narrowing the spectral widths of the two input beams (at least until one approaches the Raman linewidth). Resolution of 0.3 to 0.03 cm^{-1} is readily achieved and, with single mode pulsed and cw lasers, a resolution of 0.01 to 0.001 cm^{-1} has recently been attained [7.69,88,94,95].

At low pressures collisional broadening can be minimized and only the Doppler width limits the ultimate resolution. It is noteworthy that this is significantly reduced for forward scattering (CARS) versus 90° scattering (Raman) [7.126][4]. In addition, KHOO and YEUNG suggest that Doppler-free linewidths can be determined if one spectrally analyzes the CARS signal with an interferometer [7.96,97].

At present, vibration-rotation CARS spectra have been obtained for H_2 [7.29,61-64, 66,85,89,92], D_2 [7.92,95,120], N_2 [7.38,68,70,93,126], O_2 [7.68,82], HF [7.71,119], CO [7.70], CH_4 [7.38,46,47,59,69,85], C_2H_2 [7.94], C_6H_6 [7.38,59,127], and C_3H_4

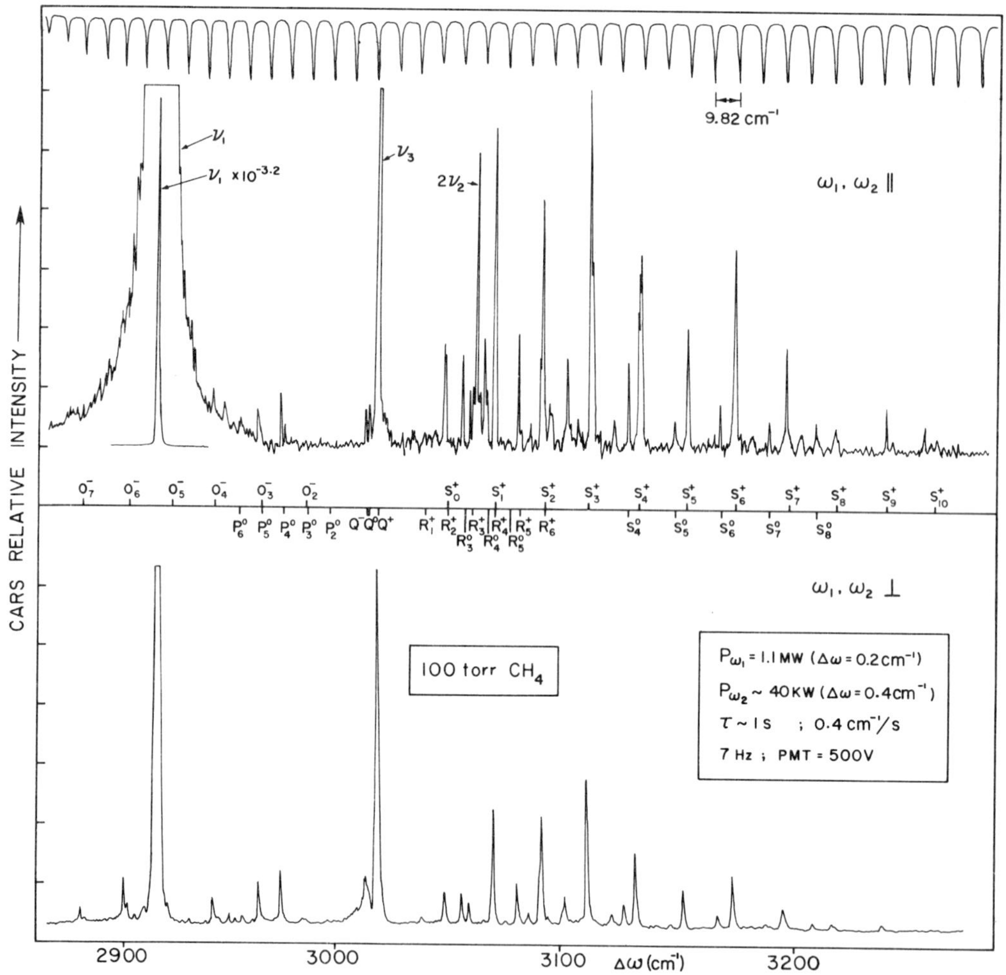

Fig. 7.17. CARS spectra of the ν_1 and ν_3 vibrational modes of CH_4 for two polarization arrangements

[4]See WEBER Ref.3.1, pp.729-733 for the angular dependence of the Doppler width of Raman lines.

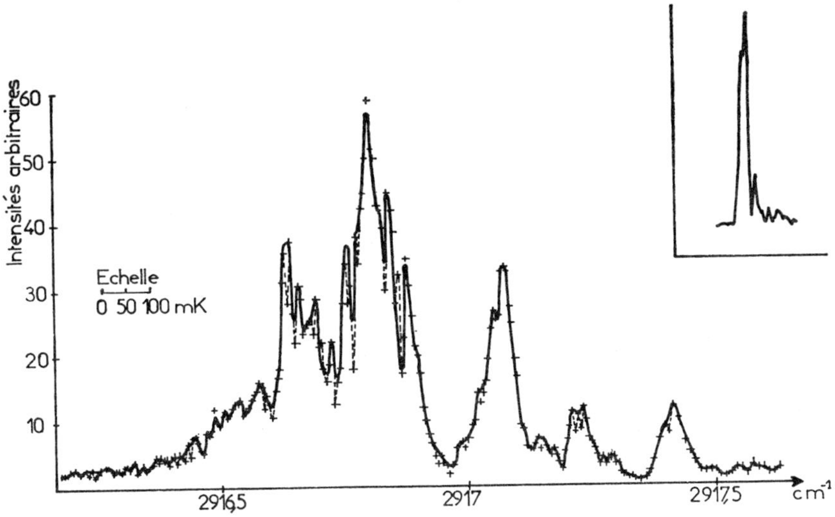

Fig. 7.18. High-resolution scan of the ν_1 Q-branch for CH_4 at 60 torr. A pulsed ruby laser - dye laser system was used to record this spectrum at about 0.005 cm^{-1} resolution. The inset is a normal Raman spectrum by H. Berger: Thesis, Dijon (1976). From [7.69] with permission

[7.127]. The only pure rotational CARS spectrum reported is for the $J=3\rightarrow5$ rotational transition of H_2 at 1033 cm^{-1} [7.48,127]. The most extensive spectrum yet recorded is that shown in Fig.7.17 for the ν_1 and ν_3 CH stretches of CH_4 [7.38,59]. The large decrease in the intensity of the ν_1 symmetric stretch for the perpendicular scan is apparent; the residual intensity is attributed to the fact that all beams were not 100% polarized. A much higher resolution scan of the ν_1 Q-branch was recently obtained with a pulsed ruby laser by BOQUILLON et al. [7.69], Fig.7.18. This spectrum and that of C_2H_2 shown in Fig.7.10 give a good indication of the potential of high-resolution CARS spectroscopy. Very precise frequency measurements should be possible since this information is inherently contained in the intense ω_1, ω_2 beams. An accurate relative frequency mark can thus be obtained by passing the ω_2 beam through an analyzing etalon (the upper peaks in Fig.7.17 were generated in this manner). In addition, accurate linewidth measurements are possible with CARS and pressure broadening effects have recently been determined for H_2 [7.64,92], D_2 [7.92,95], and C_2H_2 [7.94].

7.4.4 Solutions

Because the CARS signal depends on the bulk susceptibility of a sample, interference effects between $\chi^{(3)}_{RES}$ for a solute and χ_{NR} for the solvent are measurable. As discussed earlier, the frequency shift between the minimum and maximum of a resonance permit the determination of the real part of χ_{NR} and this has been exploited in a number of solution studies [7.33-35]. It is also found that any imaginary contribution to χ_{NR} will increase the width of the minimum and LEVENSON and BLOEMBERGEN have shown from such measurements that, for example, χ''_{NR} is less than 5% of χ'_{NR} for benzene [7.34,35]. For 2,5-diphenyloxazole (1 M. in benzene) $\chi''_{NR} \simeq \chi'_{NR}$, an increase largely attributed to two photon absorption processes, e.g., Fig.7.2c [7.33]. In fact, with CARS, it is possible to use known Raman cross sections as a calibration standard in measuring two photon cross sections more accurately than by previous methods [7.33-37].

From an analytical standpoint, χ_{NR} is undesirable since it limits the detection of small amounts of solutes by CARS. Studies of benzene-toluene mixtures show that the 992 cm^{-1} benzene resonance follows a squared dependence on concentration only down to ~10%. At lower concentration the dependence is linear due to the cross term with χ_{NR} for the solvent, see (7.70), and the signal becomes lost in the background for benzene concentrations below ~1% [7.50]. Similarly, it is found that the 1005 cm^{-1} resonance of sodium benzoate is comparable to the background in 0.3 M aqueous solutions [7.75].

One way to improve the sensitivity of CARS is to use resonance enhancement by choosing ω_1 or ω_3 to be near a one photon resonance, see (7.56,57). Similar enhancement occurs in conventional Raman spectroscopy but resonant CARS has the advantage that fluorescence interference is avoided because of the anti-Stokes position of the signal. CHABAY et al. were the first to demonstrate this enhancement effect in a study of 1.4×10^{-3} M diphenyloctatetraene in benzene [7.73,75]. Figure 7.19 compares their CARS and Raman spectra and the improvement in S/N and fluorescence background for the CARS result is apparent. They noted preferential enhancement of the 1580 cm^{-1} mode compared to the 1571 cm^{-1} feature as λ_1 was shifted from 545 to 480 nm, nearer to the first sample absorption at 404 nm. The need for some care in optimizing experimental parameters (conc., λ_1, and focal position) was also discussed and a detectivity limit for the 1170 cm^{-1} band was estimated at 5×10^{-5} M.

Resonance enhancement of CARS signals of biological samples in aqueous solutions has recently been observed by NESTOR et al. [7.76]. With optimization of sample concentration, cell pathlength and ω_1, ω_2 frequencies, resonance CARS spectra were obtained for 10^{-3} M solutions of ferrocytochrome and vitamin B_{12}. For ω_1 frequencies chosen to be off-resonance with sample absorptions, the CARS spectra of both compounds showed "negative" peaks attributed to an inverse Raman process induced by the ω_1 photons in the presence of the CARS nonresonant background of the solvent. This interpretation is disputed by HUDSON et al. [7.9,75] and LYNCH et al. [7.128] who argue that such negative peaks arise from a cross term between the background susceptibility with χ''_{RES} which is enhanced by the proximity of a one photon resonance. CARREIRA et al. present experimental evidence for the latter interpretation in a power dependence study of a 5×10^{-3} M benzene solution of N, N-diethyl-p-nitrosoaniline, another compound which shows "negative" peaks in the CARS spectrum [7.78]. These workers have

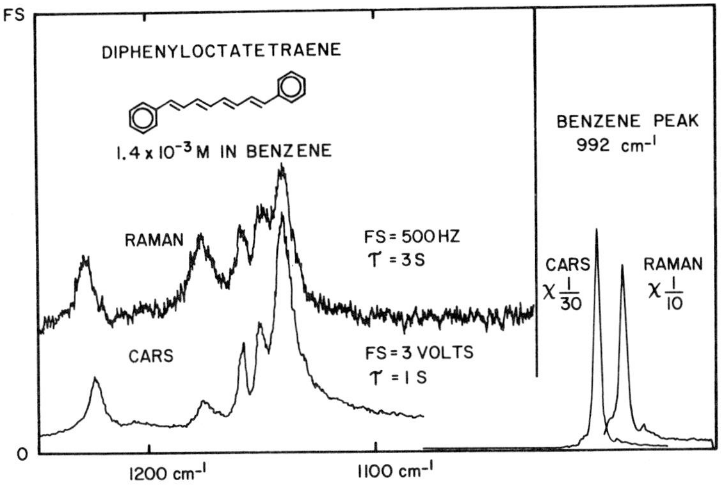

Fig. 7.19. CARS and Raman spectra of DPOT in benzene. The Raman spectrum is on a wave number scale; the CARS spectrum is on wavelength scale. The spectra are aligned at the 1140 cm^{-1} peak of DPOT. A 1 cm^{-1} bandwidth applies to the Raman spectrum and 0.3 cm^{-1} to the CARS. For CARS, ω_1 was 480.2 nm, as compared to 488.0 nm for the Raman. Non-resonant susceptibility, rather than fluorescence as for Raman, is the primary source of background signal for CARS. FS denotes full scale which for the Raman signal was 500 counts/s with a 10% quantum efficiency; for CARS it was 3 V from a photodiode. From [7.73] with permission

also obtained enhanced CARS spectra of β-carboline, a strongly fluorescent compound, and of 10^{-3} M β-carotene in benzene. It was found that the CARS signal saturated for ω_1 powers in excess of ~1 kW due to sample absorption. A detection limit of $\sim 5 \times 10^{-7}$ M was estimated for β-carotene and the use of melting point capillaries as a small volume cell was demonstrated [7.79]. NITSCH and KIEFER have also obtained CARS spectra of β-carotene in 10^{-4} M benzene solutions [7.77]. Resonance spectra of 10^{-4} M solutions of several dyes and of chlorophyll in methanol have been recorded recently by LAU and co-workers [7.91]. HUDSON has recently reviewed the application of CARS to biophysical problems [7.129]. No gas phase resonance enhanced CARS spectra have been reported but DRUET has considered the theory in this case and estimates a detection capability of a few ppm for the halogens [7.130,131]. CARREIRA and co-workers have obtained CARS spectra in the ultraviolet region below 300 nm, a region which should make resonance enhancement feasible for a wider variety of compounds [7.78].

7.4.5 Background Suppression in CARS

Several schemes have been proposed to reduce the nonresonant CARS contribution to improve the detection limits for mixtures. For a two component sample, the bulk susceptibility can be written

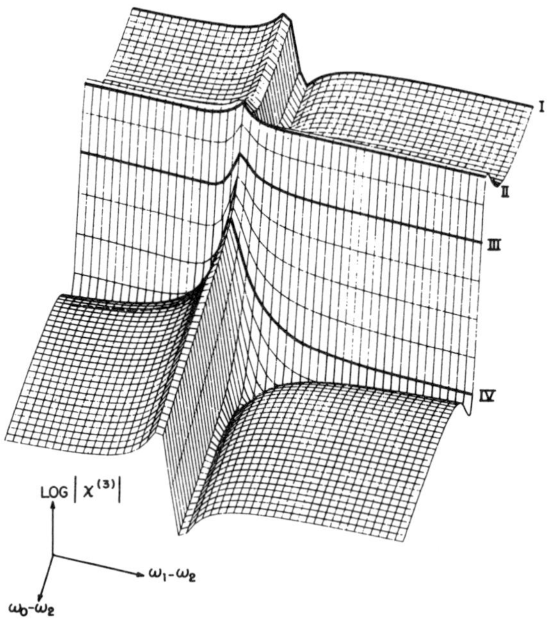

Fig. 7.20. Three-dimensional plot of $\log|\chi^{(3)}(\omega_3=\omega_0+\omega_1-\omega_2)|$ versus $\omega_0-\omega_2$ and $\omega_1-\omega_2$. From [7.104] with permission

$$|\chi^{(3)}|^2 = (\chi_{NR}+\chi'_A+\chi'_B)^2 + (\chi''_A+\chi''_B)^2$$

(we assume that χ_{NR} is real). Component A contains a Raman resonance at ω_A while χ_B becomes large at a different frequency ω_B. This second frequency could arise from another Raman resonance or from a one or two photon resonance in A, in the solvent or in some other added material. Figure 7.20 shows the general form of the susceptibility surface, with the two resonances occurring at $\omega_1 - \omega_2 = \omega_A$ and $\omega_0 - \omega_2 = \omega_B$. LYNCH et al. have used three input laser frequencies, ω_0, ω_1, ω_2, to map such a surface for a benzene-cyclohexane mixture [7.37,104]. They obtained spectra with shapes much like cuts I-IV of Fig.7.20 and they noted that scans in $\omega_1-\omega_2$ along a valley of minimum background (produced by cancellation of χ_{NR} by the negative part of χ'_B) can enhance the detectivity at ω_A by at least an order of magnitude. Furthermore they demonstrate with CuCl a similar cancellation of χ_{NR} when ω_B corresponds to a two photon absorption (TPA) [7.37]. (In this latter case, additional interference effects can occur between the imaginary components.) AHKMANOV et al. [7.87] have also considered ways to cancel χ_{NR} and have employed real contributions of a broad single photon absorption of rhodamine 6G to reduce χ_{NR} of toluene by more than an order of magnitude. Interference effects between the imaginary components were not observed in a similar experiment reported in [7.111].

In the experiments described above, all beams were linearly polarized in a common direction (1) and the susceptibility element $\chi^{(3)}_{1111}(\omega_1,\omega_1,-\omega_2) = \chi_{1111}$ was probed. LEVENSON and BLOEMBERGEN have considered alternative polarization conditions for ω_1,

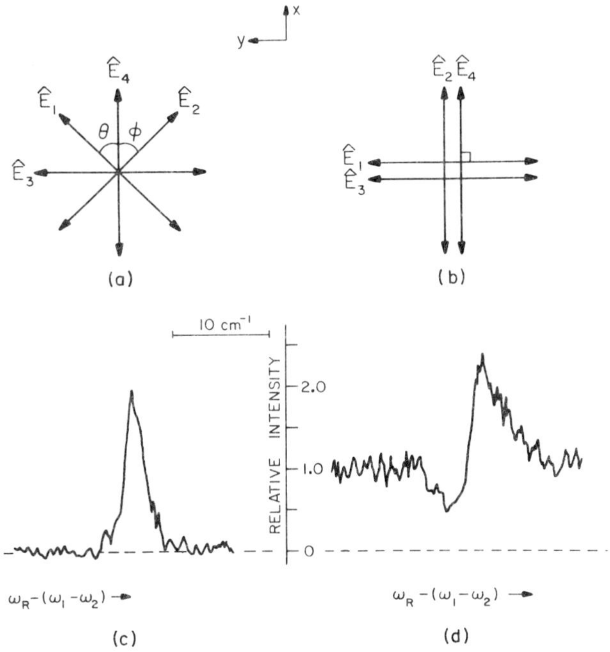

Fig. 7.21. Four-wave mixing spectra with corresponding polarization configurations. The sample is a mixture (3% C_6H_6 and 97% CS_2) and ω_R=992 cm^{-1}. From [7.132] with permission

ω_2, and ω_3 and demonstrated that an orientation with ω_1 at 45° to orthogonal ω_2, ω_3 polarizations gives a measure of the difference $|\chi_{1122}-\chi_{1221}|$ [7.34]. According to the Kleinman symmetry conjecture [7.130], $\chi^{NR}_{1122}=\chi^{NR}_{1221}$ and the nonresonant contribution to the total susceptibility cancels out. The Raman resonant terms do not cancel, however; for a symmetric mode with a Raman depolarization ratio of zero, $|\chi^{RES}_{1122}-\chi^{RES}_{1221}| = 1/2\,|\chi^{RES}_{1111}|$ and the peak CARS signal is reduced by only a factor of four from the value when all polarizations are parallel [7.34]. Unfortunately KLEINMAN symmetry is only good to 20% or so for visible wavelengths and the background is therefore only reduced by a factor of ~25 [7.133].

SONG et al. [7.133] have recently suggested a refinement to the CARS experiment by which the nonresonant terms can be cancelled exactly. They used three linearly polarized input beams ω_1, ω_2, and ω_3 with variable polarization directions as shown in Fig.7.21a. (They term this polarization condition *asterisk*.) The angles θ and ϕ are adjusted to one of many combinations for which the nonresonant background is eliminated and the resonant CARS signal ω_4 is recorded. Figure 7.21c shows such a scan of the 992 cm^{-1} mode of benzene in a 3% benzene: 97% CS_2 mixture for $\theta = 45°$, $\phi = 48 \pm 3°$. The background was reduced by more than a factor of 1000 compared to that obtained with a more conventional CARS polarization orientation (see Fig.7.21b,d) [7.133]. The lineshape was also improved significantly with the *asterisk* configuration and the peak value was only reduced by a factor of about 4.

7.5 Summary

Although CARS has many advantages, it is not likely to "replace" normal Raman spectroscopy. The experiments are not routine and are strongly dependent on the reproducible performance of the ω_1 and ω_2 lasers. Because of the nonlinear nature of the CARS process, amplitude, frequency and spatial fluctuations in ω_1 and ω_2 beams combine to yield substantial shot to shot fluctuations in CARS signals (10-50%). At very high peak powers [MW] optical damage of mirrors and other components is a problem and in addition significant pumping of the sample molecules from the desired initial state can occur. With dye lasers, a full spectral scan requires using several different dyes and changing filters in the detection system.

The nonresonant background contribution is perhaps the principal limitation of CARS. Resonance enhancement of the sample signal can be of aid in some cases but is not generally applicable. Schemes to reduce χ_{NR} by cancellation using three input beams [7.37,104] or polarization [7.133] have been discussed. An alternate approach suggested by SCEATS involves a pulse sequenced method [7.134]. This technique employs two picosecond ω_1 laser pulses with a time delay comparable to the dephasing time of molecular vibrations. Although this technique has not been experimentally demonstrated, it is proposed that this method will descriminate between χ_{NR} and χ_{RES} because the dephasing due to the latter is much slower. Like the other procedures, the experiment becomes more difficult, however, and further work will be necessary to make such variations routine for CARS.

Another nonlinear coherent Raman technique which shows promise in eliminating χ_{NR} background is Raman Induced Kerr Effect Spectroscopy (RIKES) [7.135-139]. In this experiment the polarization of a probe beam is changed due to the nonlinear Kerr effect when a pump beam passes through a common sample focus. The transmission of the probe beam through a blocking polarizer increases significantly when the frequency difference between the two beams is equal to a Raman resonance in the sample. Unfortunately, a background due to imperfect polarizations and birefringence of optical components can be troublesome in the RIKES experiments. Some gain can be had by means of an optical heterodyne technique in which this coherent probe background is mixed with the pump produced signal [7.139]. By chopping the pump pulse and using phase sensitive detection, the background can be suppressed and both the real and the imaginary susceptibility of a sample can be displayed. This approach is particularly promising for cw RIKES experiments. Optical modulation has also been proposed as a way to improve the S/N in CARS experiments [7.19,46].

Despite the above limitation of CARS, the technique promises to be of great value in a number of areas where conventional Raman and absorption spectroscopy are of little use. Very high resolution molecular spectroscopy is one logical field of endeavor. The sensitivity of CARS is such that it may find application in molecular beam studies. Because of the short duration of pulsed experiments (~10 ns), it should be possible to detect transient species and to monitor their reaction and decay within a few collisions. Its application as both a pump and a probe for energy transfer research is likely. Atmospheric chemical and photo-chemical reactions can be simulated and conveniently monitored by CARS. Diagnostic applications involving flames and electrical discharges have been mentioned and research in these areas will continue to grow. The use of CARS for other luminescing and fluorescing systems is probable. Most of this research will concentrate on gas phase systems but, with resonance enhancement, CARS will continue to be useful in studies of colored molecules of biological interest.

Acknowledgments. We thank the National Science Foundation for research support during the writing of this chapter. The assistance of Mr. Dennis Guthals in the calculations and plots for several figures is greatly appreciated. We also wish to thank many individuals active in CARS research who kindly supplied preprints of papers and permitted us to reproduce figures in this chapter.

References

7.1 Shen, Y.R.: Rev. Mod. Phys. *48*, 1 (1976)
7.2 Bloembergen, N.: *Nonlinear Optics*, 3rd printing (W.A. Benjamin, New York 1977)
7.3 Wang, C.S.: "The Stimulated Raman Process", in *Quantum Electronics: A Treatise*, Vol.1, ed. by Rabin,H., Tang,C.L. (Academic Press, New York 1975) Chap.7
7.4 Shen, Y.R.: "Stimulated Raman Scattering", in *Light Scattering in Solids*, ed. by M. Cardona, Topics in Applied Physics, Vol.8 (Springer, Berlin, Heidelberg, New York 1975) Chap.7
7.5 Lallemand, P.: In *The Raman Effect*, ed. by A. Anderson, Vol.1 (Marcel Dekker, New York 1971) p.287
7.6 Maier, M., Kaiser, W., Giordmaine, J.A.: Phys. Rev. *177*, 580 (1969)
Maier, M.: Appl. Phys. *11*, 209 (1976)
7.7 Terhune, R.W.: Bull. Am. Phys. Soc. *8*, 359 (1963)
7.8 Maker, P.D., Terhune, R.W.: Phys. Rev. *137*, A801 (1965)
7.9 Andersen, H.C., Hudson, B.S.: In *Molecular Spectroscopy*, ed. by D.A. Long, Vol.5 (Chemical Society, New York 1977)
7.10 DeWitt, R.W., Harvey, A.B., Tolles, W.M.: "Theoretical Development of Third-Order Susceptibility as Related to Coherent Anti-Stokes Raman Spectroscopy", NRL Memor. Rpt. No.3260 (1976)
7.11 Armstrong, J.A., Bloembergen, N., Ducuing, J., Pershan, P.S.: Phys. Rev. *127*, 1918 (1962)
7.12 Terhune, R.W., Maker, P.D.: In *Lasers*, ed. by A.K. Levine, Vol.2 (Marcel Dekker, New York 1968) p.295
7.13 Butcher, P.N.: "Nonlinear Optical Phenomena", Ohio State University Engineering Publications, Columbus (1965)
7.14 Flytzanis , C.: In *Quantum Electronics: A Treatise*, ed. by Rabin,H., Tang,C.L., Vol.1 (Academic Press, New York 1975) p.9
7.15 Minck, R.W., Terhune, R.W., Wang, C.C.: Appl. Opt. *5*, 1595 (1966)
7.16 Pantell, R.H., Puthoff, H.E.: *Fundamentals of Quantum Electronics* (Wiley and Sons, New York 1969) Chaps.6,7
7.17 Hauchecorne, G., Kerherve, F., Mayer, G.: J. Physique *32*, 47 (1971)
7.18 Regnier, P., Moya, F., Taran, J-P.E.: AIAA J. *12*, 826 (1974)
7.19 Akhmanov, S.A., Koroteev, N.I.: Sov. Phys. JETP *40*, 650 (1975)
7.20 Krochik, G.M., Khronopulo, Yu.G.: Sov. J. Quant. Electron. *5*, 917 (1975)

7.21 Yariv, A.: *Quantum Electronics*, 2nd ed. (Wiley and Sons, New York 1975)
7.22 Yuratich, M.A., Hanna, D.C.: Mol. Phys. *33*, 671 (1977)
7.23 Tolles, W.M., Nibler, J.W., MacDonald, J.R., Harvey, A.B.: Appl. Spectrosc. *31*, 253 (1977)
7.24 Kogelnik, H., Li, T.: Proc. IEEE *54*, 1312 (1966)
7.25 Jackson, J.D.: *Classical Electrodynamics*, 2nd ed. (Wiley and Sons, New York 1975)
7.26 Bjorklund, G.C.: IEEE J. QE-*11*, 287 (1975)
7.27 Bjorklund, G.C.: "Effects of Focusing on Third-Order Nonlinear Processes in Isotropic Media", Bell Lab., Tech. Memo. 74-1313-23 (1974)
7.28 Shaub, W.M., Harvey, A.B., Bjorklund, G.C.: J. Chem. Phys. *67*, 2547 (1977)
7.29 Regnier, P.: "Application of Coherent Anti-Stokes Raman Scattering to Gas Concentration Measurements and to Flow Visualization", Office National d'Etudes et de Recherches Aêrospatiales, Tech. Note No.215, 92320 Châtillon, France (1973); also available in English as European Space Agency technical translation ESA TT-200
7.30 Christie, J.H., Lockwood, D.J.: J. Chem. Phys. *54*, 1141 (1971)
7.31 Ward, J.F.: Rev. Mod. Phys. *37*, 1 (1965)
7.32 Wallace, R.: J. Mol. Phys. *11*, 457 (1966)
7.33 Levenson, M.D., Bloembergen, N.: J. Chem. Phys. *60*, 1323 (1974)
7.34 Levenson, M.D., Bloembergen, N.: Phys. Rev. B*10*, 4447 (1974)
7.35 Lynch Jr., R.T., Lotem, H.: J. Chem. Phys. *66*, 1905 (1977)
7.36 Anderson, R.J.M., Holtom, G.R., McClain, W.M.: J. Chem. Phys. *66*, 3832 (1977)
7.37 Lotem, H., Lynch Jr., R.T., Bloembergen, N.: Phys. Rev. A*14*, 1748 (1976)
7.38 Nibler, J.W., MacDonald, J.R., Harvey, A.B.: In *Proc. 5th Intern. Conf. on Raman Spectroscopy*, ed. by Schmid,E.D., Brandmüller,J., Kiefer,W., Schrader,B., Schrötter,H.W. (Schulz Verlag, Freiburg/Br. 1976) pp.717-725
7.39 Matsui, H., Resler Jr., E.L., Bauer, S.H.: J. Chem. Phys. *63*, 4171 (1975)
7.40 Lukasik, J., Ducuing, J.: J. Chem. Phys. *60*, 331 (1974)
7.41 Frey, R., Lukasik, J., Ducuing, J.: Chem. Phys. Lett. *14*, 514 (1972)
7.42 DeMartini, F., Ducuing, J.: Phys. Rev. Lett. *17*, 116 (1966)
7.43 Audibert, M.M., Joffrin, C., Ducuing, J.: Chem. Phys. Lett. *25*, 158 (1974)
7.44 Audibert, M.M., Joffrin, C., Ducuing, J.: Chem. Phys. Lett. *19*, 26 (1973)
7.45 Miller, R.G., Hancock, J.K.: J. Chem. Phys. *66*, 5150 (1977)
7.46 Henesian, M.A., Kulevskii, L., Byer, R.L.: J. Chem. Phys. *65*, 5530 (1976)
7.47 Barrett, J.J., Begley, R.F.: Appl. Phys. Lett. *27*, 129 (1975)
7.48 Barrett, J.J.: Appl. Phys. Lett. *29*, 722 (1976)
7.49 Begley, R.F., Harvey, A.B., Byer, R.L., Hudson, B.S.: Am. Lab. *6*, 11 (1974)
7.50 Begley, R.F., Harvey, A.B., Byer, R.L.: Appl. Phys. Lett. *25*, 387 (1974)
7.51 Levenson, M.D., Flytzanis, C., Bloembergen, N.: Phys. Rev. B*6*, 3962 (1972)
7.52 Levenson, M.D.: IEEE J. QE-*10*, 110 (1974)
7.53 Regnier, P., Taran, J-P.E.: Appl. Phys. Lett. *23*, 240 (1973)
7.54 Regnier, P., Taran, J-P.E.: In *Laser Raman Gas Diagnostics*, ed. by Lapp,M., Penney,C.M. (Plenum Press, New York 1974) p.87
7.55 Tolles, W.M., Turner, R.D.: Appl. Spectrosc. *31*, 96 (1977)
7.56 Cyvin, S.J., Rauch, J.E., Decius, J.C.: J. Chem. Phys. *43*, 4083 (1965)
7.57 Monson, P.R., McClain, W.M.: J. Chem. Phys. *53*, 29 (1970)
7.58 Yuratich, M.A., Hanna, D.C.: Opt. Commun. *18*, 134 (1976)
7.59 Nibler, J.W., Shaub, W.M., MacDonald, J.R., Harvey, A.B.: "Coherent Anti-Stokes Raman Spectroscopy", in *Vibrational Spectra and Structure*, ed. by J.R. Durig, Vol.6 (Elsevier, New York 1977) Chap.3
7.60 Nabara, A., Kubota, K.: Jpn. J. Appl. Phys. *6*, 1105 (1967)
7.61 DeMartini, F.: "High Resolution Nonlinear Spectroscopy of Molecular Vibrational Resonances in Gases", in *Fundamental and Applied Laser Physics*, ed. by Feld,M.S., Javan,A., Kurnit,N.A. (John Wiley and Sons, New York 1973) pp.549-561
7.62 DeMartini, F., Giuliani, G.P., Santamato, E.: Opt. Commun. *5*, 126 (1972)
7.63 DeMartini, F., Santamato, E., Capasso, F.: IEEE J. QE-*8*, 542 (1972)
7.64 DeMartini, F., Simoni, F., Santamato, E.: Opt. Commun. *9*, 176 (1973)
7.65 Rado, W.G.: Appl. Phys. Lett. *11*, 123 (1967)
7.66 Moya, F., Druet, S.A.J., Taran, J-P.E.: Opt. Commun. *13*, 169 (1975)

7.67 Moya, F., Druet, S.A.J., Taran, J-P.E.: In *Laser Spectroscopy*, ed. by Haroche, S., Pebay-Peyroula,J.C., Hänsch,T.W., Harris,S.E. (Springer, Berlin, Heidelberg, New York 1975) p.66
7.68 Moya, F., Druet, S.A.J., Pealat, M., Taran, J-P.E.: "Flame Investigation by Coherent Anti-Stokes Raman Scattering", AIAA Paper No.76-29, presented at the AIAA 14th Aerospace Sciences Meeting and 12th Annual Meeting Aerospace, January 26-30, 1976, Washington, D.C. (1976)
7.69 Boquillon, J.P., Moret-Bailly, J., Chaux, R.: C.R. Acad. Sc. Paris B*284*, 205 (1977)
7.70 Taran, J-P.E.: In *Tunable Lasers and Applications*, ed. by Mooradian,A., Jaeger, T., Stockseth,P., Springer Series in Optical Sciences, Vol.3 (Springer, Berlin, Heidelberg, New York 1976) pp.378-388
7.71 Taran, J-P.E.: In *Proc. 5th Intern. Conf. on Raman Spectroscopy*, ed. by Schmid, E.D., Brandmüller,J., Kiefer,W., Schrader,B., Schrötter,H.W. (Schulz Verlag, Freiburg/Br. 1976) pp.695-706
7.72 Wynne, J.J.: Phys. Rev. Lett. *29*, 650 (1972)
7.73 Chabay, I., Klauminzer, G., Hudson, B.S.: Appl. Phys. Lett. *28*, 27 (1976)
7.74 Carreira, L.A., Maguire, T.C., Malloy Jr.,T.B.: J. Chem. Phys. *66*, 2621 (1977)
7.75 Hudson, B.S., Hetherington, W., Cramer, S., Chabay, I., Klauminzer, G.: Proc. Natl. Acad. Sci. *73*, 3798 (1976)
7.76 Nestor, J., Spiro, T., Klauminzer, G.: Proc. Natl. Acad. Sci. *73*, 3329 (1976)
7.77 Nitsch, W., Kiefer, W.: In *Proc. 5th Intern. Conf. on Raman Spectroscopy*, ed. by Schmid,E.D., Brandmüller,J., Kiefer,W., Schrader,B., Schrötter,H.W. (Schulz Verlag, Freiburg/Br. 1976) pp.740-741
7.78 Carreira, L.A., Goss, L.P., Malloy,Jr., T.B.: J. Chem. Phys.*66*, 2762, 4360 (1977)
7.79 Roger, L.B., Stuart, J.D., Goss, L.P., Malloy Jr.,T.B., Carreira, L.A.: Anal. Chem. *49*, 949 (1977)
7.80 Wallan, D.J., Ritz, G.P., Morris, M.D.: Appl. Spectrosc. *31*, 475 (1977)
7.81 Begley, R.F., Harvey, A.B., Byer, R.L., Hudson, B.S.: J. Chem. Phys. *61*, 2466 (1974)
7.82 Miles, R.B., Laufer, G., Bjorklund, G.C.: Appl. Phys. Lett. *30*, 417 (1977)
7.83 Harvey, A.B., MacDonald, J.R., Tolles, W.M.: Prog. Anal. Chem. *8*, 211 (1976)
7.84 Nibler, J.W., MacDonald, J.R., Harvey, A.B.: Opt. Commun. *18*, 134, 371 (1976)
7.85 Akhmanov, S.A., Dmitriev, V.G., Kovrigin, A.I., Koroteev, N.I., Tunkin, V.G., Kholodnykh, A.I.: JETP Lett. *15*, 425 (1972)
7.86 Akhmanov, S.A., Koroteev, N.I., Kholodnykh, A.I.: J. Raman Spectrosc. *2*, 239 (1974)
7.87 Akhmanov, S.A., Bunkin, A.F., Ivanov, S.G., Koroteev, N.I., Kovrigin, A.I., Shumay, I.L.: In *Tunable Lasers and Applications*, ed. by Mooradian,A., Jaeger, T., Stockseth,P., Springer Series in Optical Sciences, Vol.3 (Springer, Berlin, Heidelberg, New York 1976) pp.391
7.88 Byer, R.L.: "Parametric Oscillators", in *Tunable Lasers and Applications*, ed. by Mooradian,A., Jaeger,T., Stockseth,P., Springer Series in Optical Sciences, Vol.3 (Springer, Berlin, Heidelberg, New York 1976) p.378
7.89 Roh, W.E., Schreiber, P., Taran, J-P.E.: Appl. Phys. Lett. *29*, 174 (1976)
7.90 Nibler, J.W., Harvey, A.B.: "Applications of CARS", in *Analytical Raman Spectroscopy*, ed. by W. Kiefer (Wiley-Interscience, New York) to be published
7.91 Lau, A., Werncke, W., Klein, J., Pfeiffer, M.: Opt. Commun. *21*, 399 (1977)
7.92 Henesian, M.A., Kulevskii, L., Byer, R.L., Herbst, R.L.: Opt. Commun. *18*, 225 (1976)
7.93 Hirth, A., Vollrath, K.: Opt. Commun. *18*, 213 (1976)
7.94 Fabelinsky, V.I., Krynetsky, B.B., Kulevsky, L.A., Mishin, V.A., Prokhorov, A.M., Savel'ev, A.D., Smirnov, V.V.: Opt. Commun. *20*, 389 (1977)
7.95 Krynetsky, B.B., Kulevsky, L.A., Mishin, V.A., Prokhorov, A.M., Savel'ev, A.D., Smirnov, V.V.: Opt. Commun. *21*, 225 (1977)
7.96 Khoo, I.C., Yeung, E.S.: Opt. Commun. *22*, 83 (1977)
7.97 Khoo, I.C., Yeung, E.S.: Phys. Rev. A , (1978)
7.98 Coffinet, J.P., DeMartini, F.: Phys. Rev. Lett. *22*, 60 (1969)
7.99 Wynne, J.J.: IEEE J. QE-*8*, 607 (1972)
7.100 Yablonovitch, E., Bloembergen, N., Wynne, J.J.: Phys. Rev. B*3*, 2060 (1971)
7.101 Yablonovitch, E., Flytzanis, C., Bloembergen, N.: Phys. Rev. Lett. *29*, 865 (1972)

7.102 Bloembergen, N., Levenson, M.D., Lynch Jr.,R.T.: In *Optical Properties of Highly Transparent Solids*, ed. by Mitra,S.S., Bendow,B. (Plenum Press, New York 1975) p.329
7.103 Kramer, S.C., Parsons, F.G., Bloembergen, N.: Phys. Rev. B*9*, 1853 (1974)
7.104 Lynch Jr.,R.T., Kramer, S.D., Lotem, H., Bloembergen, N.: Opt. Commun. *16*, 372 (1976)
7.105 Kramer, S.D., Bloembergen, N.: Phys. Rev. B*14*, 4654 (1976)
7.106 Kramer, S.D., Bloembergen, N.: In *Optical Properties of·Highly Transparent Solids*, ed. by Mitra,S.S., Bendow,B. (Plenum Press, New York 1975) p.365
7.107 Kato, Y., Takuma, H.: J. Opt. Soc. Am. *61*, 347 (1971); J. Chem. Phys. *54*, 5398 (1971)
7.108 Owyoung, A., Peercy, P.S.: J. Appl. Phys. *48*, 674 (1977)
7.109 Itzkan, I., Leonard, D.A.: Appl. Phys. Lett. *26*, 106 (1975)
7.110 Akhmanov, S.A., Koroteev, N.I., Orlov, R.Yu., Shumai, I.L.: JETP Lett. *23*, 249 (1976)
7.111 Lotem, H., Lynch Jr.,R.T.: Phys. Rev. Lett. *37*, 334 (1976)
7.112 Capasso, F., DeMartini, F.: Opt. Commun. *9*, 172 (1973)
7.113 Moya, F.: "Application de la Diffusion Raman Anti-Stokes Cohérente à des Mesures de Concentrations Gazeuses dans les Ecouillements Aérodynamiques"; Thesis, University of Paris (1976)
7.114 Moya, F., Druet, S.A.J., Taran, J-P.E.: "Gas Concentration and Temperature Measurements by Coherent Anti-Stokes Raman Scattering, in Proceedings of Project Squid Workshop on Combustion Measurements", in *Jet Propulsion Systems*, ed. by R. Goulard (Purdue Univ., Dec. 1975)
7.115 Pealat, M., Druet, S.A.J., Attal, B., Taran, J-P.E.: "Temperature and Concentration Measurements in Reactive Media by Coherent Anti-Stokes Raman Scattering", presented at 16th Intern. Symposium of Combustion, MIT, Cambridge, Mass. (1976)
7.116 Taran J-P.E.: In *Laser Spectroscopy III*, ed. by Hall,J.L., Carlsten,J.L., Springer Series in Optical Sciences, Vol.7 (Springer, Berlin, Heidelberg, New York 1977) p.315
7.117 Shaub, W.M., Lemont, S., Harvey, A.B.: Unpublished work
7.118 Roh, W.B.: "Coherent Anti-Stokes Raman Spectra of Molecular Gases", Technical Rpt. AFAPL-19-TR-77-47 (Wright Patterson Air Force Base, Dayton, Ohio)
7.119 Eckbreth, A.C., Bonczyk, P.A., Verdieck, J.F.: "Review of Laser Raman and Fluorescence Techniques for Practical Combustion Diagnostics", EPA Rpt. for contract 68-02-2176
7.120 Nibler, J.W., MacDonald, J.R., Harvey, A.B.: Opt. Commun. *18*, 371 (1976)
7.121 Shaub, W.M., Nibler, J.W., Harvey, A.B.: J. Chem. Phys. *67*, 1883 (1977)
7.122 Nibler, J.W., MacDonald, J.R., Harvey, A.B.: "Development of a Diagnostic Tool for Measuring Temperature and Number Density in the EDGL", Semiannual Technical Rpt. for DARPA order No.2062 (1976)
7.123 Calvert, J.G., Pitts Jr.,J.N.: *Photochemistry* (Wiley and Sons, New York 1967)
7.124 Goldberg, L., MacDonald, J.R., Harvey, A.B.: Unpublished work
7.125 Clements, W.R.L., Stoicheff, B.P.: J. Mol. Spectrosc. *33*, 183 (1970)
7.126 Hirth, A., Vollrath, K.: In *Proc. 5th Intern. Conf. on Raman Spectroscopy*, ed. by Schmid,E.D., Brandmüller,J., Kiefer,W., Schrader,B., Schrötter,H.W. (Schulz Verlag, Freiburg/Br. 1976) pp.734-735
7.127 Barrett, J.J.: In *Proc. 5th Inter. Conf. on Raman Spectroscopy*, ed. by Schmid, E.D., Brandmüller,J., Kiefer,W., Schrader,B., Schrötter,H.W. (Schulz Verlag, Freiburg/Br. 1976) pp.732-733
7.128 Lynch Jr.,R.T., Lotem, H., Bloembergen, N.: J. Chem. Phys. *66*, 4250 (1977)
7.129 Hudson, B.S.: Ann. Rev. Biophys. Bioeng. *6*, 135 (1977)
7.130 Druet, S.A.J.: "Coherent Anti-Stokes Raman Scattering Near Electronic Resonances and Fabrication of an Experimental Set-up, Office National d'Etudes et de Recherches Aérospatiales", Technical Note No.1976-6, Châtillon, France
7.131 Druet, S.A.J.: In *Proc. 5th Intern. Conf. on Raman Spectroscopy*, ed. by Schmid, E.D., Brandmüller,J., Kiefer,W., Schrader,B., Schrötter,H.W. (Schulz Verlag, Freiburg/Br. 1976) pp.736-737
7.132 Kleinman, D.A.: Phys. Rev. *126*, 1977 (1962)
7.133 Song, J.J., Eesley, G.L., Levenson, M.D.: Appl. Phys. Lett. *29*, 567 (1976)
7.134 Sceats, M.: Private communication

7.135 Heiman, D., Hellwarth, R.W., Levenson, M.D., Martin, G.: Phys. Rev. Lett. *36*, 189 (1976)
7.136 Levenson, M.D., Song, J.J.: J. Opt. Soc. Am. *66*, 641 (1976)
7.137 Levenson, M.D.: Phys. Today *30*(5), 44 (1977)
7.138 Song, J.J., Levenson, M.D.: J. Appl. Phys. *48*, 3496 (1977)
7.139 Eesley, C.L., Levenson, M.D., Tolles, W.M.: IEEE J. QE-*14*, 45 (1978)

Additional References with Titles

Chapter 2

Altman, K., Klöckner, W., Strey, G.: Determination of the anharmonicity constant of oxygen by Raman measurements. Chem. Phys. Lett. *46*, 461 (1977)

Brassington, N.J., Edwards, H.G.M., Farwell, D.W., Long, D.A., Mansour, H.R.: "The Pure Rotational Raman Spectra of $^3H_2$ and SO_3 and the Vibration-Rotation Raman Spectrum of $^3H_2$", in *Lasers in Chemistry*, ed. by M.A. West (Elsevier Publishing) Amsterdam 1977) p.150

Edwards, H.G.M., Long, D.A., Mansour, H.R.: Pure Rotational and Vibration-Rotational Raman Spectra of Tritium, $^3H_2$. J. Chem. Soc. Faraday Trans.II *74*, 1203 (1978)

Finsterhölzl, H., Klöckner, H.W., Srinivasan, K., Schrötter, H.W., Brandmüller, J.: Raman spectra of carbon dioxide and of its isotopic variants in the Fermi-resonance region: Part I - $^{12}C^{16}O_2$. Indian J. Pure Appl. Phys. *16*, 370 (1978)

Fletcher, W.H., Rayside, J.S., McLendon, W.B.: A computer controlled high resolution Raman spectrometer system. J. Raman Spectrosc. *7*, 205 (1978)

Hollinger, A.B., Welsh, H.L.: High resolution rotation-vibration Raman bands of benzene. I. The totally symmetric bands of C_6H_6. Can. J. Phys. *56*, 974 (1978)

Klöckner, H.W., Finsterhölzl, H. Srinivasan, K., Schrötter, H.W.: Raman spectra of carbon dioxide and its isotopic variants in the Fermi-resonance region: II. $^{12}C^{18}O_2$. Appl. Spectrosc. *32*, 401 (1978)

Ohashi, N., Kyogoku, T., Matsuoka, S.: Raman spectrum of the ν_3 band of Ethylene-h_4. J. Raman Spectrosc. *7*, 194 (1978)

Srinivasan, K., Finsterhölzl, H., Klöckner, H.W., Illig, D., Schrötter, H.W.: Raman-spectroscopic observation of the Σ-diads of $^{12}O^{16}O^{18}O$ and $^{12}C^{16}O^{17}O$. Z. Naturf. Teil A *32*, 1070 (1977)

Chapter 3

Altmann, K., Klöckner, W., Strey, G.: "Der Intensitätsverlauf im reinen Rotations-Raman-Spektrum von CO_2 and N_2O unter Berücksichtigung des 01^10-Niveaus". Z. Naturf. Teil A *31*, 1311 (1976)

Brassington, N.J., Edwards, H.G.M., Farwell, D.W., Long, D.A., Mansour, H.R.: "The Pure Rotational Raman Spectra of $^3H_2$ and SO_3 and the Vibration-Rotation Raman Spectrum of $^3H_2$", in *Lasers in Chemistry*, ed. by M.A. West (Elsevier Publishing, Amsterdam 1977) p.150

Brassington, N.J., Edwards, H.G.M., Long, D.A.: Pure Rotational Raman Spectrum of XeF_2. J. Chem. Soc. Faraday Trans.II, *74*, 1208 (1978)

Brassington, N.J., Edwards, H.G.M., Farwell, D.W., Long, D.A., Mansour, H.R.: The pure rotational Raman spectrum of sulphur trioxide. J. Raman Spectrosc. *7*, 154 (1978)

Brassington, N.J., Edwards, H.G.M., Long, D.A., Skinner, M.: The pure rotational Raman spectrum of SeO_3. J. Raman Spectrosc. *7*, 158 (1978)

Edwards, H.G.M., Farwell, D.W., Long, D.A., Najm, K.A.B.: The pure rotational Raman spectra of $^{17}O_2$, $^{16}O^{17}O$, and $^{17}O^{18}O$. Indian J. Pure Appl. Phys. *16*, 354 (1978)

Edwards, H.G.M., Long, D.A., Mansour, H.R.: Pure rotational and vibration-rotational Raman spectra of tritium, $^3H_2$. J. Chem. Soc. Faraday Trans.II, *74*, 1203 (1978)

Edwards, H.G.M., Long, D.A., Mansour, H.R.: Pure rotational Raman spectrum of $^{37}Cl_2$. J. Chem. Soc. Faraday Trans.II, *74*, 1200 (1978)
Finsterhölzl, H., Hochenbleicher, J.G., Strey, G.: Intensity distribution in pure rotational Raman spectra of linear molecules in the ground and vibrational Π-states. J. Raman Spectrosc. *6*, 13 (1977)
Hester, R.E.: A cassegrain collector with unusually high light grasp. J. Raman Spectrosc. *7*, 74 (1978)
McMullan, D., Wehiger, P.A.: Developments in Electronography for astronomical research. Endeavour *1*, 32 (1977)
Patterson, G.D.: Pure rotational Raman scattering by symmetric top molecules. J. Chem. Phys. *59*, 821 (1973)
Scherer, J.R.: Removal of I_2 absorption lines from 514-nm excited Raman spectra. Appl. Opt. *17*, 1621 (1978)
Spieweck, F.: Single-mode operation of an Ar^+ laser with an intracavity I_2 absorption cell. Appl. Phys. *1*, 233 (1973)
Spieweck, F.: Wavelength stabilization of an Ar^+ laser with an external I_2 absorption cell. Appl. Phys. *3*, 429 (1974)
Spieweck, F.: I_2-stabilisierte Argonionen-Laser mit hoher Wellenlängengenauigkeit. PTB-Mitt. *88*, 243 (1978)

Chapter 4

Abbate, S., Gussoni, M., Zerbi, G.: Raman intensities: 'Overlay' least squares calculations of Raman electro-optical parameters for simple hydrocarbons. Indian J. Pure Appl. Phys. *16*, 199 (1978)
Buckingham, A.D., Szabo, A.: Determination of the polarizability anisotropy in a diatomic molecule from relative Raman intensities. J. Raman Spectrosc. *7*, 46 (1978)
Domingo, C., Escribano, R., Orza, J.M., Montero, S.: "Absolute Raman Intensities and Electro-Optical Parameters of Chloromethanes in the Gas Phase", in *Proc. VIth Int. Conf. on Raman Spectroscopy*, Vol.2, ed. by E.D. Schmid, R.S. Krishnan, W. Kiefer, H.W. Schrötter (Heyden and Son, London-Philadelphia-Rheine 1978) pp.168-169
Gussoni, M., Abbate, S., Zerbi, G.: Raman intensities: Transferability of electro-optical parameters. J. Raman Spectrosc. *6*, 289 (1977)
Montero, S., Bermejo, D., Klöckner, W.: Electro-optical parameters and Raman intensities of $CHCl_3$ and $CDCl_3$. Mol. Phys. *33*, 1029 (1977)
Murphy, W.F.: The rovibrational Raman spectrum of water vapour ν_1 and ν_3. Mol. Phys. *36*, 727 (1978)
Schrötter, H.W., Stery, G.: "Raman Intensities in Gases and their Enhancement in Liquids", in *Proc. VIth Int. Conf. on Raman Spectroscopy*, Vol.1, ed. by E.D. Schmid, R.S. Krishnan, W. Kiefer, H.W. Schrötter (Heyden and Son, London-Philadelphia-Rheine 1978) pp.179-188

Chapter 5

Aivasova, A.A., Nosenko, B.M., Otazhonov, Sh., Printsev, A.G.: Concentration dependence of profile shape for molecular and Raman scattering of light in a benzene-carbon tetrachloride mixture. Opt. Spectrosc. *41*, 467 (1976)
Berard, M., Lallemand, P.: Influence of the use of approximate trajectories for binary collision calculations. J. Quant. Spectrosc. Radiat. Transfer *19*, 386 (1978)
Breuillard, C., Ouillon, R.: Infrared and Raman band shapes and dynamics of molecular motions for N_2O in solutions: ν_3 band in CCl_4 and liquid SF_6. Mol. Phys. *33*, 747 (1977)
Brueck, S.R.J.: Polarized vibrational Raman scattering lineshape parameters in liquid CO and liquid CO mixtures. Chem. Phys. Lett. *53*, 273 (1978)
Burshtein, A.I., Temkin, S.I.: Collapse of the rotational structure of Raman scattering spectra in dense media. Sov. Phys.-JETP *44*, 492 (1976)

Clouter, M.J., Kiefte, H., Ali, N.: Anomalous behavior in the vibrational Raman spectrum of oxygen under near-critical conditions. Phys. Rev. Lett. *40*, 1170 (1978)

Diestler, D.J., Manz, J.: Model for vibrational dephasing of diatomic molecules in liquids. Liquid N_2 and O_2. Mol. Phys. *33*, 227 (1977)

Fabelinsky, V.I., Krynetsky, B.B., Kulevskiy, L.A., Mishin, V.A., Prokhorov, A.M.: High resolution CW CARS spectroscopy of the Q-branch of the ν_2 band in C_2H_2. Opt. Commun. *20*, 389 (1977)

Ikawa, S., Yamada, M., Kimura, M.: Study of the ν_1 Raman linewidth of SO_4^{2-} in acidic solutions. J. Raman Spectrosc. *6*, 89 (1977)

Jurkowska, K.: Molecular dynamics in fluids and the shape of vibrational bands. Wiad. Chem. *30*, 641 (1976)

Korsunskii, V.I., Lavrik, N.L., Naberukhin, Yu.I.: Different positions of isotropic and anisotropic components of Raman scattering in liquids as evidence of intermolecular vibration coupling. Opt. Spectrosc. *41*, 468 (1976)

Koubaa, M., Perrot, M.: "Analyse par spectroscopie Raman des relaxations vibrationelle et orientationelle de l'ion nitrate en solution aqueuse. C.R. Acad. Sci. Ser. C *286*, 99 (1978)

Limouzi, J., Burneau, A., Perchard, J.P.: Comparison between infrared and Raman band profiles in the absence of rotational relaxation in the liquid phase: ν(OH) of methanol in acetonitrile. Chem. Phys. Lett. *47*, 346 (1977)

Lynden-Bell, R.M.: Vibrational relaxation and line widths dephasing by intermolecular forces. Mol. Phys. *33*, 907 (1977)

Lynden-Bell, R.M., Tabisz, G.C.: Vibrational relaxation and line widths in liquids: A comparison of theory and experiment. Chem. Phys. Lett. *46*, 175 (1977)

Miller, S., Clarke, J.H.R.: The effect of vibration-rotation coupling on Raman band shapes in liquids. Chem. Phys. Lett. *56*, 235 (1978)

Nienhuis, G., Schuller, F.: Collision and Doppler broadening of fluorescence and Raman scattering from atoms. Physica B and C *92*, 409 (1977)

Ouillon, R., Breuillard, C., Soussen-Jacob, J., Tsakiris, J.: Raman and infrared band shapes and dynamics of molecular motions for N_2O in solution: ν_1 and ν_3 bands in liquid solvents at various temperatures. J. Mol. Struct. *46*, 447 (1978)

Perchard, J.P., Perchard, C., Burneau, A., Limouzi, J.: Raman spectroscopy of weak hydrogen bonds in the liquid phase. J. Mol. Struct. *47*, 285 (1978)

Petrula, J.R., Strauss, H.L., Lao, K.Q.-H., Pecora, R.: Depolarized scattering at a binary liquid critical point. J. Chem. Phys. *68*, 623 (1978)

Prisiyzhnyi, V.D., Kirillov, S.A., Snezhkov, V.I.: Raman spectra of fused alkali and alkaline earth nitrites. J. App. Spectrosc. (USSR) *25*, 1058 (1976)

Schroeder, J., Schiemann, V.H., Sharko, P.T., Jonas, J.: Raman study of vibrational dephasing in liquid CH_3CN and CD_3CN. J. Chem. Phys. *66*, 3125 (1977)

Tanabe, K., Jonas, J.: Raman study of vibrational relaxation in liquid benzene-d_6 at high pressure. J. Chem. Phys. *67*, 4222 (1977)

Tanabe, K., Jonas, J.: Raman study of vibrational relaxation of benzene in solution. Chem. Phys. Lett. *53*, 278 (1978)

Tomkin, S.I., Burshtein, A.I.: "Pressure-Induced Transformation of the Shape of the Q-Band of Vibrational-Rotational Raman Spectra", in *Tesizy Dokl.-Simp. Mol. Spektrosk. Vys. Sverskhvys Razresheniya*, 3rd, ed. by V.P. Lopasov (Akad. Nauk SSSR, Sib. Otb., Inst. Opt. Atmos., Tomsk, USSR 1976) p.163

Viossat, G., Cavagnat, R., Lascombe, J.: Etude par spectroscopie Raman du mouvement orientationnel des molécules de chlore à l'état liquide. J. Raman Spectrosc. *6*, 299 (1977)

Wertheimer, R.K.: On the theory of spectral line broadening: Vibrational dephasing and resonance energy transfer in molecular liquids. Mol. Phys. *35*, 257 (1978)

Yarwood, J., Arndt, R., Döge, G.: Molecular motions and vibrational relaxation in acetonitrile. I. Raman spectral studies of the ν_1, ν_3 and $2\nu_3$ bands in the liquid and in dilute solutions. Chem. Phys. *25*, 387 (1977)

Chapter 6

Cheung, L.D., Yu, N.-T., Felton, R.H.: Resonance Raman spectrum and excitation profiles of soret-excited metalloporphyrins. Chem. Phys. Lett. *55*, 527 (1978)
Friedman, J.M., Rousseau, D.L.: Simultaneous observation of coherent and incoherent resonant re-emission in the condensed phase. Chem. Phys. Lett. *55*, 488 (1978)
Henneker, W.H., Penner, A.P., Siebrand, W., Zgierski, M.Z.: Resonance Raman excitation profiles and depolarization dispersion curves, and their use in the analysis of vibronically coupled excited states. J. Chem. Phys. *69*, 1704 (1978)
Hong, K.H.: Theory of resonant Raman scattering. III. Toward the total synthesis of molecular spectra and Raman excitation profiles within the generalized vibronic theory and from an exciton-in-molecule viewpoint. J. Chem. Phys. *68*, 1253 (1978)
Hong, K.H., Jacobsen, C.W.: Resonant Raman studies of pyrazine with tunable uv lasers: An analysis based on the generalized vibronic theory of Raman intensity. J. Chem. Phys. *68*, 1170 (1978)
Kiefer, W., Baierl, P.: Application of the resonance Raman effect to determine dephasing lifetimes of higher vibrational levels. Indian J. Pure Appl. Phys. *16*, 171 (1978)
Korenowski, G.M., Ziegler, L.D., Albrecht, A.C.: Calculations of resonance Raman cross sections in forbidden electronic transitions: Scattering of the 992 cm^{-1} mode in the $^1B_{2u}$ band of benzene. J. Chem. Phys. *68*, 1248 (1978)
Lyons, K.B., Friedman, J.M., Fleury, P.A.: Nanosecond transient Raman spectra of photolyzed carboxyhemoglobin. Nature *275*, 565 (1978)
Nishimura, Y., Hirakawa, A.Y., Tsuboi, M.: A vibronic coupling in a degenerate electronic state via a nuclear momentum and an antisymmetric Raman scattering tensor. J. Chem. Phys. *67*, 1009 (1977)
Nishimura, Y., Hirakawa, A.Y., Tsuboi, M.: The nature of the vibronic coupling in a metalloporphyrin and its Raman scattering. J. Mol. Spectrosc. *68*, 335 (1977)
Srivastava, R.B., Schuyler, M.W., Dosser, L.R., Purcell, F.J., Atkinson, G.H.: Resonance Raman spectra of cytochrome c and oxy-hemoglobin using pulsed laser excitation and optical multichannel detection. Chem. Phys. Lett. *56*, 595 (1978)
Woodruff, W.H., Farquharson, S.: Time resolved resonance Raman spectroscopy of hemoglobin derivatives: Heme structure changes in 7 nanoseconds. Science *201*, 833 (1978)
Zgierski, M.Z.: Interference between resonance and preresonance Raman scattering. J. Raman Spectrosc. *6*, 53 (1977)

Chapter 7

Akhmanov, S.A., Bunkin, A.F., Ivanov, S.G., Koroteev, N.I.: Coherent ellipsometry of Raman scattering of light. JETP Lett. *25*, 416 (1977)
Aliev, M.R., Kozlov, D.N., Smirnov, V.V.: Coherent spectroscopy of high-resolution Raman scattering of methane. JETP Lett. *26*, 27 (1977)
Attal, B., Schnepp, O.O., Taran, J.-P.E.: Resonant CARS in I_2 vapor. Opt. Commun. *24*, 77 (1978)
Azizbekyan, G.V., Badalyan, N.N., Koroteev, N.I., Nersesyan, K.A., Khurshudyan, M.A., Chilingaryan, Yu.S.: Observation of overtones of vibrational-rotational molecular transitions by coherent active spectroscopy. Sov. J. Quantum Electron. *7*, 1086 (1977)
Beatty, I.R., Black, J.D., Gilson, T.R., Greenhalgh, D., Hanna, D., Laycock, L.: "CARS in Chemistry", in *Lasers in Chemistry*, ed. by M.A. West (Elsevier Publishing, Amsterdam 1977) pp.7-12
Bloembergen, N., Lotem, H., Lynch, Jr.,R.T.: Line shapes in coherent resonant Raman scattering. Indian J. Pure Appl. Phys. *16*, 151 (1978)
Bunkin, A.F., Ivanov, S.G., Koroteev, N.I.: Coherent polarization spectroscopy of Raman scattering of light. Sov. Phys.-Dokl. *22*, 146 (1977)
Druet, S., Taran, J.-P.E.: "Coherent Anti-Stokes Raman Spectroscopy", in *Chemical and Biochemical Applications of Lasers*, Vol.IV, ed. by C.B. Moore (Academic Press, New York 1978)

Eckbreth, A.C., Bonczyk, P.A., Verdieck, J.F.: "Laser Raman Fluorescence Techniques for Practical Combustion Diagnostics", in *Applied Spectroscopic Reviews*, Vol.13, ed. by E.G. Brame (Marcel Dekker, New York 1977) pp.15-164

Eckbreth, A.C.: BOXCARS: Crossed-beam phase-matched CARS generation in gases. Appl. Phys. Lett. *32*, 421 (1978)

Harvey, A.B.: Coherent anti-Stokes Raman scattering. Anal. Chem. *50*, A905 (1978)

Henesian, M.A., Duncan, M.D., Byer, R.L., May, A.D.: Absolute Raman frequency measurement of the Q(2) line in D_2 using CW CARS. Optics Lett. *1*, 149 (1977)

Lau, A., Pfeiffer, M., Werncke, W.: The line shape of coherent anti-Stokes Raman scattering excited under resonance conditions. Opt. Commun. *23*, 59 (1977)

May, A.D., Henesian, M.A., Byer, R.L.: The CW coherent anti-Stokes Raman spectrum of the ν_1 band of CH_4 and its pressure dependence. Can. J. Phys. *56*, 248 (1978)

Morris, M.D.: "Coherent Resonance Raman Techniques in Clinical and Biochemical Analysis", in *Resonance Raman Spectroscopy as an Analytical Tool*, ed. by A. Meveger (Franklin Institute Press, 1978)

Nestor, J.R.: Polarization properties of coherent anti-Stokes Raman spectra (CARS) in isotropic liquids. J. Raman Spectrosc. *7*, 90 (1978)

Nitsch, W., Kiefer, W.: High resolution pulsed CARS spectroscopy of the Q-branches of some simple gases. Opt. Commun. *23*, 240 (1978)

Nitsch, W., Kiefer, W.: CARS-spectroscopy with a pulse dye-laser system. J. Mol. Struct. *45*, 343 (1978)

Owyoung, A., Jones, E.D.: Stimulated Raman spectroscopy using low-power CW lasers. Optics Lett. *1*, 152 (1977)

Owyoung, A.: Sensitivity limitations for CW stimulated Raman spectroscopy. Opt. Commun. *22*, 323 (1977)

Owyoung, A.: High-resolution CW stimulated Raman spectroscopy in molecular hydrogen. Optics Lett. *2*, 91 (1978)

Owyoung, A.: Coherent Raman gain spectroscopy using CW laser sources. IEEE J. QE-*14*, 192 (1978)

Taran, J.-P.E.: "Coherent Anti-Stokes Raman Spectroscopy", in *Lasers in Chemistry*, ed. by M.A. West (Elsevier Publishing, Amsterdam 1977) pp.1-6

Wiener-Avnear, E., Chandra, S., Compaan, A.: Third-order nonlinear susceptibility ratio in CARS of mixtures: CS_2 in C_6H_6. Appl. Phys. Lett. *32*, 286 (1978)

Subject Index

In addition to the usual listing of topics and some of the more frequently occurring symbols, this subject index lists all the molecules mentioned in the text, figures, and tables. Molecules are listed alphabetically and are given at the end of each alphabetical section. In organic compounds the symbol H precedes all others except C. Common names and formulas, and formulas of compounds as they appear in the text and tables are given in parentheses

Springer Series in Optical Sciences

Editor: D.L. MacAdam

A Selection

Volume 2
R. Beck, W. Englisch, K. Gürs

Table of Laser Lines in Gases and Vapors

2nd revised and enlarged edition 1978.
IX, 202 pages
ISBN 3-540-08603-X

Volume 3

Tunable Lasers and Applications

Proceedings of the Loen Conference, Norway, 1976
Editors: A. Mooradian, T. Jaeger, P. Stokseth

1976. 238 figures. VIII, 404 pages
ISBN 3-540-07968-8

Contents:
Tunable and High Energy UV-Visible Lasers. – Tunable IR Laser Systems. – Isotope Separation and Laser Driven Chemical Reactions. – Nonlinear Excitation of Molecules. – Laser Photokinetics. – Atmospheric Photochemistry and Diagnostics. – Photobiology. – Spectroscopic Applications of Tunable Lasers.

Volume 4
V.S. Letokhov, V.P. Chebotayev

Nonlinear Laser Spectroscopy

1977. 193 figures, 22 tables. XVI, 466 pages
ISBN 3-540-08044-9

Contents:
Introduction. – Elements of the Theory of Resonant Interaction of a Laser Field and Gas. – Narrow Saturation Resonances on Doppler-Broadened Transition. – Narrow Resonances of Two-Photon Transitions Without Doppler Broadening. – Nonlinear Resonances on Coupled Doppler-Broadened Transitions. – Narrow Nonlinear Resonances in Spectroscopy. – Nonlinear Atomic Laser Spectroscopy. – Nonlinear Molecular Lser Spectroscopy. – Nonlinear Narrow Resonances in Quantum Electronics. – Narrow Nonlinear Resonances in Experimental Physics.

Volume 7

Laser Spectroscopy III

Proceedings of the Third International Conference, Jackson Lake Lodge, Wyoming, USA, July 4–8, 1977
Editors: J.L. Hall, J.L. Carlsten

1977. 296 figures. XII, 468 pages
ISBN 3-540-08543-2

Contents:
Fundamental Physical Applications of Laser Spectroscopy. – Multiple Photon Dissociation. – New Sub-Doppler Interaction Techniques. – Highly Excited States, Ionization, and High Intensity Interactions. – Optical Transients. – High Resolution and Double Resonance. – Laser Spectroscopic Applications. – Laser Sources. – Laser Wavelength Measurements. – Postdeadline Papers.

Springer-Verlag
Berlin
Heidelberg
New York